T0178254

Hume-Rothery Rules for Structurally Complex Alloy Phases

Hume-Rothery Rules for Structurally Complex Alloy Phases

Uichiro Mizutani
Nagoya Industrial Science Research Institute
Japan

CRC Press
Taylor & Francis Group
Boca Raton London New York

CRC Press is an imprint of the
Taylor & Francis Group, an **informa** business
A TAYLOR & FRANCIS BOOK

CRC Press
Taylor & Francis Group
6000 Broken Sound Parkway NW, Suite 300
Boca Raton, FL 33487-2742

First issued in paperback 2019

© 2011 by Taylor and Francis Group, LLC
CRC Press is an imprint of Taylor & Francis Group, an Informa business

No claim to original U.S. Government works

ISBN-13: 978-1-4200-9058-1 (hbk)
ISBN-13: 978-0-367-38337-4 (pbk)

This book contains information obtained from authentic and highly regarded sources. Reasonable efforts have been made to publish reliable data and information, but the author and publisher cannot assume responsibility for the validity of all materials or the consequences of their use. The authors and publishers have attempted to trace the copyright holders of all material reproduced in this publication and apologize to copyright holders if permission to publish in this form has not been obtained. If any copyright material has not been acknowledged please write and let us know so we may rectify in any future reprint.

Except as permitted under U.S. Copyright Law, no part of this book may be reprinted, reproduced, transmitted, or utilized in any form by any electronic, mechanical, or other means, now known or hereafter invented, including photocopying, microfilming, and recording, or in any information storage or retrieval system, without written permission from the publishers.

For permission to photocopy or use material electronically from this work, please access www.copyright. com (http://www.copyright.com/) or contact the Copyright Clearance Center, Inc. (CCC), 222 Rosewood Drive, Danvers, MA 01923, 978-750-8400. CCC is a not-for-profit organization that provides licenses and registration for a variety of users. For organizations that have been granted a photocopy license by the CCC, a separate system of payment has been arranged.

Trademark Notice: Product or corporate names may be trademarks or registered trademarks, and are used only for identification and explanation without intent to infringe.

Library of Congress Cataloging-in-Publication Data

Mizutani, U.
    Hume-Rothery rules for structurally complex alloy phases / author, Uichiro Mizutani.
        p. cm.
    Includes bibliographical references and index.
    ISBN 978-1-4200-9058-1 (hardcover : alk. paper)
    1. Alloys--Structure. 2. Physical metallurgy. 3. Phase rule and equilibrium. I. Title.

TN690.M553 2011
669'.9--dc22
                                               2010030832

Visit the Taylor & Francis Web site at
http://www.taylorandfrancis.com

and the CRC Press Web site at
http://www.crcpress.com

Table of Contents

Preface

There is no doubt that in the design and use of materials, the studies and the understanding of the factors that enhance, or hinder, the stability of phases play a crucial and time-tested role. In the present book, I have focused on one particular aspect of such studies, the Hume-Rothery rules for metals and alloys. Over a span of some 90 years Hume-Rothery rules have been used to focus attention on three particularly important aspects of alloying, the electrochemical effects, the size factor effects, and the change of electron concentration. Each of these concepts has been the subject of detailed examination and elucidation. It is my hope and belief that an additional attempt to enhance further the understanding of the fundamental bases of the Hume-Rothery rules is likely to help disentangle at least some of the remaining puzzles evident in the stability mechanisms of the increasingly complex alloys, particularly with respect to the most prominent of the Hume-Rothery parameters used in alloy design, the electron concentration.

I have been involved during the past 40 years in research on topics related to the Hume-Rothery rules, starting from the work on the noble metal alloys in the 1970s, and amorphous alloys in the 1980s, initiated from the collaboration with Professor T.B. Massalski, in whose research group I worked during an extended visit to Pittsburgh, USA, during the early 1970s. A growing interest in the Hume-Rothery rules has been revived and became intensified soon after the discovery of the quasicrystals in 1984. Such structures are characterized by the symmetry of an icosahedron, but lacking the translational symmetry, and they can be regarded as structurally complex aperiodic, but ordered, electron compounds.

During the last decade or two, much research interest has occurred in this new field. In connection with the needed understanding of the more basic physics behind the stabilizing effects related to the electron concentration, my colleagues and I embarked on theoretical studies to gain a deeper insight into the origins of the Hume-Rothery electron concentration rule, especially by exploring the opportunities provided by the use of the first-principles band calculations. The results of these efforts are the basis of the present monograph. My aim here is to review in a 'teaching format' the most recent developments in the interpretation of the Hume-Rothery electron concentration rule with a particular emphasis on the stability of structurally complex metallic alloys (CMAs).

The present book starts in Chapter 1 with a review of the historical developments on the Hume-Rothery electron concentration rule, followed by the chemical bonding and phase diagram approaches to alloy phase stability in Chapter 2, a brief description of early theories advanced by Mott and Jones in Chapter 3, and the introduction of the two first-principles band calculation methods in Chapter 4: LMTO-ASA (Linear Muffin-Tin Orbital-Atomic Sphere Approximation) and FLAPW (Full-potential Linearized Augmented Plane Wave) methods, on which the present work is based. Chapter 5 is devoted to the interpretation of the α/β-phase transformation in the Cu-Zn alloy system, which has been known as one of the Hume-Rothery electron concentration rules. Since a series of gamma-brasses and several 1/1-1/1-1/1 approximants were chosen as working substances in this volume, the atomic structure of these CMAs is briefly described in Chapter 6. The first six chapters constitute an introductory part of the present volume. Chapters 7 to 10 focus on the main subjects to interpret the Hume-Rothery electron concentration rule for the CMAs and the **e/a** determination for alloys containing transition metal elements. The gist of the Hume-Rothery stabilization mechanism is discussed as clearly as possible.

I hope that the present book provides a fundamental knowledge on phase stability aspects in general, and also provides useful ideas and data in the development of new functional metals and alloys. I hope that researchers and engineers in universities and industries may benefit from this book. I trust that the writing style and approach that I adopted also makes it suitable as an advanced textbook to those who are now graduate students in the Departments of Materials Science, Metallurgy, Physics and Chemistry and will enable them to make early contact with original research papers in this field. A number of these are listed in the Reference section at the end of each Chapter and suitable review articles and more advanced textbooks are also included.

My writing efforts have benefited very substantially from the support, advice, and help of many friends and colleagues. I am also grateful for permissions to reproduce some figures and data.

I wish to express my deep thanks to Professor Jean-Marie Dubois, Institut Jean Lamour, Ecole des Mines de Nancy, France, and Dr. Esther Belin-Ferré, LCPMR-UMR, Paris, France, for critically reading the manuscript and providing me many comments and suggestions. The present work could not be completed without a strong collaboration with my associates in Japan: Professor Hirokazu Sato, Aichi University of Education, Professor Tsunehiro Takeuchi, Nagoya University, Professor Yoichi Nishino, Nagoya Institute of Technology, and Dr. Ryoji Asahi and Tatsuo Noritake, the Toyota Central Research and Development Laboratories, Inc., to whom I wish to express my sincere thanks on this occasion.

I would like to thank the Toyota Physical & Chemical Research Institute, which had provided me an opportunity to concentrate on this research work during my stay over 2005 to 2009 immediately after my retirement from Nagoya University. Thanks are also due to the Nagoya Industrial Science Research Institute for offering me a comfortable office since April 2009. The present work has been financially supported by the Grant-in-Aid for Scientific Research (Contract No.20560620) from the Japan Society for the Promotion of Science.

To my wife Kazuyo Mizutani I owe much felt gratitude not only for making this monograph possible, but also for her patience, support, advice, and encouragement over the many years of my research career.

Finally, very special and personal thanks are due to my good friend and mentor Dr. Thaddeus B. Massalski, Emeritus Professor at Carnegie-Mellon University, Pittsburgh, for carefully reading the whole manuscript at its various stages and for suggesting many improvements. His support, advice and encouragement are very much appreciated. Professor Massalski has actually guided me to this fascinating research field, which I found so challenging, at the time in early 1970s when I worked as a postdoctoral fellow at Carnegie-Mellon. It is my pleasure to dedicate this book to him.

Uichiro Mizutani

CHAPTER 1

Introduction

1.1 WHAT IS THE DEFINITION OF THE HUME-ROTHERY RULES?

As a set of basic rules describing the conditions under which an element could dissolve in a metal to form a solid solution, the Hume-Rothery rules have earned a great reputation in the field of materials science as simple but powerful guides to be considered when designing a new alloy. The rules were established in the 1920s and 1930s by the efforts of Hume-Rothery (1899–1968) and his associates, as well as other crystallographers and physicists, as will be described in the following sections. Hume-Rothery, a British metallurgist, was definitely a leading contributor in developing and exploiting such concepts.

What is the definition of the Hume-Rothery rules? At the Hume-Rothery Symposium held in St. Louis, in 2000, Massalski [1] summarized the existing confusion in the definition of the Hume-Rothery rules. He noted that many authors in different books and publications have presented the number and description of the Hume-Rothery rules (or effects), and their significance, in different ways. For example, Hume-Rothery himself listed five factors affecting the stability of alloy phases, without explicitly referring to them as the "Hume-Rothery rules" [2]. These are

1. The difference between the electronegativities, $\Delta\chi$, of the elements involved. The larger the $\Delta\chi$, the higher is the tendency for the atoms to unite in either liquid or solid phases. This is often called the *electrochemical effects*.

2. A tendency for atoms of elements near the ends of the short periods and B subgroups to complete their octets of electrons.

3. Size factor effects, that is, effects related to the difference in the atomic diameters of the elements.

4. A tendency for definite crystal structures to occur at characteristic numbers of electrons per unit cell, which, if all atomic sites are occupied, is equivalent to saying that similar structures occur at characteristic electrons per atom ratio **e/a** or the electron concentration.*

5. Orbital-type restrictions.

On the other hand, Kittel in his textbook on solid-state physics [3] discussed only the factor (4) mentioned above as the Hume-Rothery rule. Thus, there seems to exist no consensus on how many rules there are and what is their more precise significance.

Among factors listed by Hume-Rothery, the size factor, the electronegativity difference $\Delta\chi$, and the electron concentration effects appear to be the most pertinent three factors that affect the stability of metallic phases. Consider first the size factor effect. During alloying, if a solute differs in its atomic size by more than about 15% from the host atomic size, then it is likely to have a low solubility in that host. Here, the size factor is said to be unfavorable. Alternatively, in terms of the ratio of the Goldschmidt radii of two constituent atoms, a favorable alloy formation may be expected when the ratio is between 0.8 and 1.2. This effect is often referred to as the *Hume-Rothery size factor rule*. Among the three rules, the size factor rule seems to be the least controversial, since its quantitative evaluation had been made by evaluating the elastic energy of a solid solution [4–6] (see Appendix 1). The size factor rule is important in the sense that the formation of a wide primary solid solution is prohibited when the size factor is unfavorable.

The situation for the latter two rules is more complex. We consider the difficulties to originate largely from the fact that they may be related in a complex way to the electronic interactions among the constituent elements in a solid. Electrochemical effects serve a key feature in the description

---

* Valence electrons are important in determining how an element reacts chemically with other elements in an alloy. So is the parameter **e/a** in an alloy, which is defined as the average number of valence electrons of constituent elements per atom, in which the number of valence electrons in an element is determined by the group (vertical column) in the periodic table. However, it is less clear in transition metal (TM) elements. We will discuss a new clear-cut determination for the **e/a** value of elements including TM elements in Chapters 7 to 10.

of the electronic interactions pictured in terms of covalent bonding. Its degree in a binary alloy system has been quantitatively evaluated by taking a difference in the Pauling electronegativity χ between the two constituent elements [7]. If a solute has a large difference in electronegativity relative to the host, then the resulting charge transfer is more likely to favor compound formation. Its solubility in the host would therefore be limited. Thus, we may say that the formation of a wide primary solid solution range is again prohibited when the electrochemical effect is large.

The electron concentration effects are clearly evident when the size factor and the $\Delta\chi$ are of minor importance. It can be singled out by analyzing the electronic structure from the viewpoint of itinerant-electron picture and is the central issue in this monograph. Let us consider what the Hume-Rothery electron concentration rule means. Figure 1.1 shows the equilibrium phase diagrams for the Cu-Zn, Cu-Ga, and Cu-Ge alloy systems [8]. As is clear from Figure 1.1, different phases appear successively with increasing concentration of the partner elements Zn, Ga, and Ge. The face-centered cubic (fcc) phase extending from Cu forms a primary solid solution and is called the α-*phase*. Its maximum solubility limit is found to be 38.3 at.%Zn, 19.9 at.%Ga, and 11.8 at.%Ge in the respective alloy systems. Obviously, the maximum solubility limit decreases with increasing valency of the partner element: two for Zn, three for Ga, and four for Ge.

The β- and β′-phases appear next to the α-phase in the neighborhood of 50 at.%Zn in the Cu-Zn system. The β-phase is disordered bcc and stable at high temperatures but transforms into the β′-phase at low temperatures. The β′-phase has the CsCl-type ordered structure or the B2-structure. Similarly, the β-phase exists at around 25 at.%Ga at high temperatures in the Cu-Ga system. We see again that the concentration range over which the β- and β′-phases are stable moves to lower concentrations with increasing valency of the solute elements. Further increase in the solute concentration leads to the formation of the γ-phase in both Cu-Zn and Cu-Ga systems. As will be described in more details in Chapter 6, the structure of the γ-phase is constructed by stacking three bcc cells in x-, y-, and z-directions and subsequently removing the center and corner atoms with slight displacements of the remaining atoms. Its unit cell contains a total of 52 atoms. The γ-phase is often referred to as the *gamma-brass phase* and will be treated in this volume as the representative of structurally complex metallic alloy (CMA) phases.

It is evident from the previous description that the concentration range of different phases, appearing successively and systematically with an

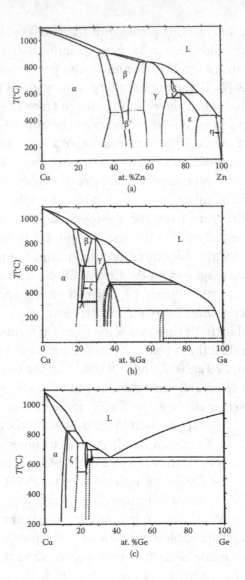

FIGURE 1.1 Phase diagram of (a) Cu-Zn, (b) Cu-Ga, and (c) Cu-Ge alloy systems. [From H. Okamoto, *Phase Diagrams for Binary Alloys* (ASM International, OH, 2000).]

increasing amount of a solute element added to the noble metal Cu, is shifted to lower concentrations with an increase in the valency of the solute element. These features are quite regularly observed not only in the Cu-based alloys but also in Ag- and Au-based alloys, as long as the partner element is chosen from polyvalent elements in the periodic table. Figure 1.2 schematically represents a composite phase diagram of the noble metals alloyed with polyvalent metal elements, represented in terms of **e/a**. As is clear, the α-, β-, γ-, ε-, and η-phases successively appear at particular **e/a** ranges, regardless of the solute element added to the noble metal [9]. This is the *Hume-Rothery electron concentration rule*. How can we explain this unique **e/a**-dependent phase stability? What happens if the transition metal (TM) element is involved as a partner element? What about the **e/a** value for the TM element? All these searching questions are addressed as the major topics in this monograph.

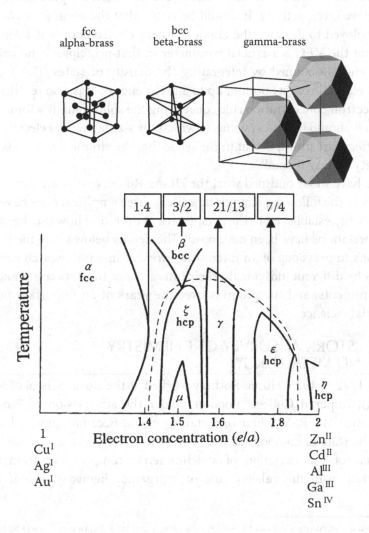

FIGURE 1.2 Schematic phase diagram as a function of electron concentration **e/a** in noble metal alloys showing the Hume-Rothery electron concentration rule.

As will be discussed in later chapters, in order to consider the Hume-Rothery electron concentration rule as rigorously as possible, we exclusively rely on the electronic structure derived by performing first-principles electronic structure calculations. Here it should be kept in mind that there are two different parameters that may be used to define the electron concentration: one is an average number of itinerant electrons per atom, e/a, and the other the number of total electrons, including d-electrons accommodated in the valence band, which is hereafter referred to as **VEC** in the present chapter.* For example, the e/a and **VEC** for pure Cu is unity and eleven, respectively. It should be noted that the parameter e/a must be employed to describe the Hume-Rothery electron concentration rule, but that the **VEC** is a crucial parameter in first-principles band calculations and is obtained by integrating the density of states (DOS) of the valence band from its bottom up to a given energy. It is also recalled that the electron concentration rule concerning the solid solubility limit of the α-phase should hold in systems, in which the size factor and electrochemical effects are unimportant in the sense that the atomic size ratio is close to unity and $\Delta\chi$ is small.

We have so far outlined what the Hume-Rothery rules are. Let us now survey in the following sections how these three empirical rules have been historically established in the perspective of time, and how the theoretical interpretations have been developed. The details below relate the present author's impressions of an interesting story of intense research commitments by different individuals, their interactions, their successes and disappointments, and the progress over the years of an important field in materials science.

1.2 HISTORICAL SURVEY OF CHEMISTRY AND METALLURGY

From 1922 to 1925, Hume-Rothery worked at the Royal School of Mines (RSM), Imperial College, London, under the supervision of Professor Carpenter, the RSM Chair of metallurgy, and became interested in the metallic state of compounds [10]. This field was then in a confused state because of the deviation of stoichiometric compositions from those expected from the valency rule of inorganic chemistry. People knew

---

* The e/a deduced from the Hume-Rothery plot (see Chapter 7, Section 7.4) refers to an average number of itinerant electrons per atom dominant outside the muffin-tin (MT) sphere in an alloy.

that inorganic compounds do not conduct electricity because they use up all the electrons to form stable orbits. However, there exists another class of compounds that are composed of definite proportions of two or more elemental metals and conduct electricity. They have been referred to as *intermetallic compounds* and received great attention at that time, as they apparently involve loosely bound electrons. In 1925, x-ray studies by Westgren and Phragmén [11] revealed the structure analogies of successively appearing phases, including α-, β-, γ-, ε-, and η-phases in Cu-Zn, Ag-Zn, and Au-Zn systems, all of which were indeed typical of intermetallic compounds. The collapse of the ordinary valency rules was mentioned. Nevertheless, no one in the mid-1920s could think of the style of metallic bonding, since it was before the birth of Schrödinger's equation in 1926 [12] and was certainly much earlier than the establishment of metallic cohesion by Wigner and Seitz [13] based on quantum mechanics in 1933.

In 1926, Hume-Rothery published the historic paper reporting the synthesis of various intermetallic compounds of Sn with reactive alkali and alkaline earth metals and pointed out that there was regularity in the occurrence of the structure in these alloys, though valency rules were not obeyed [14]. He noted with his own words in its PART III: "At first sight, there appears to be little connection between the formulae Cu_5Sn, Cu_3Al and CuZn, but examination showed them to possess one characteristic." His finding can be now restated that, in spite of different stoichiometric ratios involved, these three compounds crystallize into a common structure of the body-centered cubic (bcc) with the possession of the same number of electrons per atom ratio e/a of 3/2. We know that this has become the basis for the later establishment of the Hume-Rothery electron concentration rule. The preceding work led him to receive his Ph.D. degree from the University of London in 1925.

Structure information was of course of vital importance in his conclusions above. However, Hume-Rothery simply noted in [14] that the x-ray investigations had shown the Cu-Zn β-phase to possess the bcc structure. Regarding Cu-Al and Cu-Sn β-phases, he speculated them to be also bcc from indirect evidence on the relevant phase diagrams. Westgren and Phragmén were much stimulated by Hume-Rothery's proposal and extended their x-ray studies in 1926–1928 to examine structure analogies in Cu-Zn, Cu-Al, and Cu-Sn systems [15,16]. Moreover, they were the first to point out that the three gamma-brasses Cu_5Zn_8, Cu_9Al_4, and $Cu_{31}Sn_8$ appear to be stabilized at the common electron concentration e/a = 21/13, though the solute concentration, expressed in atomic percent, is quite different among them [16–18]. This is one of the most significant contributions to the establishment of the Hume-Rothery electron concentration rule over 1925 to 1929. More details about structural studies will be reviewed in the next section.

Hume-Rothery published another seminal paper in 1934 in collaboration with his two research students, Mabbott and Channel-Evans [19]. They revealed that the solid solution is limited, if the atomic diameter of the solute metal differs from that of the solvent metal by more than ~15%. When a difference in atomic diameter is less than this, a wide solid solution formation is possible and subject to the influence of factors of an electronic origin. It is also shown that the maximum solid solubilities of B-subgroup elements in copper and silver alloys, when the size relationship is favorable, are limited not by the solute concentration but by the **e/a** ratio of 1.4. This work demonstrated for the first time that details of alloy formation in certain solvents can be rationalized in terms of both atomic diameters and valencies. Both the size factor and electron concentration rules are summarized by him as the Hume-Rothery rules in the monograph *The Structure of Metals and Alloy* originally published in 1936 [20], which covers not only the maximum solubility limit of α-phase at **e/a** = 1.4 but also stability of β-, γ-, and ε-phases in the **e/a** range centered at 3/2, 21/13, and 7/4, respectively, in noble metals alloyed with B-subgroup elements in the periodic table. Its schematic illustration was already shown in Figure 1.2.

The preceding work attracted considerable attention not only among metallurgists but also among physicists in the field of solid-state physics. Here, at this stage, we must mention the initiation of theoretical works on solid-state physics in England, which dates back to 1930, when Jones was employed to conduct the national project initiated at Bristol [21,22]. Jones in collaboration with Mott [23] could successfully interpret the empirical Hume-Rothery electron concentration rule on the basis of the free-electron model in the presence of the Brillouin zone.* The application of the Brillouin zone concept, being often cited as the Fermi surface–Brillouin zone (FsBz) interaction throughout the present volume, and the free-electron model appeared for a long time to account for empirical relationships discovered by Hume-Rothery very satisfactorily. More details about the theoretical developments will be described in Section 1.4.

---

* The concept of the Brillouin zone was established in 1930 by a French physicist L. Brillouin.

Hume-Rothery continued to study the intermediate phases in great detail in different alloy systems, which occur at approximate e/a value of 1.5. He revealed that the bcc structure is not the only one to occur at this electron concentration but close-packed hexagonal phases, referred to as the ζ-phase, frequently occur, and phases with β-manganese structure or the μ-phase are also found. Some alloy systems based on the noble metals contain two of these phases, whereas the Ag-Al system contains all three types. In 1940, Hume-Rothery and his collaborators [24] showed how the electronic factor was modified by the atomic size difference between the constituent elements and also the difference in electronegativity, Δχ. Raynor, who followed Hume-Rothery ideas, could reveal the existence of many e/a-dependent phases in both binary and ternary alloy systems in 1940s [25].

The theoretical basis of the experimental relationships has become somewhat clouded in late 1950s, when Pippard demonstrated in 1957 [26] that the Fermi surface of copper at e/a = 1.0 already makes a contact with the {111} zone planes of the Brillouin zone of the face-centered cubic lattice. This was because Jones [27] assumed the Fermi surface of copper to be fully contained within the first Brillouin zone. So, the basis of the Jones interpretation was rendered doubtful. More details about the model of Jones will be described in Chapter 3. Hume-Rothery thought deeply about this and, in his papers in the 1960s, sought to reconcile the new knowledge with the experimental relationships [28–31]. He noted [28] that "this later experimental work by Pippard could be regarded as a striking confirmation of Jones's calculations but the price paid was a very heavy one, because the solubility limit of α-phase could no longer be associated with the region immediately after a peak on an N(E) curve. Further, the occurrence of the β-phase at an electron concentration of 1.5 could no longer be correlated with the peak on the body-centered cubic curve at e/a = 1.22. The whole position was one of great confusion."

Throughout his academic career, Hume-Rothery had always tried to interpret experimental data concerning the features of phase diagrams in terms of the basic properties such as the atomic size and the electron concentration. Hume-Rothery died on September 27, 1968, at the age of 69, and the problem his earlier work confronted and the difficulty of a proper interpretation of the electron concentration rule remained unsettled.

In 1960s–1980s, various physical properties, including the lattice constant [32], electronic specific heat coefficient [9], and positron annihilation angular correlation curves [33,34] of noble metal alloys, were systematically studied as a function of **e/a**. For example, the electronic specific heat coefficient, which is proportional to the density of states (DOS) at the Fermi level, was found to fall on universal curves for given phases only when plotted against **e/a**. Encouraged by this finding, Massalski and Mizutani [9] attempted to construct the Fermi surface of highly concentrated noble metal alloys on the basis of the FsBz interaction consistent with the observed **e/a**-dependent electronic specific heat coefficients. Their proposed Fermi surface was later confirmed by the positron annihilation angular correlation curves.

Concerning developments after 1980s till the present in relation to the Hume-Rothery electron concentration rule, we must point to works accumulated in the field of quasicrystals. Following the discovery of a quasicrystal in Al-Mn alloy system by Shechtman et al. [35] in 1984, Tsai and his collaborators discovered a series of thermally stable quasicrystals in Al–Cu–TM (TM = Fe, Ru, Os) and Al–Pd–TM (TM = Mn and Re) systems over 1988 to 1991 by using the empirical Hume-Rothery electron concentration rule as a guide [36,37]. They assigned negative valencies to the TM constituent elements, as proposed earlier by Raynor [25] and pointed out that new quasicrystals could be synthesized by searching for alloys, whose average valency falls into **e/a** values in the vicinity of 1.8. Their works certainly stimulated both metallurgists and physicists not only to search further for new quasicrystals along this line but also to reexamine the physics behind the Hume-Rothery electron concentration rule.

At this stage, we need to use up one paragraph to define briefly what quasicrystals and their approximants mean. A *quasicrystal* is defined as a solid satisfying the following conditions [38–40]: (1) the diffraction intensities consist of an infinite number of the δ-functions, (2) the number of basic vectors is larger than that of its dimension, and (3) rotational symmetry forbidden in crystals is manifest in quasicrystals. To realize the preceding conditions, the introduction of the concept of a quasi-lattice has been successful, which is constructed on the basis of the so-called cut-and-projection method from a higher-dimensional space to three- or two-dimensional real space [38–40]. This method can generate not only its basic structure but also a crystalline structure having nearly the same local structure as that of a quasicrystal. The latter is specifically called an *approximant* to the quasicrystal. Approximants are, therefore, crystals containing generally many atoms over 100 to 1000 atoms in the giant unit cell and characterized by the possession of the local rotational symmetry similar to that in quasicrystals. Many intermetallic compounds having been regarded as a CMA phase before the discovery of the quasicrystal have been newly identified as approximants described by the cut-and-projection method from a higher-dimensional space. Readers are asked to consult excellent textbooks on the introduction to quasicrystals [39,40]. More details about approximants found in various alloy systems will be described in Chapter 6.

In 1988, Belin-Ferré and her associates [41] pointed out for the first time from the soft x-ray emission spectra for the Al-Mn quasicrystal that the DOS at the Fermi level is substantially depressed relative to not only pure Al but also both the Al-Mn crystalline and amorphous alloys and suggested that this is a possible reason for a large enhancement in its resistivity. Independently, Fujiwara [42] theoretically revealed the presence of a depression in the DOS across the Fermi level by performing the linearized muffin-tin orbital (LMTO) band calculations for the Al-Mn approximant containing 138 atoms in its unit cell and suggested it to contribute to the stabilization of a quasicrystal. A depression in the DOS at the Fermi level has been called a *pseudogap* (see more details in Chapter 2, Section 2.3; Chapter 4, Section 4.2; Chapter 7, Section 7.2; etc.).* First-principles band calculations are not feasible for quasicrystals, in which the size of the unit cell is infinitely large and, hence, the presence of a pseudogap in quasicrystals has only been conjectured on the basis of experimentally observed data. However, first-principles band calculations have been performed for approximants, where the lattice periodicity is assured. It has now been generally established that the family of the CMA phases, including quasicrystals, is characterized in most cases by the existence of a pseudogap at the Fermi level in the corresponding electronic band structure.

Recent developments on various topics directly or indirectly related to the Hume-Rothery rules were well reviewed by outstanding researchers in physics and materials science and compiled in the two volumes published from TMS (The Minerals, Metals & Materials Society) in 2000 [1] and 2005 [43].

1.3 HISTORICAL SURVEY OF CRYSTALLOGRAPHY

In 1912, Friedrich, Knipping, and Laue showed the first diffraction effect by a copper sulfate single crystal. Shortly after this, key papers on the explanation of the zinc-blende diffraction pattern and the formulation of

---

* The concept of the "pseudogap" was first introduced by Mott [see N.F.Mott, Phil.Mag. 19 (1969) 835] to discuss metal–insulator transition of expanded liquid mercury.

the diffraction condition known as the *Bragg law* were published by W. L. Bragg (son) and William Henry Bragg (father) in 1913. Both W. H. Bragg and W.L.Bragg were jointly awarded the Nobel Prize in Physics in 1915 for their invention of the x-ray spectrometer and substantial contributions to the foundation of a new science of x-ray analysis of a crystal structure [44].

The powder diffraction method developed by Debye and Scherrer in Switzerland in 1916 and Hull in America in 1917 had been proved to be valuable in studies of elements and other materials. This approach had been taken up by Westgren in Sweden, whose laboratory had become one of the centers for the study of metals and alloys. In 1921, Westgren succeeded in obtaining funds for expanding the x-ray diffraction equipment with cameras for powders in collaboration with Phragmén. Since then, they could substantially contribute to the establishment of the Hume-Rothery electron concentration rule, as described in detail in the following text.

In 1924, Jette, Phragmén, and Westgren [45] studied the Cu-Al alloys, in which a phase with the gamma-brass structure was revealed for the first time. In 1925, Westgren and Phragmén published the paper entitled "X-ray Analysis of Copper-Zinc, Silver-Zinc and Gold-Zinc Alloys" [11], in which three important conclusions were drawn: first, structure analogies are revealed such that, with increasing Zn concentration, the face-centered cubic α-phase is followed by β-, γ-, and ε-phases before ending with the Zn primary solid solution called the η-phase; second, owing to a significant difference in diffractive powers between constituent elements, the β-phase in Ag-Zn and Au-Zn was safely proved to crystallize in the CsCl-type structure; and, third, all the gamma-phases contain 52 atoms in the cubic unit cell. The CsCl-type structure was also suggested for the Cu-Zn β-phase, though its identification was difficult because of a negligibly small difference in diffractive power between Cu and Zn.

At the Annual General Meeting in the Institute of Metals in London on March 11, 1926, Hume-Rothery [14] had called attention to the similarities between the β-phases of Cu-Zn, Cu-Al, and Cu-Sn systems, and had suggested that these similarities were connected to the fact that the ratio of the number of valency electrons to the number of atoms in all three phases is 3:2. In the paper received by the *Journal of the Institute of Metals* on November 1925, Hume-Rothery [14] mentioned that the Cu-Zn beta-brass is body-centered cubic without any reference to the work by Westgren and Phragmén published on February 1925 [11]. We see that Hume-Rothery was apparently unaware of the x-ray diffraction studies by Westgren and Phragmén at that time.

The gamma-brass, which is known as being typical of a CMA phase subjected to the Hume-Rothery electron concentration rule, has played a considerable role in the development of modern solid-state physics. As already mentioned in Section 1.2, Westgren and Phragmén pointed for the first time in 1926 to structural analogies among the Cu-Zn, Cu-Al, and Cu-Sn gamma-brasses [15]. In particular, they revealed that the Cu_4Sn gamma-brass forms a super-lattice constructed by doubly stacking the unit cell of the Cu-Zn or Cu-Al gamma-brass along the x-, y-, and z-directions and contains 416 atoms per unit cell with the lattice constant of 1.791 nm.

It was in 1926 that Bradley and Thewlis [46] were able to determine all 52 atom positions of the gamma-brass in the Cu-Zn system, using the x-ray diffraction data taken by Westgren and Phragmén [11]. The chemical formulae Cu_5Zn_8, Ag_5Zn_8, and Au_5Zn_8 were established by their work (see Appendix 2, Section A2.2). They also succeeded in determining 58 atom positions in its unit cell of the α-manganese possessing a similarly complex structure even as an element [47]. In 1929, Bradley [48] further determined all 52 atom positions of the Cu-Al gamma-brass and described its atomic structure in terms of two different 26-atom clusters forming the CsCl-type structure with the chemical formula Cu_9Al_4. However, as already noted in Section 1.2, the first proposal on the e/a=21/13 rule for Cu_5Zn_8, Cu_9Al_4, and $Cu_{31}Sn_8$ gamma-brasses was made by Westgren and Phragmén one year earlier in 1928 [16–18].* Bernal in 1928 also analyzed the diffraction data of Cu-Sn gamma-brass single crystals and led to the chemical formula $Cu_{41}Sn_{11}$ by dividing 416 atoms in the unit cell into 328 Cu and 88 Sn atoms to fit with the measured density [49]. This corresponds to 21.15 at.%Sn alloy, which is slightly richer in Sn than 20.5 at.%Sn for $Cu_{31}Sn_8$ suggested by Westgren and Phragmén [16,17].

It was in the 1926–1940 period that Bradley and his associates made significant contributions to studies of alloys and alloy phase diagrams in a series of papers [46–48,50,51]. Bragg saw the importance of his work and discussed it with a series of lecturers and visitors in the department

---

* At that time, Bradley was apparently one of the very few researchers, who were able to refine the atomic structure of such CMAs with the accuracy that has hardly been surpassed even in present times. Westgren and Phragmén simply speculated the chemical formula Cu_9Al_4 from the composition they studied when they proposed the e/a = 21/13 rule for the gamma-brasses in 1928. A further confusion seems to exist regarding the e/a=21/13 rule. This is because Mott and Jones stated on page 170 in their book [23] as if Hume-Rothery were the first to claim the e/a = 21/13 rule for the gamma-brass in 1931.

at Manchester [21]. Many physicists, including H. Jones, N. F. Mott, H. A. Bethe, and R. E. Peierls as well as the metallurgist W. Hume-Rothery, were among such visitors. The knowledge so gained enabled the empirical electron concentration rule to receive an explanation by Jones by introducing the concept of the Brillouin zones into the free-electron model. The models of Jones will be described as a main topic in Chapter 3.

It seems unwarranted to review here numerous numbers of crystallographic works on metals and alloys from 1940s up to present. Since the present monograph focuses on the stability mechanism of structurally complex metallic alloys, and primarily deals with the gamma-brasses and approximants as representatives, we shall only briefly mention in this section how crystallographic works on these CMAs have been making progress. From late 1960s to 1970s, Westman in Stockholm was guided by Westgren to the field of alloy chemistry and published with his coworkers a large number of detailed structural data on many kinds of gamma-brasses. Independently, Pearson and his group in Waterloo, Ontario, Canada, also carried out atomic structure determination of various gamma-brasses over 1970s. Thanks to great efforts by these two research groups, we can reliably use the atomic structure data for a large number of gamma-brasses. The details in individual systems will be surveyed in Chapter 6 and Appendix 2.

Similarly, the atomic structure of various approximants has been experimentally determined for the last two decades. The data are again of vital importance to study the stabilization mechanism of such highly complex compounds in relation to the Hume-Rothery electron concentration rule and will be discussed in Chapter 6.

1.4 HISTORICAL SURVEY OF PHYSICS

In the late 1920s, Carpenter at the Royal School of Mines in London, one of the foremost metallurgists and also known as a supervisor of Hume-Rothery, strongly supported the view that theoretical work on the nature of metals and alloys should be encouraged in England. In February 1930, Lindemann, as head of the Clarendon Laboratory, Oxford, proposed that the theoretical research on metals and alloys should be undertaken as an alternative to the empirical metallurgical studies and that such pioneering works had already been attempted by Lennard-Jones in Bristol. The fund was finally awarded in 1930 to Lennard-Jones, who decided to employ Harry Jones as a research assistant to carry out a theoretical investigation of metals and alloys [21,22].

Jones obtained a Ph.D. degree in experimental physics from Leeds but was also a former student of Professor Fowler, a theoretician working on statistical physics at the Cavendish Laboratory, Cambridge. Jones was selected as the best young physicist for the project, since he was familiar with both experimental techniques and the knowledge of mathematical theory. In the autumn of 1930, he became an assistant to Professor Lennard-Jones at the University of Bristol and immediately began to study various papers on the electron theory of metals published by Bloch, Peierls, Brillouin, and Wilson. He was attracted by the concept of energy zones in the reciprocal space and considered the relation between the planes of energy discontinuity and the Bragg law to provide an opportunity of relating the theory to the properties of actual metals and alloys.

However, Jones encountered serious difficulty in the summer of 1932, when Lennard-Jones left Bristol to accept the Chair of Theoretical Chemistry at Cambridge. In spite of this development, he remained at Bristol. In 1933, Jones happened to attend a talk given by W. L. Bragg, a director of the physics laboratory at Manchester, on the structure of alloys. Bragg had been giving lectures to many visitors, emphasizing the possibilities opened up for the study of metals and alloys by x-ray structural crystallography. Jones was greatly impressed by his talk, in particular, about the structure of the gamma-brass and the empirical Hume-Rothery rules and began to think on whether any connection could be made with his approach based on the new quantum mechanics and electron theory of metals developed by Sommerfeld in 1928.

Jones realized that the most intense line in the x-ray powder diffraction spectrum of the gamma-brass should correspond to zone planes in the reciprocal space all at the same distance from the origin. By this, Jones had hinted that the zone formed by these planes might be related to the stability of the phase and its physical properties. In 1934, he successfully explained the observed large diamagnetic susceptibility in the Cu-Zn and Cu-Sn gamma-brasses by assuming that the Fermi surface must lie very close to the surface of discontinuity in the energies, and relying on the theory by Peierls, which was constructed on the basis of the Landau diamagnetism in metals [52]. Jones already noted in this paper that the number of electrons as given by Hume-Rothery's e/a ratio of 21/13 occupies a volume somewhere between the volume of the inscribed sphere and the total volume of the Brillouin zone.

In the autumn of 1933, Mott joined the Department of Theoretical Physics at Bristol as a successor of Lennard-Jones [22,53]. In his first month,

Mott understood all that Jones told him about metals and mentioned to Bragg: "I think that metals are exciting and lots are to be done from the theoretical point of view." Mott further recalled that "I need hardly say that my interest became even greater when, soon afterwards, I discussed these problems with Hume-Rothery himself, filled up with ideas" and admitted to say that "it was Jones who, by providing a quantum mechanical explanation of the Hume-Rothery rules, essentially vindicated the application of the new ideas to metallurgical problems, thereby convincing others including Hume-Rothery of their superiority over classical models" [21]. It may be also noted that Mott and Jones were deeply influenced by the comprehensive report in the *Handbuch der Physik* by Sommerfeld and Bethe on the electron theory of metals published in 1933 [54].

In 1934, Jones used his ideas to account for the semimetal properties of bismuth. At the same time, he continued to work closely with Mott, who was stimulated by practical rather than purely theoretical considerations. For example, a query by Bragg about the high reflecting power of alloys in the gamma-brass led Mott and Jones to study the whole subject of the energy distribution of conduction electrons in metals. Indeed, Jones, Mott, and Skinner [55] interpreted the soft x-ray emission spectra from the light metals Li, Be, Na, Mg, and Al measured by Skinner during his stay in the United States. This paper not only demonstrated the possibility of wave-functions changing type from s-like to p-like within a single valence band but also confirmed implicitly the physical reality of the Fermi distribution of conduction electrons.

The collaboration between Mott and Jones culminated in the publication of their highly influential textbook *The Theory of the Properties of Metals and Alloys* in 1936 [23], in which the stabilization mechanism of the gamma-brass was elegantly described in terms of the FsBz interaction. The publication of this book also marked the end of the first phase in the formation of the Bristol solid-state theory group. Jones took a major role in the first initiatives at Bristol in metal physics. However, he completed his last work at Bristol in 1937 on the electron theory of phase formation by calculating the electronic energies of the α- and β-phase Cu-Zn alloy system [27], the details of which will be described in Chapter 3, and following a brief appointment at Cambridge, he accepted a readership in mathematics at Imperial College, London.

The development in electron theory of metals and alloys from 1940s up to now cannot be summarized only by a few paragraphs. Readers may be encouraged to consult the textbook published by the present author in 2001 in case they encounter here unfamiliar terms and concepts while reading this monograph [38]. Before ending this section, a few words may be added. We consider a true understanding of the Hume-Rothery electron concentration rule to owe its origin to the development of self-consistent first-principles band calculations, the efficiency and reliability of which have been dramatically improved by the progress in computer science. In the present monograph, we have employed two first-principles band calculation methods: the LMTO-ASA (linear-muffin-tin orbital-atomic sphere approximation) and the FLAPW (full-potential linearized augmented plane wave). The outline of these two methods will be described in Chapter 4, since they are definitely an indispensable tool to analyze the Hume-Rothery electron concentration rule in realistic metals and alloys.

REFERENCES

1. T.B. Massalski, *The Science of Alloys for the 21st Century: A Hume-Rothery Symposium Celebration*, edited by P.E.A. Turchi, R.D.Shull and A. Gonis (TMS, PA, 2000), pp. 55–70.
2. W. Hume-Rothery, *Phase Stability in Metals and Alloys*, edited by P.S. Rudman, J. Stringer, and R.I. Jaffee, Series in *Materials Science and Engineering* (McGraw Hill, New York, 1967), pp. 3–23.
3. C. Kittel, *Introduction to Solid State Physics*, 2nd edition (John Wiley & Sons, New York, 1956), pp. 325–327.
4. J. Friedel, *Adv.Phys.* 3 (1954) 446.
5. J.D. Eshelby, *Solid State Physics*, edited by F. Seitz and D. Turnbull (Academic Press, New York, 1956), vol. 3, pp. 115–119.
6. T. Egami and Y. Waseda, *J. Non-Cryst. Solids* 64 (1984) 113.
7. L. Pauling, *Nature of the Chemical Bonds and the Structure of Molecules and Crystals: An Introduction to Modern Structural Chemistry*, Cornell University Press, N.Y. 1960, 3rd ed.
8. H. Okamoto, *Phase Diagrams for Binary Alloys* (ASM International, OH, 2000).
9. T.B. Massalski and U. Mizutani, *Prog. Mat. Sci.* 22 (1978) 151.
10. G.V. Raynor, *J. Inst. Metals*, 98 (1970) 321.
11. A. Westgren and G. Phragmén, *Phil. Mag.* 50 (1925) 311.
12. E. Schrödinger, *Ann. Physik*, 79 (1926) 361, 489, 734.
13. E. Wigner and F. Seitz, *Phys. Rev.* 43 (1933) 804.
14. W. Hume-Rothery, *J. Inst. Metals*, 35 (1926) 295.
15. A. Westgren and G. Phragmén, *Z. Metallkunde* 18 (1926) 279.
16. A. Westgren and G. Phragmén, *Z. Anorg. Chemie*, 175 (1928) 80.
17. A. Westgren and G. Phragmén, *Metallwirtschaft* 7 (1928) 700.
18. A. Westgren and G. Phragmén, *Trans. Farad. Soc.*, 25 (1929) 379.
19. W. Hume-Rothery, G.W. Mabbott, and K.M. Channel-Evans, *Phil. Trans. Roy. Soc.* A 233 (1934) 1.

20. W. Hume-Rothery and G.V. Raynor, *The Structure of Metals and Alloys*, Institute of Metals Monograph and Report Series, No.1 (1962) pp. 210–217.
21. S.T. Keith and P.K. Hoch, *Br. J. Hist. Sci.* 19 (1986) 19.
22. H. Jones, *Proc. R. Soc. Lond.* A371 (1980) 52–55; N.F. Mott, ibid., pp. 56–66.
23. N.F. Mott and H. Jones, *The Theory of the Properties of Metals and Alloys* (Clarendon Press, Oxford, 1936; Dover, 1958).
24. W. Hume-Rothery, P.W. Reynolds, and G.V. Raynor, *J. Inst. Metals*, 66 (1940) 191.
25. G.V. Raynor, *Prog. Met. Phys.* 1 (1949) 1.
26. A.B. Pippard, *Phil. Trans. R. Soc.* A250 (1957) 325.
27. H. Jones, *Proc. Phys. Soc.* A49 (1937) 250.
28. W. Hume-Rothery, *J. Inst. Metals* 9 (1961–62) 42.
29. W. Hume-Rothery, *Atomic Theory for Students of Metallurgy*, 4th edition, Institute of Metals, London, Monograph and Report Series No. 3, 1962, p. 312.
30. W. Hume-Rothery, *Application of X-ray Diffraction to Metallurgical Science*, edited by P.P. Ewald *Fifty Years of X-ray Diffraction* (Utrecht, 1962), pp. 190–211.
31. G.V. Raynor, *Biographical Memoirs of Fellows of the Royal Society*, 15 (1969) 109: p. 122.
32. T.B. Massalski and H.W. King, *Prog. Mat. Sci.* 10 (1961) 1.
33. M. Haghgooie, S. Berko, and U. Mizutani, *Proc. 5th Int. Conf. Positron Annihilation* (The Japan Institute of Metals, Japan, 1979). p. 291.
34. T. Suzuki, M. Hasegawa, and M. Hirabayashi, *J. Phys.* F6 (1976) 779.
35. D. Shechtman, I. Blech, D. Gratias, and J.W. Cahn, *Phys. Rev. Letters* 53 (1984) 1951.
36. A.P. Tsai, A. Inoue, and T. Masumoto, *Jpn. J. Appl. Phys.* 27 (1988) L1587; A.P. Tsai, A. Inoue, Y. Yokoyama, and T. Masumoto, *Mater. Trans. Jpn. Inst. Met.* 31 (1990) 98.
37. Y. Yokoyama, A.P. Tsai, A. Inoue, T. Masumoto, and H.S. Chen, *Mater. Trans. Jpn. Inst. Met.* 32 (1991) 421.
38. U. Mizutani, *Introduction to the Electron Theory of Metals* (Cambridge Univ. Press, 2001) (see Chapter 15).
39. C. Janot, *Quasicrystals-A primer* (Oxford Science Publications, Oxford, 1994).
40. J.M. Dubois, *Useful Quasicrystals* (World Scientific, Singapore, 2005).
41. A. Traverse, L. Dumoulin, E. Belin, and C. Sénémaud, in *Quasicrystalline Materials*, edited by Ch. Janot and J.M. Dubois (World Scientific, Singapore, 1988), pp. 399–408.
42. T. Fujiwara, *Phys. Rev.* B40 (1989) 942.
43. *The Science of Complex Alloy Phases*, edited by T.B. Massalski and P.E.A. Turchi (TMS, PA, 2005).
44. D. Phillips, *Fifty Years of X-Ray Diffraction* (dedicated to the International Union of Crystallography on the occasion of the commemoration meeting in Munich, July 1962), pp. 75–143; W.L. Bragg, ibid., pp. 120–135; W.

Hume-Rothery, ibid., pp. 190–211; G. Hägg, ibid., "VI. Schools and Regional Development, 22.Scandinavia" pp. 479–483; A. Hultgren, B. Kalling and A. Westgren, ibid., pp. 360–364.

45. E.R. Jette, G. Phragmén, and A.F. Westgren, *J. Inst. Metals* 31 (1924) 193.
46. A.J. Bradley and J. Thewlis, *Proc. Roy. Soc.* (A) 112 (1926) 678.
47. A.J. Bradley and J. Thewlis, *Proc. Roy. Soc.* A115 (1927) 456.
48. A.J. Bradley, *Phil. Mag.* 6 (1929) 878.
49. J.D. Bernal, *Nature, Lond.* 122 (1928) 54.
50. A.J. Bradley and P. Jones, *J. Inst. Metals.* 51 (1933) 131.
51. A.J. Bradley, H.J. Goldschmidt, and H.J. Lipson, *J. Inst. Metals*, 63 (1938) 149.
52. H. Jones, *Proc. Roy. Soc. Lond.* A144 (1934) 225.
53. N.F. Mott, *A Life in Science* (Taylor & Francis, London, 1986).
54. A. Sommerfeld and H. Bethe, Handbuch der Physik (Geiger/Scheel, 1933), Band 24/2, *Aufbau der Zusammenhängen der Materie*, S.333–622; *Elektronentheorie der Metalle* (Springer-Verlag, Berlin, 1967).
55. H. Jones, N.F. Mott, and H.W.B. Skinner, *Phys. Rev.* 45 (1934) 379.

CHAPTER 2

Chemical Bonding and Phase Diagrams in Alloy Phase Stability

2.1 COHESIVE ENERGY OF A SOLID

The cohesive energy of a solid refers to the energy required to separate constituent atoms from each other and bring them to an assembly of neutral free atoms. Cohesive energies for elements in the periodic table are listed in Table 2.1 [1].* Among the elements, tungsten is known to possess the highest melting point as well as the highest cohesive energy, of 837 kJ/mol. Alkali metals generally possess rather low cohesive energies, for instance, that of Na is 108 kJ/mol. The cohesive energy of solid Ar is merely 7.7 kJ/mol.

Elements with metallic and covalent bonding are important to us. We see from Table 2.1 that their cohesive energies are distributed over the range 150–400 kJ/mol (Mg = 148, Al = 322, Si = 448, Ti = 469, Fe = 414, Cu = 338, Ag = 286, Au = 365, and Bi = 208 kJ/mol). There are typically three bonding types for a solid: ionic, covalent, and metallic bondings. In an ideally ionic crystal, its cohesive energy can be calculated by summing up the electrostatic energy for an assembly consisting of ions with unlike charges. This is known as the calculation of the Madelung constant

---

* The cohesive energy, total-energy, free energy of formation, and enthalpy of formation are expressed in units of kJ/mol, cal/mol, or eV/atom. We use units of either eV/atom or kJ/mol (1 eV/atom = 96.44 kJ/mol) in the present book.

TABLE 2.1 Cohesive Energies of Elements in the Periodic Table

1	2	3	4	5	6	7	8	9	10	11	12	13	14	15	16	17	18
Li 159	Be 322											B 561	C 711	N 477	O 251	F 84	Ne 2.1
Na 108	Mg 148											Al 322	Si 448	P 332	S 277	Cl 135	Ar 7.7
K 91	Ca 176	Sc 379	Ti 469	V 511	Cr 396	Mn 288	Fe 414	Co 424	Ni 428	Cu 338	Zn 130	Ga 269	Ge 374	As 289	Se 206	Br 118	Kr 11
Rb 83	Sr 164	Y 423	Zr 610	Nb 720	Mo 658	Tc	Ru 639	Rh 555	Pd 380	Ag 286	Cd 112	In 247	Sn 301	Sb 259	Te 193	I 107	Xe 15
Cs 80	Ba 179	La 434	Hf 611	Ta 781	W 837	Re 782	Os 783	Ir 670	Pt 565	Au 365	Hg 67	Tl 181	Pb 197	Bi 208	Po 144	At	Rn
Fr	Ra	AC															

Ce 460	Pr 373	Nd 323	Pm	Sm 203	Eu 174	Gd 399	Tb 393	Dy 297	Ho 293	Er 322	Tm 247	Yb 151	Lu 427
Th 572	Pa 527	U 522	Np 440	Pu 385	Am 251	Cm	Bk	Cf	Es	Fm	Md	No	Lr

Note: The values are in units of kJ/mol converted from values in units of cal/mol listed in Reference 1.

[1]. Since ionic bonding originates from electrostatic interaction, cohesive energy can arise even without overlap of wave functions between the neighboring atoms and, hence, in the absence of formation of the valence band. However, any realistic ionic crystals are certainly not in such an ideal state but form the valence band through overlap of wave functions of the outermost electrons on neighboring atoms. For example, Gellat et al. (1983) [2] discussed the characteristic features of ionic crystals by performing LMTO band calculations for PdF, PdB, and PdLi as compounds exhibiting both strong ionicity and covalency.

Degenerate orbital levels in the assembly of neutral atoms are lifted and split into the valence band through the orbital hybridization among neighboring atoms upon the formation of a solid. The cohesive energy in both metallic and covalent bonding types is gained by lowering the total-energy relative to that of the assembly of free atoms. In these cases, it is meaningless to evaluate potential energies classically in the same manner as in ionic crystals discussed earlier. We need to calculate the total-energy of a system in the presence of periodically arranged ionic potentials of a solid. First-principles band calculations are indispensable to evaluate the cohesive energy in a solid, where metallic and covalent bondings are essential.

We will review metallic bonding briefly at this stage [3]. We take as a starting point that the free electron gas with negative charges is uniformly distributed in an array of potentials due to periodically arranged positive ions and that the total charge of the electrons is just large enough to cancel that of the ions. The electrostatic energy per atom, when the array of potentials forms a bcc lattice, reduces to

$$\varepsilon_0 = -\frac{2348}{(r_s/a_0)} \text{ kJ/mol} \tag{2.1}$$

where r_s represents the average radius of the sphere that each free electron occupies in the space and a_0 is the Bohr radius, which equals 0.0529 nm [4]. In the case of Na, for example, the electrostatic energy amounts to $\varepsilon_0 = -600$ kJ/mol by inserting its appropriate ratio $r_s/a_0 = 3.91$ into Equation 2.1.*

---

* The value of r_s turns out to be 0.207 nm for Na by inserting its lattice constant a (= 0.422 nm) into the relation

$$\frac{4\pi}{3}r_s^3 = \frac{a^3}{2}$$

There is another well-known method, due to Wigner–Seitz [5], to extract the potential energy contribution to the cohesive energy. The wave function of an electron propagating in a periodic lattice must be of the form $\psi_k(r) = e^{ik \cdot r} u_k(r)$ as a Bloch wave [3]. Here, the wave vector k specifies a quantized Bloch state, and $u_k(r)$ is a periodic function satisfying the relation $u_k(r) = u_k(r + l)$, where l is the lattice vector. The electronic state with $k = 0$ refers to the energy state having a vanishing kinetic energy. Wigner and Seitz assumed the lowest energy state of the valence band to represent the contribution solely from the potential energy. The wave function obviously becomes $\psi_0(r) = u_0(r)$, possessing the same symmetry as the lattice. A normal derivative of $u_0(r)$ on the surface of the Wigner–Seitz cell of the bcc lattice must be zero, since $u_k(r) = u_k(r + l)$ holds. The Wigner–Seitz cell was approximated by a sphere of an equal volume, and the Schrödinger equation was solved under the condition

$$\frac{d\psi_0(r)}{dr} = 0$$

on its surface. Instead, the condition above for a free atom is satisfied only at an infinitely large distance. In this way, the difference in the energy eigenvalue between an electron at the bottom of the valence band in a metal and that in the free atom was calculated as a function of distance r. The result is shown as $\varepsilon_o(r)$ in Figure 2.1.

Let us suppose 1 mole of Na metal contains the Avogadro number of valence electrons. No more than two electrons can occupy the same quantum state due to the Pauli exclusion principle. As a result, the Fermi sphere is formed in the reciprocal space, and the kinetic energy is inevitably increased. According to the free electron model, an average kinetic energy per electron is given by

$$\varepsilon_{kin} = \frac{3}{5} E_F = \frac{2903}{\left(r_s / a_0\right)^2} \text{ kJ/mol} \tag{2.2}$$

where E_F is the Fermi energy [3,4].* It amounts to $\varepsilon_{kin} = 190 \text{ kJ/mol}$ for Na metal. The kinetic energy is a quantity with a positive sign. Hence,

---

* The Fermi energy refers to the energy of the highest-occupied electronic states at absolute zero. The Fermi level or chemical potential at absolute zero coincides with the Fermi energy.

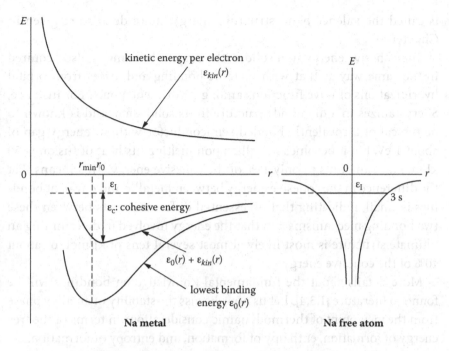

FIGURE 2.1 Cohesive energy in metallic bonding. Na metal is used as an example. The curve $\varepsilon_0(r)$ represents the lowest energy of electrons of the wave vector $\mathbf{k} = 0$, while the curve ε_{kin} represents an average kinetic energy per electron. ε_I represents the ionization energy needed to remove the outermost 3s electron in a free Na metal to infinity, and ε_c is the cohesive energy. The position of the minimum in the cohesive energy gives an equilibrium interatomic distance r_0. [From U. Mizutani, *Introduction to the Electron Theory of Metals* (Cambridge Univ. Press, 2001).]

we see that the smaller the r_s, the more unfavorable the metallic bonding. The minimum appears at a particular distance r_0 in the total-energy given by the sum of Equations 2.1 and 2.2. The system is stabilized when the total-energy is lower than that of the assembly of free atoms. This is the mechanism of metallic bonding. The difference between the two energies is called the cohesive energy. The value r_0 at the minimum corresponds to an equilibrium interatomic distance.* A variety of metallic phases are stabilized in nature by a mechanism unique to an individual system to lower the kinetic energy of electrons as much as possible (it

---

* Rigorously speaking, one needs to take into account the electron–electron interaction properly in the evaluation of cohesive energy (see Chapters 4 and 5, and Reference 4).

is called the valence-band structure energy; more details are given in Chapter 5).

The cohesive energy in a solid with covalent bonding is also acquired in the same way as that with metallic bonding and arises from orbital hybridizations of wave functions among constituent atoms. For instance, Si crystallizes in a diamond structure in its solid state, and is known to be typical of a covalently bonded semiconductor with an energy gap of about 1 eV, but it becomes metallic upon melting. Its heat of fusion is 50 kJ/mol and amounts to only 11% of the cohesive energy. This means that the difference in the cohesive energy between metallic and covalent bondings is small, indicating that no essential difference exists between these two bonding mechanisms and that the energy involved in determining an ultimate structure is most likely at most several tens of kJ/mol, or about 10% of the cohesive energy.

More details about the fundamental knowledge on bondings will be found in literature [1,3,4]. Let us now discuss the stability of an alloy phase from the viewpoint of thermodynamic considerations in terms of the free energy of formation, enthalpy of formation, and entropy of formation.

2.2 FREE ENERGY OF FORMATION AND ENTHALPY OF FORMATION

Suppose we prepare a solid solution (x_A, x_B) at absolute temperature T under a constant pressure P by mixing x_A of component A and x_B (= 1 − x_A) of component B in an A-B alloy system. The free energy of formation or free energy of mixing ΔG_m is given by the relation

$$\Delta G_m = \Delta H_m - T\Delta S_m \qquad (2.3)$$

where ΔH_m is the enthalpy of formation or heat of mixing and ΔS_m is the entropy of formation or entropy of mixing [6]. Instead of the cohesive energy, we discuss now the energy and entropy relative to those of pure metals A or B. In other words, thermodynamic quantities obtained by a composition-weighted average of relevant values of constituent pure elements are taken as a reference state in Equation 2.3. An equilibrium phase diagram can be understood by studying both temperature and composition dependences of these thermodynamic quantities. An alloy will be formed by mixing the atomic species A and B in proportion to respective concentrations x_A and x_B, provided that ΔG_m is negative under given temperature and composition.

Figures 2.2 and 2.3 show the composition dependence of thermodynamic quantities at 1173 K in Au-Ni and at 1625 K in Ni-Pt alloy systems, respectively [7,8]. All thermodynamic quantities ΔG_m, ΔH_m, and ΔS_m become zero at both ends of the composition axis. The enthalpy of formation ΔH_m is positive in the Au-Ni system. At low temperatures, the entropy term in Equation 2.3 is so small that ΔG_m becomes positive.

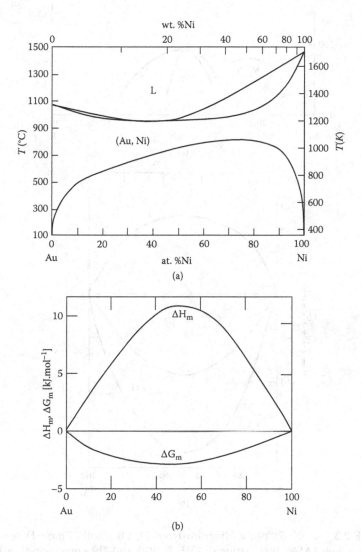

FIGURE 2.2 (a) Au-Ni phase diagram [from H. Okamoto, *Phase Diagrams for Binary Alloys* (ASM International, OH, 2000)] and (b) heat of mixing ΔH_m and free energy of mixing ΔG_m at 1173 K. [From B.L. Averbach, A. Flinn and M. Cohen, *Acta Met.* 2 (1954) 92].

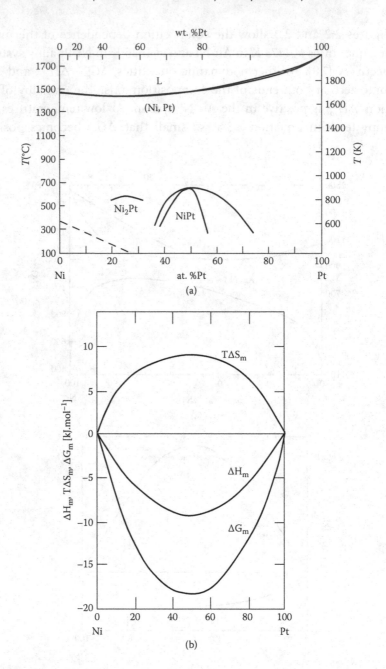

FIGURE 2.3 (a) Ni-Pt phase diagram from H. Okamoto, *Phase Diagrams for Binary Alloys* (ASM International, OH, 2000)] and (b) corresponding heat of mixing ΔH_m, entropy of mixing ΔS_m and free energy of mixing ΔG_m at 1625K. [From R.W. Swalin, *Thermodynamics of Solids* (Second edition, John Wiley & Sons, New York, 1972).]

Hence, as can be seen in Figure 2.2a, the decomposition into two phases takes place at low temperatures. But a solid solution is formed when $\Delta G_m <$ 0 at high temperatures as a result of an increase in the second term $T\Delta S_m$ in Equation 2.3. The value of ΔH_m at x = 0.5 in Figure 2.2b takes its maximum of 11.7 kJ/mol, being merely 3% of the cohesive energy of pure elements such as Au. In contrast, as shown in Figure 2.3, a complete solid solution is formed over a whole temperature range in the Ni-Pt alloy system, since $\Delta H_m < 0$.

As far as stability at absolute zero is concerned, we can discuss it in terms of ΔH_m, since no entropy term exists.* The value of ΔH_m has been evaluated by a large number of theoretical and experimental methods and distributed over a wide range covering from only ± a few up to ±100 kJ/mol, depending on the alloy system chosen. In the pair potential approach, ΔH_m of an A-B alloy is given by

$$\Delta H_m = P_{(AB)} \left[H_{AB} - \frac{1}{2} \left(H_{AA} + H_{BB} \right) \right] \qquad (2.4)$$

where H_{AA}, H_{AB}, and H_{BB} represent mean potentials associated with three types of atom pairs A-A, A-B, B-B, respectively, and $P_{(AB)}$ is the number of the atom pair A-B [6]. If H_{AB} is smaller than the average of H_{AA} and H_{BB}, $\Delta H_m < 0$ and, hence, a solid solution will be formed.

Miedema and his collaborators constructed the macroscopic atom (MA) model and evaluated the heat of mixing ΔH_m for hypothetical intermetallic compounds AX_5, AX_4, AX_2, A_3X_5, A_2X_3, AX, A_3X_2, A_5X_3, A_2X, A_3X, and A_5X in an A-X binary alloy system containing the 3d-, 4d-, and 5d-transition metal element [9]. In the MA model, the value of ΔH_m is calculated by using two parameters: one associated with the difference in electronegativities between elements A and X and the other with the difference in the charge density on the boundary of the Wigner–Seitz cell. The Miedema model is known to provide fairly accurate data for ΔH_m not only in sign but also in absolute value. However, it is not based on first-principles electronic structure calculations, so one cannot pursue band structure effects, such as the effect of a pseudogap on its stability.

---

* In principle, the formation of an alloy is possible at absolute zero when $\Delta H_m < 0$. According to the third law of thermodynamics, however, entropy, including the configuration entropy, should vanish at absolute zero. Hence, any alloy with chemical disorder or structural disorder remains metastable at T = 0.

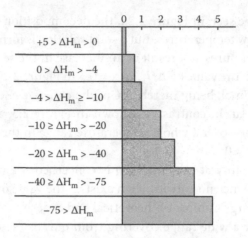

FIGURE 2.4 The number of intermetallic compounds existing in the phase diagram increases with increasing the heat of mixing ΔH_m to form an AB intermetallic compound in an A-B intertransition metal alloy system. [From F.R. de Boer, R. Boom, W.C.M. Mattens, A.R. Miedema, and A.K. Niessen, *Cohesion in Metals* (North-Holland, 1988).]

Miedema and his coworkers examined binary phase diagrams consisting of two transition metal (TM) elements and pointed out that the number of intermetallic compounds existing in the resulting phase diagram increases from zero to five when the heat of mixing of a 1:1 stoichiometric compound is increased in a negative direction. As is clearly shown in Figure 2.4, we see that the more negative the enthalpy of formation of the 1:1 compound, the more intermediate phases can exist [9]. Figure 2.5 is another demonstration, showing that the number of intermediate phases increases from one to three and to five, with increasing heat of mixing in a negative direction as $\Delta H_m = 0$, −37, and −75 kJ/mol at $x_B = 0.5$. The most important message in this argument is that the difference in the heat of mixing amounts to only 5–20 kJ/mol when several intermediate phases compete with one another in a given alloy system, indicating that the energy difference involved in the competition between the two neighboring phases is fairly small.

2.3 KINETIC ENERGY OF ELECTRONS AND THE ROLE OF THE PSEUDOGAP

The average kinetic energy per electron for pure Cu becomes $\varepsilon_{kin} = 405$ kJ/mol, if its Fermi energy of 7.0 eV is inserted into Equation 2.2. Instead, a rough estimate of the potential energy using Equation 2.1 amounts to −880 kJ/mol for pure Cu. Hence, we see that the kinetic

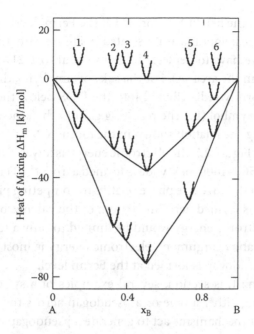

FIGURE 2.5 The number of intermediate phases increases from unity up to five as the heat of mixing ΔH_m increases in a negative direction from zero, −37 to −75 kJ/mol in an A-B binary system. The abscissa represents the concentration of the element B. [From F.R. de Boer, R. Boom, W.C.M. Mattens, A.R. Miedema, and A.K. Niessen, *Cohesion in Metals* (North-Holland, 1988).]

energy of electrons should play a critical role in cohesive energy. Since the kinetic energy, positive in sign, acts against cohesive energy increase, nature often devises a mechanism to lower the kinetic energy of a given system as much as possible (see Section 5.2).

We have so far compared various energies relevant to phase stability and are ready to discuss the mechanism that lowers the kinetic energy of electrons by forming a pseudogap across the Fermi level. We consider it most important to learn how much energy can be lowered through the formation of a pseudogap at the Fermi level. To grasp its essence, we assume the valence band to be approximated by a density of states (DOS) of a rectangle, as shown in Figure 2.6a [10]. The Fermi level E_F and the DOS at the Fermi level, $N(E_F)$, are set to 7.0 eV and $N(E_F) = 0.21$ states/eV.atom appropriate to pure Cu in the free electron model, respectively. First-principles band calculations performed for various approximants have revealed that the depth of a pseudogap at the Fermi level is generally about 20–60% that of the free electron DOS and that the width of the pseudogap is extended over the range $\Delta E = 500\text{--}1500$ meV (see Section 7.2). Let us assume that a pseudogap with the height H is formed at the Fermi level on the rectangular DOS with the height H_0, as illustrated in Figure 2.6b. In order to create a pseudogap, we have to deplete electrons equal to $0.21 \times (1 - H / H_0) \cdot \Delta E$ states/atom from the area (A). For the sake of simplicity, the depleted electrons are uniformly redistributed into the DOS below the pseudogap, as marked by the symbol (B). The resulting gain in the kinetic energy is calculated by using the relation shown in the footnote.*

As shown in Figure 2.7, the electronic energy is lowered by 30–60 kJ/mol if a pseudogap 500–1000 meV wide is formed across the Fermi level. This is large enough to stabilize one phase relative to competing phases. However, if a pseudogap is formed near the bottom of the valence band, the reduction in the electronic energy would be limited to only a few kJ/mol. As is clear from the above argument, electronic energy is most effectively lowered when a pseudogap is formed at the Fermi level.

Before ending this section, several examples of a system being stabilized by forming either a true or a pseudogap across the Fermi level are given. Different mechanisms act to generate a pseudogap below.

1. The presence of four outermost electrons in the free atom of Si and Ge facilitates formation of covalent bonding upon solidification into a diamond structure. That is, half-filled outermost electrons around a given atom tend to share the orbital with those of neighboring atoms. The resulting orbital hybridizations split the band into bonding and antibonding states and allow electrons to fill only the bonding states, resulting in an energy gap of about 1 eV across the Fermi level in Si. The formation of a true gap contributes to lowering the heat of mixing and stabilizing the diamond structure. As mentioned above, a gain in the cohesive energy relative to that of liquid Si in the metallic state is about 50 kJ/mol.

2. Some metals such as Al, V, Pb, etc., in the periodic table undergo the superconducting state at low temperatures by opening an energy

---

* A reduction in the electronic energy ΔU is calculated by using the relation:

$$\Delta U = \int\limits_{0}^{7-\Delta E} \{N_0(E) + \Delta N(E)\} + \int\limits_{7-\Delta E}^{7} N_{pg}(E) E dE - \int\limits_{0}^{7} N_0(E) E dE$$

where $N_0(E) = H_0 = 0.21$ states/eV.atom, $\Delta N(E)$ is an increment in DOS due to transfer of depleted electrons (marked with (B)) and $N_{pg}(E) = 0.21(H/H_0)$ states/eV.atom. A simple manipulation leads to the relation ΔU [kJ/mol] = 70.88 ΔE [eV] × (1 − (H/H_0)).

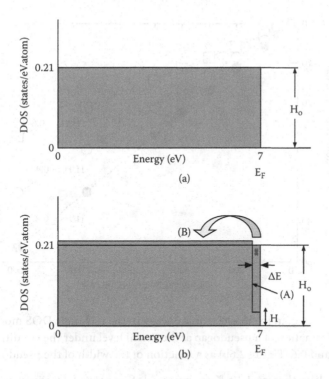

FIGURE 2.6   (a) DOS of Cu approximated by a rectangle with a height H_0. Free electron values are assigned to the Fermi level and the height of DOS at the Fermi level [from U. Mizutani, *MATERIA* (in Japanese), 45 No. 8 (2006) 605–610]. As a result, 1.4 electrons per atom are accommodated below the Fermi level. (b) A pseudogap with ΔE in width and H in height is created at the Fermi level. Depleted electrons in (A) are assumed to be uniformly redistributed over a whole valence band, as marked with (B).

gap of 0.5–1 meV at the Fermi level. This is brought about by forma-tion of the Cooper pair electrons. The superconducting energy gap is much smaller than the gap shown in Figure 2.6. According to the BCS theory, a gain in the cohesive energy for Al upon superconduct-ing transition amounts to

$$\frac{\left(k_B T_c\right)^2}{E_F} = 2.4 \times 10^{-9} \text{ eV/electron}$$

or 7.7×10^{-8} kJ/mol [3]. A small energy involved in the transforma-tion explains why the superconducting transition is generally very

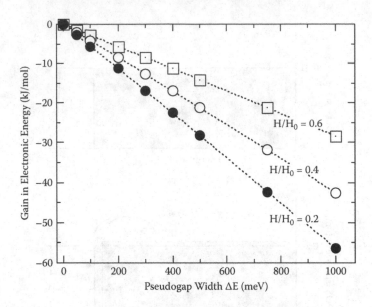

FIGURE 2.7   A gain in the electronic energy in the rectangle DOS model for Cu upon the formation of a pseudogap at the Fermi level under the conditions H/H_0 = 0.2, 0.4 and 0.6 (Figure 2.6b) as a function of the width of the pseudogap ΔE.

low, like that of 1.19 K for pure Al. See more basic information on superconductivity in Reference 3.

3. A highly anisotropic organic molecular metal characterized as a pseudo-one-dimensional conductor is known to be stabilized through the so-called Peierls transition [11]. Charge-density waves or spin-density waves are excited by formation of a periodic modulation through lattice deformation. As a result, a pseudogap is formed across the Fermi level and contributes to lowering the electronic energy. For example, $K_2Pt(CN)_4Br_{0.3} \cdot 3.2H_2O$, an inorganic compound known as KCP, opens a gap of 100 meV at the transition temperature of 189 K [11].

REFERENCES

1. C. Kittel, *Introduction to Solid State Physics* (Third edition, John Wiley & Sons, New York, 1967), chap. 3.
2. C.D. Gellat, Jr., A.R. Williams, and V.L. Moruzzi, *Phys. Rev.* B27 (1983) 2005.
3. U. Mizutani, *Introduction to the Electron Theory of Metals* (Cambridge Univ. Press, Cambridge, 2001).
4. N.W. Ashcroft and N.D. Mermin, *Solid State Physics* (Saunders College, West Washington Square, Philadelphia, PA 19105 (1976)), chap. 20, p. 410.

5. E. Wigner and F. Seitz, *Phys. Rev.* 43 (1933) 804.
6. R.W. Swalin, *Thermodynamics of Solids* (Second edition, John Wiley & Sons, New York, 1972).
7. H. Okamoto, *Phase Diagrams for Binary Alloys* (ASM International, OH, 2000).
8. B.L. Averbach, A. Flinn and M. Cohen, *Acta Met.* 2 (1954) 92.
9. F.R. de Boer, R. Boom, W.C.M. Mattens, A.R. Miedema, and A.K. Niessen, *Cohesion in Metals*, (North-Holland 1988).
10. U. Mizutani, *MATERIA* (in Japanese), 45 No. 8 (2006) pp. 605–610.
11. G. Gruner, *Density Waves in Solids* (Addison-Wesley Longmans, MA., 1994).

CHAPTER 3

Early Theories of Alloy-Phase Stability

3.1 MOTT-JONES MODEL FOR ALPHA-, BETA-, AND GAMMA-BRASSES

In the book published by Mott and Jones in 1936, they discussed how to approach the Hume-Rothery electron concentration rule by referring to the gamma-brass phase, which occurs at the ratio of 21 valence electrons to 13 atoms: the value of **e/a** is obviously equal to 21/13 for both Cu_5Zn_8 and Cu_9Al_4 gamma-brasses [1]. They assumed that the free energy against solute concentration would suddenly increase, as the concentration passes across the boundary of the phase and that it would be most likely caused by an increase in the electronic energy at absolute zero. Though they admitted that no precise calculation had yet been carried out, they proposed its critical concentration to be estimated from the FsBz interaction for a given phase. They tried to explain its mechanism by using a schematic DOS curve, as reproduced from [1] in Figure 3.1. A round maximum "A" with a subsequent rapid declining slope in the DOS was attributed to the FsBz interaction. When the Fermi surface approaches and touches the Brillouin zone planes, they could naturally assume a *grad* E in the energy dispersion to become very small and, hence, the DOS to be sharply enhanced. They considered the electronic energy of a system to rise rapidly, once electrons fill up the band to just beyond the point A, as shown by the shaded area.

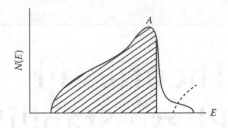

FIGURE 3.1 DOS in the Mott-Jones model [from N.F. Mott and H. Jones, *The Theory of the Properties of Metals and Alloys* (Oxford University Press, England, 1936)]. A symbol "A" indicates the peak in the DOS.

Note that Figure 3.1 was schematically drawn without specifying any particular FsBz interaction.*

The critical number of electrons per atom, $(e/a)_c$, was simply calculated in the free electron model under the assumption that it is given by the number of electrons filling a sphere inscribed to the Brillouin zone. The values of $(e/a)_c$ for the alpha-, beta-, and gamma-brasses are 1.362, 1.480, and 1.538, respectively, as illustrated in Figure 3.2. These values are indeed not too far from values of 1.4, 3/2, and 21/13 (=1.615) in the empirical Hume-Rothery electron concentration rule. This is one of the most important conclusions advanced by Mott and Jones in 1936. Obviously, these numbers were simply calculated in the free electron model in the presence of the Brillouin zone having a vanishing energy gap for the three crystal structures. This is a very simple model but did grasp in essence the physics involved. Its importance will be emphasized in Chapter 11.

3.2 THE MODEL OF JONES (I)

In 1937, Jones attempted to interpret the phase competition between the α- and β-phases in the Cu-Zn system within the framework of the two-wave approximation in the nearly free electron (NFE) model [2]. He essentially made a comparison of the valence-band structure energies between the face-centered cubic (fcc) and body-centered cubic (bcc) Cu as a function of electron concentration **e/a** in the context of the rigid-band model, which assumes that the addition of Zn to increase **e/a** does not change the DOS of the parent Cu with the fcc and bcc structures. The DOSs of

---

* Mott and Jones [1] did not explicitly mention how the round maximum "A" in the DOS was derived. As will be discussed in Chapter 3, Section 3.2 (see Figures 3.3 and 3.4), Jones later replaced it by a sharp cusp derived from the nearly-free electron (NFE) model, which was formulated by Bethe in 1928. See Figure 5.10 (a), where more realistic DOSs calculated from first-principles band calculations are shown.

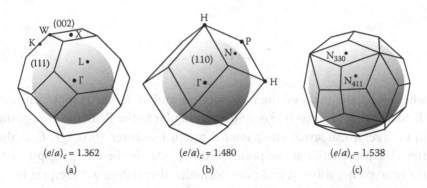

$(e/a)_c = 1.362$

(a)

$(e/a)_c = 1.480$

(b)

$(e/a)_c = 1.538$

(c)

FIGURE 3.2 The FsBz interaction in the Mott and Jones theory. A critical value of $(e/a)_c$ is obtained when the spherical Fermi surface touches the zone plane of the respective Brillouin zones: (a) the principal symmetry points L in the fcc Brillouin zone, (b) the principal symmetry points N in the bcc Brillouin zone, and (c) the symmetry points N_{330} and N_{411} in the Brillouin zone for the gamma-brass structure. Note that the distance from the origin to the points N_{330} and N_{411} is the same and equal to $\sqrt{18}/2$ in units of $2\pi/a$, where a is the lattice constant.

the valence band for the fcc- and bcc-Cu calculated by Jones are repro-duced in Figure 3.3a, along with the **e/a** dependence of the valence-band structure energy difference in Figure 3.3b, which will be described later. A large cusp found in the respective DOSs represents the so-called van Hove singularity, which Jones believed to be responsible for the interpretation of the Hume-Rothery electron concentration rule.* Note that the parameter **e/a** but not **VEC** can be directly obtained by integrating the DOS shown in Figure 3.3 in the model of Jones (I).

By applying the NFE model, Jones could express the energy of the valence electron in the neighborhood of the center of the zone boundary by equations:

$$E = \left(\hbar^2/2m\right)\left[k_x^2 + k_y^2 + k_o^2 + \alpha\left(k_z - k_o\right)^2\right] - \Delta E/2$$

$$= E_o\left[x^2 + y^2 + 1 + \alpha\left(z-1\right)^2\right] - \Delta E/2 \qquad (3.1)$$

$$= E_o\left[x^2 + y^2 + 1 + \left(1-z\right)^2 - 2\sqrt{\left(1-z\right)^2 + \frac{1}{\left(1-\alpha\right)^2}}\,\right]$$

---

* The anomaly in the DOS caused by the FsBz interaction is referred to as the van Hove singularity.

and

$$\alpha = 1 - 4E_o / \Delta E \qquad (3.2)$$

where m is the mass of the free electron and $\hbar$ is the Planck constant divided by 2π, k_z-axis is chosen perpendicular to the Brillouin zone plane in the reciprocal space and passes through its center $(0,0,k_o)$, E_o is the free electron energy at the point $(0,0,k_o)$, ΔE is the energy gap across the zone plane, and x, y, and z are normalized variables with respect to k_o. The point $(0,0,k_o)$ refers to the principal symmetry points L and N at the center of the {111} and {110} zone planes in the Brillouin zone of the fcc and bcc lattices, respectively. Equation 3.1 indicates that the free electron behavior is maintained, as far as the electronic states (x,y,z) are away from $(0,0,1)$, but, as z approaches unity, the effect of the zone boundary is progressively increased through the term α connected to the energy gap. This is a well-known feature in the NFE model [3].

In the case of the fcc alpha-phase, Jones took into account only the effect of the {111} zone in Equation 3.1. Hence, one can immediately understand the cusp at about 6.6 eV in Figure 3.3a to be caused by the contact of the Fermi surface with the {111} zone planes of the fcc Brillouin zone. Jones

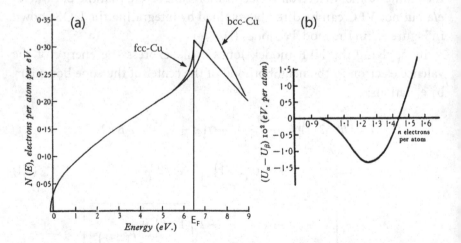

FIGURE 3.3 (a) DOS derived from the model of Jones (I) for the fcc- and bcc-Cu. (b) e/a-dependence of the valence-band structure energy difference between these two phases in the Cu-Zn alloy system. [From H. Jones, *Proc. Phys. Soc.* 49 (1937) 250.]

adopted the energy gap across the {111} zone planes to be 4.1 eV for pure Cu and assumed it to remain unchanged throughout the α-phase. It happened that the Fermi level is located just at the {111} peak for pure Cu by filling one electron per atom into the DOS thus calculated.

The determination of the DOS for the hypothetical bcc-Cu is inevitably uncertain. As can be seen in Figure 3.3a, the two DOSs are identical to each other in the parabolic region, indicating the validity of the free-electron model. The DOS in the free-electron model is expressed in the form [3]:

$$D(E) = \left(\frac{V_a}{2\pi^2}\right)\left(\frac{2m}{\hbar^2}\right)^{3/2}\sqrt{E} \qquad (3.3)$$

where $D(E)$ represents the DOS and V_a is the volume per atom. A perfect coincidence of two parabola over the energy range below about 5 eV for the fcc- and bcc-Cu shown in Figure 3.3a indicates that Jones implicitly assumed $V_a^{fcc} = V_a^{bcc}$ in Equation 3.3.

The volumes per atom V_a in fcc and bcc lattices are obviously given by $a_{fcc}^3/4$ and $a_{bcc}^3/2$, respectively, where a is the lattice constant. The preceding condition immediately leads to the relation:

$$a_{fcc} = 2^{1/3} a_{bcc} \qquad (3.4)$$

However, there is no guarantee in validating relation 3.4, though its assumption is critically important to allow Jones to attribute the valence-band structure energy difference solely to the difference in the van Hove singularities. As discussed in Chapter 5, first-principles band calculations generally do not assume relation 3.4 and, instead, the lattice constant is determined so as to minimize the total-energy. Indeed, the FLAPW band calculations for fcc- and bcc-Cu reveal that the relation (3.4) is fairly well satisfied but is not absolutely obeyed (see Chapter 5, Section 5.2).

The van Hove singularity in the bcc-Cu is caused by the contact of the Fermi surface with the {110} zone planes. As can be seen in Figure 3.3a, the {110} peak in the DOS of the bcc-Cu is located about 1 eV at higher energy than the {111} peak in the fcc-Cu. This is essential in the model of Jones to account for the e/a-dependent stability of the α- and β-phases. The reason why the {110} peak of the β-phase appears at higher energy relative to the

{111} peak of the α-phase can be easily understood by using Equation 3.4. First of all, Jones assumed the energy gap across {110} zone planes in bcc-Cu to be the same as that across the {111} zone planes in fcc-Cu, though this is merely an assumption. Now, the distances from the origin to the center of {111} and {110} zone planes are given by $\pi\sqrt{3}/a_{fcc}(=5.441/a_{fcc})$ and $\pi\sqrt{2}/a_{bcc}(=4.442/a_{bcc})$, respectively, [3]. Since we have the relation (3.4), the latter in the bcc-Cu is rewritten as $5.596/a_{fcc}$. This is obviously longer than that in the fcc-Cu. We see that the relative position of the van Hove singularities for the fcc- and bcc-Cu shown in Figure 3.3a was derived under such crude assumptions.

Jones tried to explain the α/β phase transformation by calculating the valence-band structure energy U of the respective phases. The value of U is calculated by integrating the DOS shown in Figure 3.3a times energy E up to the energy corresponding to the highest occupied states. The valence-band structure energy difference between these two phases, ΔU, is given as a function of e/a by

$$\Delta U(e/a) = \int_0^{E_{fcc}} E D_{fcc}(E) dE - \int_0^{E_{bcc}} E D_{bcc}(E) dE \qquad (3.5)$$

where the upper limit of the integral E_{fcc} or E_{bcc} is the energy for fcc or bcc phases when the same number of electrons is filled into the respective DOSs shown in Figure 3.3a. Hence, $E_{fcc\ or\ bcc}$ is directly linked with the variable e/a through the relation:

$$e/a = \int_0^{E_{fcc}} D(E)_{fcc} dE \qquad (3.6a)$$

and

$$e/a = \int_0^{E_{bcc}} D(E)_{bcc} dE \qquad (3.6b)$$

The resulting e/a dependence of the valence-band structure energy difference $\Delta U = U_\alpha - U_\beta$ is plotted in Figure 3.3b. The difference is negative up to about $e/a = 1.45$ as a result of a rapid increase in the fcc-DOS due to the {111} peak. However, when e/a is increased above about 1.2, the bcc-DOS can accommodate more electrons than the fcc-DOS, since the fcc-DOS begins to sharply drop, whereas the bcc-DOS is still increasing due to the approach to the {110} peak. As can be seen in Figure 3.3b, the valence-band structure energy of the β-phase becomes lower than that of the α-phase above about $e/a = 1.45$. This beautifully explained the Hume-Rothery electron concentration rule concerning the α/β phase transformation. We see from Figure 3.3b that the magnitude of ΔU involved is merely of the order of $\pm 1 \times 10^{-2}$ eV/atom or ± 1 kJ/mol.

As mentioned in Chapter 1, the difficulty in the model of Jones was seriously recognized only after Pippard discovered the neck in the Fermi surface of pure Cu in 1957 [4]. As described earlier, there exist many unjustified assumptions in the model of Jones, which were made to conveniently account for the Hume-Rothery rule. Among them, we must note here only the most fundamental difficulties in the model of Jones:

1. The model is not consistent with the possession of the neck in the Fermi surface contour of pure Cu. The drawback originated from the choice of e/a as the electron concentration parameter. In first-principles band calculations discussed after Chapter 4, the integration of the DOS naturally results in **VEC**, which includes not only s- and p-electrons but also d-electrons forming the valence band. The parameter **VEC** instead of e/a should be employed in realistic electronic structure calculations to take into account the d-electron contribution.

2. The application of the rigid-band model for alpha- and beta-phase Cu-Zn alloys is too naïve to be justified. As we discussed in Chapter 2, we have to discuss a very subtle energy difference of only a few to a few tens kJ/mol or a few hundreds meV/atom between two competing phases. At present, we can say that the accuracy in first-principles band calculations for a perfectly periodic system has reached the level of satisfaction. In the case of disordered alloys, the coherent-potential approximation (CPA) and average t-matrix approximation (ATA) have been developed as a tool available to describe their electronic structures. Indeed, the composition dependence of the Fermi surface properties, optical properties and angle-resolved photoemission spectra in noble metal alloys has been successfully explained [5–8]. However, to the best of knowledge, the application of the CPA

technique to the discussion on the phase stability like that between the alpha- and beta-phase Cu-Zn alloys with the accuracy of detecting ΔU of only a few kJ/mol has not been successfully made. This is partly because an imaginary part of the Bloch wave vector cannot be ignored in a disordered alloy, which certainly gives rise to the smearing of fine structures in the DOS and masks the van Hove singularities. At present, a practical compromise to work on a concentrated alloy would be a choice of a rigid-band model or the virtual crystal approximation (VCA) or the super-cell approximation.*

3.3 THE MODEL OF JONES (II)

It may be worthwhile mentioning at this stage another model of Jones for the competition of the two relevant phases. As shown in Figure 3.4, Jones (1962) discussed the phase competition in terms of the DOS curve characterized by the van Hove singularity near the Fermi level relative to the free-electron-like monotonic DOS [9]. He again assumed the range of stability to be determined solely by the difference between the respective valence-band structure energies of two phases 1 and 2 in the same way as given by Equations 3.5, 3.6a, and 3.6b:

$$\Delta U = U_1 - U_2 = \int_0^{E_1} D_1(E)E\,dE - \int_0^{E_2} D_2(E)E\,dE \qquad (3.7)$$

where E_1 and E_2 are linked with a given **e/a** value through the relation:

$$e/a = \int_0^{E_1} D_1(E)\,dE; \qquad e/a = \int_0^{E_2} D_2(E)\,dE \qquad (3.8)$$

---

* The VCA assumes a perfectly periodic array of ionic potentials given by $V_{alloy} = c_A V_A + (1 - c_A)V_B$ for an A-B alloy, where c_A and V_A are the concentration of the atom A and its potential, respectively. The VCA model is claimed to be reasonable when the elements A and B are close to each other in the periodic table. The super-cell approximation for an A-B alloy distributes c_A% of the atom A and c_B% of the atom B randomly in a super-cell, say, containing 100 atoms per unit cell. Now a system having the super unit cell is perfectly periodic and, hence, first-principles band calculations based on the Bloch theorem in the reciprocal space can be performed.

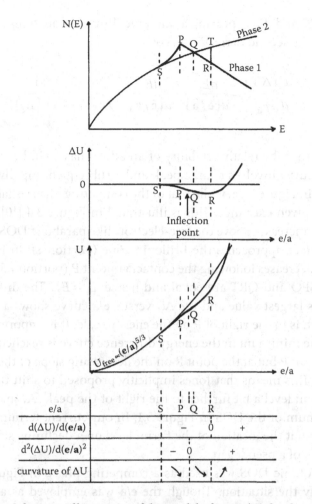

FIGURE 3.4 Schematic illustration of the model of Jones (II) for the competition between phases 1 and 2 [from T.B. Massalski and U. Mizutani, *Prog. Mat. Sci.* 22 (1978) 151]. The van Hove singularity in Figure 3.3a derived from the NFE model for the fcc or bcc phase is assumed for phase 1, whereas the free electron-like monotonic DOS is assumed for phase 2.

The first derivative of Equation 3.7 with respect to **e/a** is given by

$$\frac{d(\Delta U)}{d(e/a)} = E_1 D(E_1) \frac{dE_1}{d(e/a)} - E_2 D(E_2) \frac{dE_2}{d(e/a)} = E_1 - E_2 \qquad (3.9)$$

which shows that, at any given **e/a**, ΔU between the two competing phases changes at a rate equal to the difference in their respective maximum energies E_1 and E_2 appearing as an upper limit of Equations 3.7 and 3.8. Similarly, the second derivative becomes

$$\frac{d^2(\Delta U)}{d(e/a)^2} = \frac{dE_1}{d(e/a)} - \frac{dE_2}{d(e/a)} = \frac{1}{D_1(E_1)} - \frac{1}{D_2(E_2)} \qquad (3.10)$$

It follows that the relative stability of an alloy phase will be enhanced if the DOS curve involves a large peak and a subsequent rapidly declining slope like in Figure 3.3a, while that of the competing phase is fairly monotonic in a given **e/a** range. This is illustrated in Figure 3.4 [10]. The DOS in phase 1 increases above the free-electron-like parabolic DOS, when the Fermi surface approaches the Brillouin zone (portion SP in Figure 3.4), and then decreases following the contact at point P (portion PR). At point R, areas SPQ and QRT are equal and hence $E_1 = E_2$. The difference ΔU reaches its largest value and the ΔU versus **e/a** curve shows a minimum at R, which is to the right of P on the energy scale. It is important to realize that the minimum in the energy difference curve is reached not at the contact point P but at the point R on the decreasing slope of the DOS past the peak. This means that Jones implicitly proposed to shift the position of the Fermi level a bit further to the right of the peak "A," namely, near the minimum of the DOS in Figure 3.1. In our current terminology, this is nothing but the location of the Fermi level on a declining slope toward the bottom of a pseudogap.

A monotonic DOS was used as a competing phase in Figure 3.4 just to simplify the situation. Though the **e/a** was employed as an electron concentration parameter in the model of Jones (II), we can equally apply Equations 3.7 to 3.10 by using the **VEC** as an electron concentration parameter for any two realistic phases involving a d-band. In Chapter 5, we will try to interpret the α/β-phase transformation in the Cu-Zn alloy system by making full use of the electronic structure derived from the FLAPW band calculations for fcc- and bcc-Cu and the model of Jones (II).

REFERENCES

1. N.F. Mott and H. Jones, *The Theory of the Properties of Metals and Alloys* (Oxford University Press, England, 1936).
2. H. Jones, *Proc. Phys. Soc.* 49 (1937) 250.
3. U. Mizutani, *Introduction to the Electron Theory of Metals* (Cambridge University Press, Cambridge, 2001).

4. A.B. Pippard, *Phil. Trans. R. Soc.* A250 (1957) 325.
5. A. Bansil, H. Ehrenreich, L. Schwartz, and R.E. Watson, *Phys. Rev.* B 9 (1974) 445.
6. R. Prasad, S.C. Papadopoulos, and A. Bansil, *Phys. Rev.* B 23 (1981) 2607.
7. R.S. Rao, R. Prasad, and A. Bansil, *Phys. Rev.* B 28 (1983) 5762.
8. E. Arola, R.S. Rao, and A.Bansil, *Phys. Rev.* B 41 (1990) 7361.
9. H. Jones, *J. Phys. Radium*, Paris, 23 (1962) 637.
10. T.B. Massalski and U. Mizutani, *Prog. Mat. Sci.* 22 (1978) 151.

CHAPTER 4

First-Principles Band Calculations Using the Muffin-Tin Potential

4.1 FIRST-PRINCIPLES BAND CALCULATIONS VERSUS THE HUME-ROTHERY ELECTRON CONCENTRATION RULE

As discussed in Chapter 3, the pioneering theory put forward by Mott and Jones in 1936 [1] certainly served as a milestone in establishing a basic idea for interpreting the Hume-Rothery electron concentration rule, which relates phase stability to the Fermi surface. However, since the discovery of the neck in the Fermi surface contours of pure Cu by Pippard in 1957, one could not help but recognize that the free-electron model Mott and Jones relied on is far from being satisfactory. Needless to say, the presence of the neck in pure Cu cannot be reproduced from the free-electron model.*

Moreover, recent research after the 1990s on the stability of quasicrystals and their approximants, discovered by using the Hume-Rothery electron concentration rule as a guide, has gradually built up a general consensus such that the stability of such structurally complex metallic alloys (CMAs) is also most likely a consequence of lowering the electronic energy brought

---

* This does not mean that the free-electron model is of no use. Instead, we will emphasize at the very end of conclusions in Chapter 11 that the Mott and Jones theory based on the free electron model did grasp the essence behind the Hume-Rothery electron concentration rule in many CMAs.

about by the development of a deep pseudogap at the Fermi level. Since a pseudogap cannot be generated from the free-electron model, first-principles band calculations must be employed as a key tool to evaluate quantitatively the valence band structure and to show why a pseudogap is formed near the Fermi level in CMAs.

There are two possible mechanisms for the formation of a pseudogap at the Fermi level: orbital hybridizations and Fermi surface-Brillouin zone (FsBz) interactions. We will explain in this Chapter 4 why the LMTO-ASA (Linear Muffin-Tin Orbital-Atomic Sphere Approximation) method is best suited for extracting orbital hybridizations, while the FLAPW (Full-Potential Linearized Augmented Plane Wave) method is best suited to elucidate the FsBz interactions. These approaches involve many mathematical formulations to explain how these two first-principles band calculation methods can be utilized for our purpose. Readers who are not acquainted with advanced treatments of quantum mechanics may wish to skip Sections 4.4 to 4.12. It may also be helpful to start with textbooks on electron theory of metals [2,3], whenever needed.

4.2 ORIGIN OF THE PSEUDOGAP: ORBITAL HYBRIDIZATIONS VERSUS FsBz INTERACTIONS

The origin of the pseudogap at the Fermi level can be discussed from two different approaches: one from covalent bonding and the other from metallic bonding. Let us consider the former approach by considering an Al-Mn alloy. Here, we assume a situation such that both Al and Mn atoms are placed a few one-tenth of a nanometer apart, corresponding to an average atomic distance in the alloy. If the Al-3p and Mn-3d energy levels are close to each other, the two atomic wave functions will overlap with one another. This is called the orbital hybridization and results in bonding and antibonding levels, as illustrated in Figure 4.1a. This also holds true upon forming a solid (i.e., an Al-Mn alloy phase). The bonding and antibonding levels will naturally be broadened into the respective bands, leaving a pseudogap in between bonding and antibonding subbands formed by Mn-3d states mixed with Al-3p states. Figure 4.1b shows the DOS for the Al-Mn approximant containing 138 atoms in its cubic unit cell, which was calculated by Fujiwara in 1989 [4]. Gaussian curves are roughly drawn in Figure 4.1b as eye guide to represent the resulting Al-3p/Mn-3d bonding and antibonding subbands. The Fermi level is found to fall inside the pseudogap created between the bonding and antibonding Mn-3d subbands.

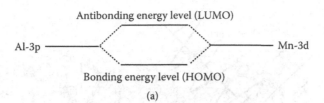

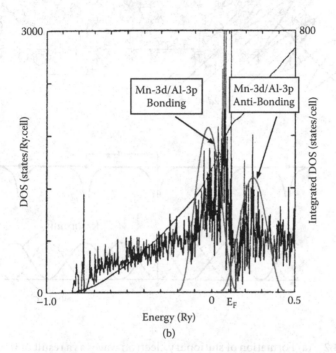

FIGURE 4.1 (a) Formation of bonding and antibonding levels due to orbital hybridization between neighboring Al-3p and Mn-3d atomic wave functions. (b) Formation of a pseudogap between bonding and antibonding subbands due to orbital hybridizations between Al and Mn atoms in an Al-Mn approximant [from T. Fujiwara, *Phys.Rev.* B 40 (1989) 942]. Gaussian curves are roughly fitted to the subband profiles as eye guide.

The Al-Mn approximant mentioned above is favorably stabilized, since the bonding band is almost fully filled by electrons, whereas the antibonding band remains almost empty. This is the stabilization mechanism due to orbital hybridizations. Here it is kept in mind that it has little to do with the interference phenomenon described below and, hence, does not involve a parameter directly pertaining to the electron concentration **e/a** (see Chapter 10, Section 10.7). Instead, a parameter determining the Fermi level must be a dominant factor. This is the total number of electrons per

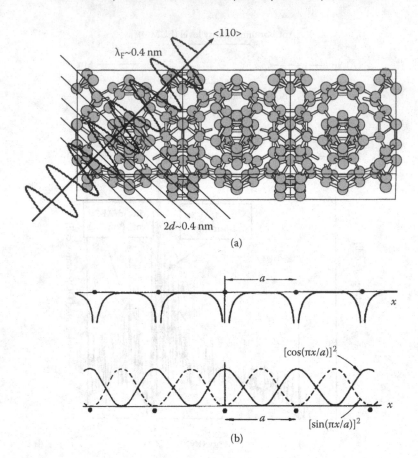

FIGURE 4.2 (a) Formation of stationary electron waves as a result of interference with periodically arranged ionic potentials. (b) Formation of either cosine- or sine-type stationary waves in one-dimensional periodic ion potential field. [From C. Kittel, *Introduction to Solid State Physics* (Third edition, John Wiley & Sons, New York, 1967).]

atom, **VEC**, in the valence band (see the difference between **e/a** and **VEC** in Section 1.1 and more details in Chapter 10.).

We have an alternative mechanism to generate a pseudogap at the Fermi level. This is due to the FsBz interaction, which is approached from the metallic bonding picture. Let us consider itinerant electrons propagating throughout a periodically arranged ionic potential field. Stationary waves will be formed when the wavelength of the electron wave matches the period of the ionic potential. The situation in real space is illustrated in Figure 4.2a. This is called the *interference phenomenon* and is equivalent to the fulfillment of the Bragg law. For the sake of simplicity, consider an

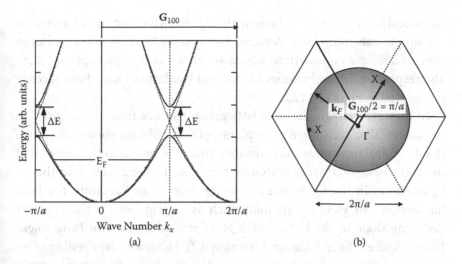

FIGURE 4.3 (a) Formation of an energy gap at the wave number $k = \pm a/\pi$ obtained by bisecting the reciprocal lattice vector $\mathbf{G}_{100}$ in a simple cubic lattice. The Fermi level is located well below the energy gap. (b) The Brillouin zone for a simple cubic lattice. The Fermi surface with the Fermi radius $\mathbf{k}_F$ corresponding to the electronic structure in (a) is schematically drawn.

electron propagating through a one-dimensional periodic potential field. Either a sine- or cosine-type stationary wave is formed, when the wave number of electron reaches $k = \pm\pi/a$, as shown in Figure 4.2b [2,3].* A closer look into Figure 4.2b indicates that the electronic energy of the cosine-type stationary wave must be lowered relative to the free electron value, since its charge density is the highest at the center of the ion, where the potential is the lowest. The opposite is true for the sine-type stationary wave. This is the mechanism for the formation of an energy gap at the wave number $k = \pm\pi/a$ in the energy dispersion relation.

Figure 4.3a illustrates the opening of an energy gap at the wave vector obtained by bisecting the reciprocal lattice vector $\mathbf{G}_{100}$ (i.e., $k = \pm|\mathbf{G}_{100}|/2 = \pm\pi/a$ in a simple cubic lattice). A polyhedron is constructed in the reciprocal space by perpendicularly bisecting six equivalent

---

* The Bloch wave function in one-dimensional periodic array of ionic potentials is given by a linear combination of the unperturbed plane wave $A_0 e^{ikx}$ and the wave $A_1 e^{i[k-(2\pi/a)]x}$ perturbed by the set of lattice planes: $\psi(x) = \exp(ikx)[A_0 + A_1 \exp\{-i(2\pi/a)x\}]$. At $k = \pi/a$, the relation $A_0 = \pm A_1$ holds and the wave function is reduced to either $\sin(\pi x/a)$ or $\cos(\pi x/a)$. The wave of $k = \pi/a$ is reflected to the wave of $k' = -\pi/a$ by receiving a crystal momentum $G = -2\pi/a$ from the lattice planes and the reflected wave of $k = -\pi/a$ is again reflected to the wave of $k' = \pi/a$ by receiving a crystal momentum $G = 2\pi/a$ from the lattice planes. This process is infinitely repeated, resulting in a cosine- or sine-type stationary wave [2].

reciprocal lattice vectors. A cube with edge length of $2\pi/a$ is formed for the simple cubic lattice, as depicted in Figure 4.3b. This is the Brillouin zone of a simple cubic lattice, across which an energy gap appears [2,3]. The respective Brillouin zones for fcc and bcc lattices have been already introduced in Figures 3.2a,b.

In the case of a simple cubic lattice, the distance from the origin Γ to the center X of six square zone planes of its Brillouin zone is the same (Figure 4.3b). In contrast, the distance from the origin to the center L of the eight hexagonal faces is shorter than that to the center X of the six square faces in the fcc-Brillouin zone (Figure 3.2a). Certainly, the FsBz interaction can yield an anomaly such as a cusp, with a subsequently declining slope in the DOS, which is referred to as the *van Hove singularity* (Figure 3.3a in Chapter 3, Section 3.2). However, the opening of an energy gap across different sets of zone planes would not necessarily result in an energy gap in the DOS.* Only a small cusp followed by a rather shallow dip is generally created in the DOS in structurally simple fcc, bcc, or hcp metals and alloys. In CMAs, however, the number of zone planes of the Brillouin zone increases, as exemplified, for example, for gamma-brass in Figure 3.2c. Now we often encounter a rather unique situation in which a deep valley is formed in the DOS, but the electronic states remain finite along this minimum. This has been already referred to as a *pseudogap* (see Chapter 1, Section 1.2 and Chapter 2, Section 2.3). As emphasized in Chapter 2, Section 2.3, its contribution to the stability of a solid would become the most effective, when it is formed across the Fermi level. A FsBz-induced pseudogap plays a key role in stabilizing a complex metallic structure, in which the diffraction spectrum consists of a series of Bragg peaks. Included are not only crystalline metals and alloys but also quasicrystals, which exhibit a more marked pseudogap the better ordered they are [2].

Another key issue must be addressed in relation to a FsBz-induced pseudogap. An interference phenomenon of electrons with a given set of lattice planes discussed above does not necessarily occur at the Fermi level. In other words, it is possible to have a pseudogap off from the Fermi level (see Chapter 8, Section 8.4). But we are most interested in a system, where a pseudogap is formed just across the Fermi level. This is what we call a FsBz-induced pseudogap system.

---

* See more detailed discussion in Section 5.10 in [2].

Let us suppose the interference phenomenon above to occur at the Fermi level and to be strong enough to cause a sizable pseudogap. Now the situation is envisaged such that the effective Fermi sphere with the diameter $2k_F$ is in contact with the relevant Brillouin zone plane characterized by the reciprocal lattice vector $|\mathbf{G}|$. Figure 3.2 illustrates this situation in the free electron model. This immediately leads to the condition:

$$2k_F = |\mathbf{G}| \tag{4.1}$$

Equation 4.1 has been often referred to as the *Hume-Rothery matching condition*. It is important to keep in mind that Equation 4.1 is no longer based on the free electron model. In Chapters 7 to 10, we will introduce an elegant technique on the basis of first-principles FLAPW band calculations to determine the effective Fermi diameter $2k_F$ and the *critical* reciprocal lattice vector $|\mathbf{G}|$ involved in Equation 4.1 in alloys, regardless of whether the transition metal (TM) element is involved or not. A rigorous test of the matching condition given by Equation 4.1 for alloys, including many CMAs, will be discussed in Chapter 10.

In the past two decades, expressions such as the "Hume-Rothery stabilization mechanism" and/or "Hume-Rothery-type stabilization" have been frequently employed, particularly upon discussing the stability of quasicrystals [5–7]. However, many people have used these phrases, whenever a pseudogap is experimentally found, or theoretically predicted, in a CMA without differentiating between the two origins: the FsBz and the orbital-hybridizations. A very careful discussion is needed to extract the role of the FsBz interactions on the formation of a pseudogap in systems, in which orbital hybridizations are predominant. Detailed discussions will be made on this critical issue in Chapters 8 to 10.

4.3 WHAT ARE FIRST-PRINCIPLES BAND CALCULATIONS?

As discussed in Chapter 2, Section 2.1, degenerate 3s-levels in an assembly of molar Na free atoms are "lifted" or are split into slightly different energies upon formation of solid Na. This is the formation of the valence band in metallic bonding. Itinerant electrons in the valence band move in a lattice while interacting with each other via the Coulomb force. Rigorously speaking, we must treat the electron-electron interaction in the context of the "many-body problem," which cannot be analytically solved. The motion of an electron in a metal has been treated in the so-called one-electron approximation, under which each electron independently propagates in an effective averaged potential.

Both Hartree and Hartree–Fock approximations had been employed to construct an effective one-electron potential until 1964, when the density functional theory (DFT) has been established [8–10]. This brought us a substantial progress in the reliability of one-electron band calculations. According to the DFT, the total-energy of an electron running in an effective potential field is given as the functional of electron density. Indeed, Kohn and Sham [9] provided the method of calculating the total-energy of a system by treating the exchange-correlation energy of the electron in the local density approximation (LDA). Since details about the DFT-LDA theory will be found in the literature [8–10], we simply note that the total-energy of a system at absolute zero [9] is expressed as

$$
U_{total} = \sum_i \varepsilon_i - \frac{1}{2} \iint \frac{n(\mathbf{r})n(\mathbf{r}')}{|\mathbf{r}-\mathbf{r}'|} d\mathbf{r}d\mathbf{r}'
$$
$$
+ \int n(\mathbf{r})\Big[\varepsilon_{XC}\big(n(\mathbf{r})\big) - \mu_{XC}\big(n(\mathbf{r})\big)\Big]d\mathbf{r}
$$

(4.2)

where ε_i is the solution of an effective one-electron Schrödinger equation given by

$$
\left[-\frac{\hbar^2}{2m}\nabla^2 + v_{eff}(\mathbf{r})\right]\psi_i(\mathbf{r}) = \varepsilon_i\psi_i(\mathbf{r})
$$

(4.3)

The first term in Equation 4.2 represents the one-electron band structure energy due to both valence and core electrons, whereas the second term is often referred to as the *Hartree term*, representing the Coulomb potential energy due to the nucleus-electron interaction plus an average electron-electron interaction energy, and the third term represents the exchange-correlation energy derived in the LDA. More recently, the exchange-correlation energy functional of electrons has been calculated within the generalized gradient approximation coupled with the Perdew–Burke–Ernzerhof hybrid scheme (GGA-PBE) [11]. Obviously, the first term due to the contribution from valence electrons in Equation 4.2 is essentially equivalent to the electronic energy of valence electrons given by the first or second term in Equation 3.5.

The expression "first-principles" in first-principles band calculations refers to the method of calculating the electronic structure by solving a one-electron Schrödinger equation without relying on any experimental results under the assumption that an electron experiences an effective potential within the framework of the DFT-LDA theory. It is also called *ab initio band calculations*, since the Latin term *ab initio* means "from the beginning" in English. There are two different approaches in first-principles band calculations: one is the pseudopotential method, which allows us to employ the set of plane waves as basis functions, and the other an all-electron method, which treats all electrons, including core states, explicitly. In the pseudopotential approach, the potential becomes very smooth so that plane waves can be safely used as basis functions. But ignorance of the core states poses difficulties not only in discussing issues related to core states, such as core excitations and core level shifts, but also in accurately treating d-electrons having a higher tendency of localization in the valence band.

The all-electron method employs a one-electron effective potential determined by both nuclei and all electrons and, hence, s-, p-, and d-electrons are treated on the same ground. There are several all-electron first-principles band calculation methods, all of which basically employ a spherically symmetric muffin-tin (MT) potential within a sphere centered at a given nucleus with radius a, and a constant potential outside it to approximate an ionic potential in a given system. The sphere is called the *muffin-tin (MT) sphere*. The wave function inside the MT sphere can be rigorously solved as the product of the radial wave function and spherical harmonics. As will be described below, Augmented Plane Wave (APW), Korringa-Kohn-Rostoker (KKR), and muffin-tin orbital (MTO) methods are typical of all-electron first-principles band calculation methods. In particular, we shall focus on LMTO-ASA and FLAPW methods with an emphasis on what information can be extracted from them to analyze the Hume-Rothery electron concentration rule. Both methods employ an effective potential constructed within either the DFT-LDA or DFT-GGA-PBE theory to treat the electron-electron interaction.

4.4 ALL-ELECTRON BAND CALCULATIONS WITHIN THE MUFFIN-TIN APPROXIMATION

Wigner and Seitz [12] divided a bcc crystal into the smallest volume enclosed by planes bisecting the interatomic distances. Their operation is illustrated in Figure 4.4a, using atoms in the plane. A resulting polyhedron is called the *truncated octahedron* and can fill the space without any

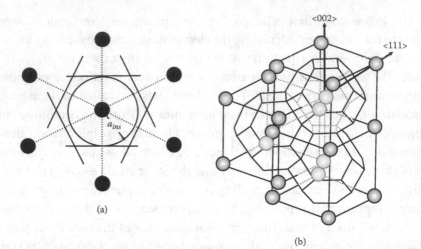

FIGURE 4.4 (a) Construction of the Wigner–Seitz cell and the MT sphere inscribed in the cell. Solid circle symbolizes the atom on the lattice. (b) The Wigner–Seitz cell for a bcc lattice.

overlap or void (see also Chapter 6, Section 6.3). Obviously, it contains a single atom at its center. This is called the *Wigner–Seitz cell* of a bcc lattice. Its structure is shown in Figure 4.4b. They assumed spherical symmetry of an ionic potential in a crystal to hold up to the boundary of the Wigner–Seitz cell and expressed the wave function of an electron as

$$\psi(\mathbf{k},\mathbf{r}) = \sum_{\mathbf{R}} e^{i\mathbf{k}\cdot\mathbf{R}} \sum_{\ell m} b_{\ell m}^{j\mathbf{k}} \theta(\mathbf{r}-\mathbf{R}) i^\ell Y_{\ell m}(\widehat{\mathbf{r}-\mathbf{R}}) R_\ell\left(E, \left|\mathbf{r}-\mathbf{R}\right|\right) \quad (4.4)$$

where $\mathbf{r}$ and $\mathbf{R}$ are coordinates of an electron and a nucleus, respectively, $\mathbf{k}$ is the wave vector of the Bloch electron, ℓ and m are azimuthal and magnetic quantum numbers of partial waves,* $R_\ell(E, \left|\mathbf{r}-\mathbf{R}\right|)$ is the radial wave function of the electron having an energy E, $Y_{\ell m}(\widehat{\mathbf{r}-\mathbf{R}})$ is a spherical harmonic, and the function θ takes unity inside the Wigner–Seitz cell and otherwise zero.† In the Wigner–Seitz method, the coefficient $b_{\ell m}^{j\mathbf{k}}$ is determined in such a way that Equation 4.4 is continuous and differentiable

---

* A spherical coordinate representation is best suited to describe the motion of an electron in a spherically symmetric potential field. Its quantized motion can be described in terms of four quantum numbers: principal quantum number n, azimuthal quantum number ℓ, magnetic quantum number m, and spin quantum number s. A spherical wave specified by azimuthal quantum number ℓ and magnetic quantum number m is called the partial wave.

† A symbol ⌢ over $\mathbf{r}$–$\mathbf{R}$ represents an angular variable (θ, φ) of the vector $\mathbf{r}$–$\mathbf{R}$.

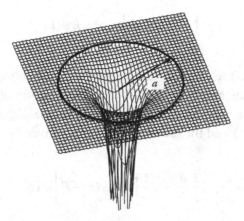

FIGURE 4.5 A single MT potential located at **R=R**. A spherically symmetric potential is assumed inside the MT sphere with radius a and a constant potential outside it. [From U. Mizutani, *MATERIA* (in Japanese), 45, No. 9 (2006) 677.]

across the cell boundary. However, it is almost impossible to apply these boundary conditions rigorously. To avoid this difficulty, Slater [13] introduced the concept of the MT potential and developed the Augmented Plane Wave (APW) method in 1937.

For example, an ionic potential is assumed to be spherically symmetric within a sphere with radius a_{ins} inscribed in the Wigner–Seitz cell, as illustrated in Figure 4.4a. The potential outside the sphere is assumed to be constant. However, the choice of a radius of the MT sphere is rather arbitrary. Its radius in the Atomic Sphere Approximation (ASA) discussed in Section 4.7 will be given by that of an atomic sphere whose volume is equal to that of the Wigner-Seitz cell. A much smaller radius is employed in APW and FLAPW methods. For this reason, the radius of the MT sphere is hereafter simply denoted as a, unless otherwise stated.

To begin with, we consider a single ionic potential located at a lattice site $\mathbf{R} = \mathbf{R}$, as illustrated in Figure 4.5 [14]. Now a spherically symmetric potential at the position $\mathbf{r}$ may be explicitly expressed as

$$v(\mathbf{r})=\begin{cases} v(|\mathbf{r}-\mathbf{R}|) & |\mathbf{r}-\mathbf{R}|\leq a \\ V_{MTZ} & |\mathbf{r}-\mathbf{R}|>a \end{cases} \tag{4.5}$$

where V_{MTZ} is a constant called the muffin-tin zero outside the MT sphere. An electron moving in the potential field $v(\mathbf{r})$ centered at $\mathbf{R} = 0$ obeys the Schrödinger equation given by

$$\left[-\nabla^2 + v(\mathbf{r})\right]\psi(\mathbf{r}) = E\psi(\mathbf{r}) \tag{4.6}$$

in atomic units. The electron can propagate freely between spheres with a constant kinetic energy $\kappa^2 = E - V_{MTZ}$. The solution of Equation 4.6 is expressed as the product of a spherical harmonic $Y_{\ell m}(\hat{\mathbf{r}})$ with angular variable $\hat{\mathbf{r}} = (\theta, \phi)$ and the radial wave function given by

$$\left[-\frac{d^2}{dr^2} - \frac{2}{r}\frac{d}{dr} + \frac{\ell(\ell+1)}{r^2} + v(r) - E\right]R_\ell(E,r) = 0 \tag{4.7}$$

where the radial wave function $R_\ell(E,r)$ is normalized inside the MT sphere to satisfy the relation:

$$\int_0^a R_\ell^2(E,r)r^2 dr = 1 \tag{4.8}$$

The spherical harmonic $Y_{\ell m}(\hat{\mathbf{r}})$ satisfies the relation

$$\int Y_{\ell m}^*(\hat{\mathbf{r}})Y_{\ell' m'}(\hat{\mathbf{r}})d\hat{\mathbf{r}} = \delta_{\ell\ell'}\delta_{mm'} \tag{4.9}$$

and, hence, $Y_{\ell' m'}(\hat{\mathbf{r}})$ and $Y_{\ell m}(\hat{\mathbf{r}})$ are orthonormal to each other, unless $\ell = \ell'$ and $m = m'$. If we introduce a new radial wave function defined as $P_\ell(E,r) = rR_\ell(E,r)$, Equation 4.7 is simplified as

$$\left[-\frac{d^2}{dr^2} + \frac{\ell(\ell+1)}{r^2} + v(r) - E\right]P_\ell(E,r) = 0 \tag{4.10}$$

and the orthogonality condition becomes

$$\int_0^a P_\ell^2(E,r)dr = 1$$

4.5 WAVE FUNCTIONS INSIDE AND OUTSIDE THE MT SPHERE

By displacing the depth of the potential $v(r)$ by V_{MTZ}, the MT potential at the position $\mathbf{R}=0$ is rewritten as

$$V_{MT}(r) = \begin{cases} v(r) - V_{MTZ} & r \le a \\ 0 & r > a \end{cases} \tag{4.11}$$

Now, the Schrödinger Equation 4.6 can be rewritten as

$$\begin{aligned} \left[H - E\right]\psi_{\ell m}(E,r) &= \left[\left(-\nabla^2 + v(r)\right) - E\right]\psi_{\ell m}(E,r) \\ &= \left[-\nabla^2 + v(r) - \left(\kappa^2 + V_{MTZ}\right)\right]\psi_{\ell m}(E,r) \tag{4.12} \\ &= \left[-\nabla^2 + V_{MT}(r) - \kappa^2\right]\psi_{\ell m}(E,r) = 0 \end{aligned}$$

where $E = \kappa^2 + V_{MTZ}$ is used [15]. Equation 4.12 describes the motion of an electron with the kinetic energy κ^2 in a space, where an isolated spherically symmetric potential is located at the origin in the range $r \le a$ and an otherwise constant potential equal to zero.* The wave function inside the MT sphere is expressed as

$$\psi_{\ell m}(E,\mathbf{r}) = i^\ell Y_{\ell m}(\hat{\mathbf{r}}) R_\ell(E,r) \tag{4.13}$$

whereas the electron propagating outside it should be treated as a free electron so that the Schrödinger equation in the Cartesian coordinate system is reduced to the form

$$\left[-\nabla^2 - \kappa^2\right]\psi(E,\mathbf{r}) = 0 \tag{4.14}$$

and its solution is obviously given by

---

* κ represents the wave number of an electron moving in a space, where only a single MT potential is embedded. This is different from the wave vector $\mathbf{k}$ in Equation 4.4. Note that the latter is derived from the Bloch theorem. In the free electron model, the relation $|\mathbf{k}+\mathbf{G}|^2 = \kappa^2$ holds, where $\mathbf{G}$ is the reciprocal lattice vector [2].

$$\psi(E,\mathbf{r}) = \Omega^{-1/2} \exp\left[i\mathbf{\kappa}\cdot\mathbf{r}\right] \tag{4.15}$$

where Ω is the volume of the system. In the spherical coordinate system, the Schrödinger equation in the range $r > a$ is reduced to

$$\left[-\frac{d^2}{dr^2} + \frac{\ell(\ell+1)}{r^2} - \kappa^2\right] P_\ell(\kappa^2, r) = 0 \tag{4.16}$$

The solution of Equation 4.16 representing a spherical wave of the wave number κ is given by a linear combination of spherical Bessel function $j_\ell(\kappa r)$ and spherical Neumann function $n_\ell(\kappa r)$. When $\kappa r \ll 1$, asymptotic relations hold

$$\left. \begin{aligned} j_\ell(\kappa r) &\rightarrow (\kappa r)^\ell / (2\ell+1)!! \\[2mm] n_\ell(\kappa r) &\rightarrow \frac{-(2\ell-1)!!}{(\kappa r)^{\ell+1}} \end{aligned} \right\} \quad \kappa r \rightarrow 0 \tag{4.17}$$

where $!! = 1\cdot3\cdot5\ldots$ and $-1!! = 1$. Both $(\kappa r)^\ell$ and $(\kappa r)^{-\ell-1}$ will appear frequently in the following sections. When κr is large, both functions take the following asymptotic forms:

$$\left. \begin{aligned} j_\ell(\kappa r) &\rightarrow \frac{\sin(\kappa r - \ell\pi/2)}{\kappa r} \\[2mm] n_\ell(\kappa r) &\rightarrow -\frac{\cos(\kappa r - \ell\pi/2)}{\kappa r} \end{aligned} \right\} \quad \kappa r \rightarrow \infty \tag{4.18}$$

The solution of Equation 4.12, in which an ionic potential $V_{MT}(r)$ is embedded, is now expressed as [15]:

$$\psi_{\ell m}(E,\kappa,r) = i^\ell Y_{\ell m}(\hat{\mathbf{r}}) \begin{cases} R_\ell(E,r) & r \le a \\ \kappa\left[n_\ell(\kappa r) - \cot\eta_\ell \cdot j_\ell(\kappa r)\right] & r > a \end{cases} \tag{4.19}$$

where the constant term $\cot \eta_\ell$ is determined from the condition that the wave function is continuous and differentiable across the MT sphere and is given by

$$\cot\left(\eta_\ell\left(E,\kappa\right)\right) = \frac{n_\ell(\kappa r)}{j_\ell(\kappa r)} \cdot \frac{D_\ell(E) - \kappa r n_\ell'(\kappa r)/n_\ell(\kappa r)}{D_\ell(E) - \kappa r j_\ell'(\kappa r)/j_\ell(\kappa r)}\bigg|_{r=a} \tag{4.20}$$

where $D_\ell(E)$ is defined as

$$D_\ell(E) = \frac{a}{R_\ell(E,a)} \frac{\partial R_\ell(E,r)}{\partial r}\bigg|_{r=a} \tag{4.21}$$

and represents the nondimensional logarithmic derivative of the radial wave function at the MT sphere.* In the region far away from the origin, the wave function is described by a spherical wave in the free space but with a finite phase shift η_ℓ as a result of the presence of the MT potential at $\mathbf{R}=0$:

$$\psi_\ell(E;\kappa,r) \rightarrow -\frac{\sin(\kappa r + \eta_\ell - \ell\pi/2)}{r \sin \eta_\ell} \tag{4.22}$$

where $E=\kappa^2$ is its kinetic energy.

4.6 ORTHOGONALITY CONDITION WITH CORE ELECTRON STATES

We show below that the wave function inside the MT sphere is orthogonal to any electronic states of core electrons. Since Equation 4.9 holds, we need to consider only the case, where the radial wave function $P_{n\ell}^c(E,r)$ of a core electron has the same azimuthal quantum number ℓ as that in Equation 4.10. The Schrödinger equation for the core electron is written as

---

* Since $\kappa r j_\ell'(\kappa r)/j_\ell(\kappa r)\big|_{r=a} = \ell$ and $\kappa r n_\ell'(\kappa r)/n_\ell(\kappa r)\big|_{r=a} = -\ell-1$ hold, Equation 4.18 involves the term $\{D_\ell(E) + \ell + 1\}\{D_\ell(E) - \ell\}^{-1}$, which will appear as the potential function in Equation 4.27.

$$\left[-\frac{d^2}{dr^2} + \frac{\ell(\ell+1)}{r^2} + v(r) - E_{n\ell}^c \right] P_{n\ell}^c(E,r) = 0 \qquad (4.23)$$

where $E_{n\ell}^c$ is its energy eigenvalue. $P_{n\ell}^c(E_{n\ell}^c,r)$ is multiplied from the left-hand side of Equation 4.10 and $P_\ell(E,r)$ from the left-hand side of Equation 4.23, and both are subsequently integrated over the range from 0 to a. A subtraction of the resulting two equations leads to

$$\int_0^a \left[P_{n\ell}^c \left(\frac{d^2}{dr^2} \right) P_\ell - P_\ell \left(\frac{d^2}{dr^2} \right) P_{n\ell}^c \right] dr + \left(E - E_{n\ell}^c \right)$$

$$\times \int_0^a P_{n\ell}^c(E_{n\ell}^c,r) P_\ell(E,r) dr = 0 \qquad (4.24)$$

The first term in Equation 4.24 is easily integrated to $[P_{n\ell}^c P_\ell' - P_\ell P_{n\ell}'^c]_0^a$, which vanishes, because $P_\ell(E,r) = rR_\ell(E,r)$ and $P_{n\ell}^c(E_{n\ell}^c,r)$ are zero at $r = 0$ and $P_{n\ell}^c$ and $P_{n\ell}'^c$ are zero at $r = a$. Since $E \neq E_{n\ell}^c$ holds true, we have the relation:

$$\int_0^a P_{n\ell}^c(E_{n\ell}^c,r) P_\ell(E,r) dr = 0 \qquad (4.25)$$

This proves that the wave function inside the MT sphere is orthogonal to the core electron wave function.* Thus, any electronic state in the valence band obtained in the MT sphere approximation is always orthogonal to the electronic state of the core electron. This is important in first-principles band calculations.

---

* We say that any two vectors $\mathbf{x}$ and $\mathbf{y}$ in a vector space are orthogonal to each other, provided that the relation $\langle \mathbf{x},\mathbf{y} \rangle = \langle \mathbf{y},\mathbf{x} \rangle = 0$ holds. They have no common components. Hence, any electron in the valence band and a core electron have no hybridization terms and are independent of each other. Note that two vectors are said to be orthonormal if they are orthogonal and both are of unit length.

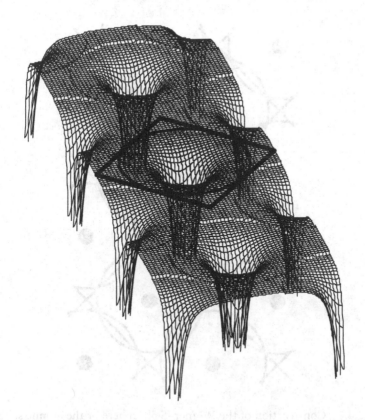

FIGURE 4.6 Periodically arranged MT potentials of pure Cu on the (100) plane. A thick square refers to the cross section of the Wigner–Seitz cell and a solid circle the inscribed MT sphere with radius a_{ins}. [From U. Mizutani, *MATERIA* (in Japanese), 45, No. 9 (2006) 677.]

4.7 KKR-ASA EQUATION

We are now ready to deal with Bloch electrons in a crystal. For the sake of simplicity, only a monatomic crystal is considered, unless otherwise stated. As shown in Figure 4.6, MT potentials are now periodically arranged around each ion [14]. As noted in Section 4.4, the ASA method approximates the Wigner–Seitz cell as given by a sphere having the same volume as that of the cell. It is clear from Figure 4.7 that they partly overlap with each other but also leave some void spaces. The radius a_{ASA} of the Wigner–Seitz sphere or an atomic sphere is obviously larger than that of an inscribed sphere a_{ins}. As will be described later, the void and overlapping regions will be ignored in the LMTO-ASA.

The "KKR" in the title of this section indicates the abbreviation of Korringa, Kohn, and Rostoker, who developed the band calculation

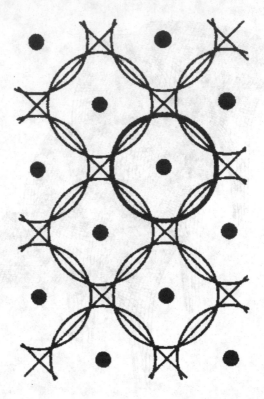

FIGURE 4.7   Construction of the Wigner–Seitz sphere or the atomic sphere. In the ASA, the radius of the MT sphere is set equal to that of the atomic sphere. The void and overlap of neighboring atomic spheres are inevitably present. [From H.L. Skriver, *The LMTO Method*, Springer series in Solid-State Sciences 41, Springer-Verlag, Berlin, 1984.]

method by ingeniously taking into account multiple-scattering of an electron propagating through the periodic MT potential field [16,17]. The boundary condition can be expressed as the condition for self-consistent multiple scattering between the MT spheres [18] or alternatively as the condition for destructive interference of tails of these waves in the core region [15,19]. We focus on the MT potential at the origin arbitrarily chosen from an array of identical atomic spheres in a monatomic crystal and consider the following trial wave function $\chi_{\ell m}(E,r)$ defined outside and inside the atomic sphere with radius a:

$$\chi_{\ell m}(E,r) = i^{\ell} Y_{\ell m}(\hat{r}) \begin{cases} R_{\ell}(E,r) + p_{\ell}(E)(r/a)^{\ell} & r \leq a \\ (a/r)^{\ell+1} & r > a \end{cases} \quad (4.26)$$

where $R_\ell(E,r)$ is the solution of the Schrödinger equation 4.7 inside the atomic sphere, $\chi_{\ell m}(E,r)$ is continuous and differentiable in all space, and $p_\ell(E)$ is called the potential function defined as

$$p_\ell(E) = \frac{D_\ell(E) + \ell + 1}{D_\ell(E) - \ell} \tag{4.27}$$

where $D_\ell(E)$ represents the logarithmic derivative of $R_\ell(E,r)$ at $r = a$, as already defined in Equation 4.21.* Equation 4.26 is essentially what is called the muffin-tin orbital or MTO [15,19].†

The term $(a/r)^{\ell+1}$ representing a tail of the MTO in Equation 4.26 has already appeared in Equation 4.17 as an asymptotic form of the spherical Neumann function. The sum of the MTO tails over all lattice sites in a crystal is expanded into series around the origin $\mathbf{R} = 0$:

$$\sum_{\mathbf{R} \neq 0} e^{i\mathbf{k}\cdot\mathbf{R}} \left(\frac{a}{|\mathbf{r}-\mathbf{R}|}\right)^{\ell+1} i^\ell Y_{\ell m}(\mathbf{r}-\mathbf{R})$$

$$= \sum_{\ell'm'} \frac{-1}{2(2\ell'+1)} \left(\frac{r}{a}\right)^{\ell'} i^{\ell'} Y_{\ell'm'}(\hat{\mathbf{r}}) S^{\mathbf{k}}_{\ell'm',\ell m} \tag{4.28}$$

where $S^{\mathbf{k}}_{\ell'm',\ell m}$ is called the *structure factor*‡ and is explicitly written as

---

* As noted in footnote on page 63, $p_\ell(E)$ is deeply related to a phase shift η_ℓ caused by the MT potential. A possible phase shift occurring outside the atomic sphere ($r > a$) is taken into account in Equation 4.19, whereas $p_\ell(E)$ is deliberately placed inside the atomic sphere in Equation 4.26.
† A symbol $\psi_{\ell m}(E, \kappa, r)$ is used in Equation 4.19, since it represents the eigenfunction in Equation 4.12. A different symbol $\chi_{\ell m}$ is intentionally used in (4.26) to emphasize a trial function.
‡ It represents the structure-dependent term in KKR and LMTO methods. Generally speaking, a quantity

$$S(\mathbf{G}) = \sum_i \exp(-i\mathbf{G}\cdot\mathbf{r}_i)$$

obtained by Fourier transforming the charge density is called the structure factor or crystal structure factor, where $\mathbf{r}_i$ is the atom position in a unit cell and $\mathbf{G}$ is the reciprocal lattice vector in the system.

$$S^{k}_{\ell'm',\ell m} = g_{\ell'm',\ell m} \sum_{R \neq 0} e^{i k \cdot R} \left(\frac{a}{|R|} \right)^{\ell''+1} \left[\sqrt{4\pi}\, i^{\ell''} Y_{\ell''m''}(\hat{R}) \right]^{*} \qquad (4.29)$$

where $\ell'' = \ell + \ell'$ and $m'' = m' - m$ hold, and the coefficient $g_{\ell'm',\ell m}$ is a constant given by [15]:

$$g_{\ell'm',\ell m} =$$

$$(-1)^{m+1} 2 \left[\frac{(2\ell'+1)(2\ell+1)}{(2\ell''+1)} \cdot \frac{(\ell''+m'')!(\ell''-m'')!}{(\ell'+m')!(\ell'-m')!(\ell+m)!(\ell-m)!} \right]^{1/2} \qquad (4.30)$$

The solution of the Schrödinger equation for an electron propagating throughout a crystal will be given by the superposition of MTOs given by Equation 4.26. To satisfy the Bloch theorem, we must take the sum over all lattice sites:

$$\sum_{\ell m} a^{jk}_{\ell m} \sum_{R} e^{i k \cdot R} \chi_{\ell m}(E, r - R) \qquad (4.31)$$

where k is the wave vector of the Bloch electron, and the superscript j of the coefficient $a^{jk}_{\ell m}$ stands for a band index. The first term $i^{\ell} Y_{\ell m}(\hat{r}) R_{\ell}(E, r)$ in the range $r \leq a$ is already a solution of the Schrödinger equation inside the atomic sphere at origin. Therefore, the following wave function must be also a solution inside any atomic spheres in a crystal:

$$\sum_{\ell m} a^{jk}_{\ell m} i^{\ell} Y_{\ell m}(\hat{r}) R_{\ell}(E, r) \qquad (4.32)$$

However, there exists an additional contribution arising from the second term in Equation 4.26. For example, it is explicitly given at $R = 0$ in the following form:

$$\sum_{\ell m} a^{jk}_{\ell m} i^{\ell} Y_{\ell m}(\hat{r}) p_{\ell}(E)(r/a)^{\ell} \qquad (4.33)$$

This extra term must vanish as a result of destructive interference with the contribution at $\mathbf{R}=0$ arising from all tails in Equation 4.28. This leads to the so-called cancellation theorem:

$$\sum_{\ell m}\left[\mathbf{P}_{\ell}(E)\delta_{\ell'\ell}\delta_{m'm}-S_{\ell'm',\ell m}^{\mathbf{k}}\right]a_{\ell m}^{j\mathbf{k}}=0 \qquad (4.34)$$

Equation 4.34 is called a secular equation. We will learn later that Equation 4.34 is quite similar to the LMTO-ASA equation. The function $\mathbf{P}_{\ell}(E)$ appearing in (4.34) differs from that in Equation 4.27 by a numerical factor and is explicitly given as

$$\mathbf{P}_{\ell}(E)=2(2\ell+1)\frac{D_{\ell}(E)+\ell+1}{D_{\ell}(E)-\ell} \qquad (4.35)$$

Equation 4.35 is again called the *potential function* and plays a key role in LMTO-ASA band calculations.*

In order to make an eigenvector $a_{\ell m}^{j\mathbf{k}}$ in secular equation 4.34 to be physically meaningful, the determinant of the coefficient must vanish:

$$\det\left[\mathbf{P}_{\ell}(E)\delta_{\ell'\ell}\delta_{m'm}-S_{\ell'm',\ell m}^{\mathbf{k}}\right]=0 \qquad (4.36)$$

Equation 4.36 is known as the KKR-ASA secular determinant [15,19], the solution of which provides $E-\mathbf{k}$ relations for an electron moving in a crystal. This is indeed the process of calculating the electronic structure. The characteristic feature of Equation 4.36 is found to consist of two separate terms: one the potential function $\mathbf{P}_{\ell}(E)$ dependent only on the MT potential inside the atomic sphere and the other the structure factor $S_{\ell'm',\ell m}^{\mathbf{k}}$ dependent only on the crystal structure. Both LMTO and LAPW methods are first-principles band calculations put forward by Andersen [20], who ingeniously developed the linearization method of solving the nonlinear determinantal equation 4.36, and applied to MTO and APW methods, respectively, to allow its fast-computation without seriously sacrificing the accuracy.

---

* Note that the potential function $\mathbf{P}_{\ell}(E)$ is a quantity entirely different from $P_{\ell}(E,r)=rR_{\ell}(E,r)$ defined in Equation 4.10. A bold character $\mathbf{P}$ is used to differentiate them.

4.8 LINEARIZATION METHOD

The KKR, MTO, and APW methods, all of which rely on the use of the MT potential and, hence, the partial-wave expansion, are devised to solve the simultaneous, linear (unknowns a_n^{kj} appear to first order), and homogeneous (right-hand side = 0) equation below:

$$\sum_n \left(H_{mn} - E_j^k O_{mn}\right) a_n^{kj} = 0 \qquad m = 0, \pm 1, \pm 2, \ldots \qquad (4.37)$$

To allow the matrix **a** of eigenfunctions to be physically meaningful, the secular equation 4.37 is reduced to solve the secular determinant $\det|H - E| = 0$.* For example, if energy-dependent MTOs like Equation 4.26 are used as basis functions, we cannot solve it by ordinary diagonalization because of the presence of energy-dependent off-matrix element H_{mn}. Andersen developed the linearization method, in which the energy involved in the MTO function is fixed [15,20]. The reason why the linearization is made without serious loss of accuracy will be described below.

Any radial wave function $\phi_\ell(D(E), r)$ with energy E may be approximated as a linear combination of $R_\ell(E_v, r)$ and its derivative $\dot{R}_\ell(E_v, r)$ with a fixed energy E_v in the following form:

$$\phi_\ell\left(D(E), r\right) = R_\ell(E_v, r) + \omega_\ell\left(D(E)\right) \dot{R}_\ell(E_v, r) \qquad (4.38)$$

where $D(E)$ is defined as

$$D(E) = a \frac{d}{dr} \ln \phi_\ell\left(D(E), r\right) \Bigg|_{r=a} \qquad (4.39)$$

and represents the logarithmic derivative on the MT sphere, as already appeared in Equation 4.21.† Our aim is to determine $\omega_\ell(D(E))$ to make $\phi_\ell(D(E), r)$ involving energy E implicitly through the logarithmic derivative D as close to a true value $R_\ell(E, r)$ as possible. As we will learn later,

---

* $O_{\ell m, \ell'm'}$ defined by $\left\langle \chi_{\ell'm'} | \chi_{\ell m} \right\rangle \equiv O_{\ell m, \ell'm'}$ is called the overlap integral.

† $\dot{X} \equiv dX/dE$ indicates the derivative with respect to energy, whereas $X'(\equiv (dX/dr))$, appearing in Equation 4.20, indicates the derivative with respect to position of an electron in real space.

the assumption that energy enters only through the term $\omega_\ell(D(E))$ in the right-hand side of Equation 4.38 is crucial.

The energy derivative $\dot{R}_\ell(E_v,r)$ in Equation 4.38 can be derived by differentiating Equation 4.7 with respect to energy:

$$\left[-\frac{d^2}{dr^2} - \frac{2}{r}\frac{d}{dr} + \frac{\ell(\ell+1)}{r^2} + v(r) - E_v \right] \dot{R}_\ell(E_v,r) - R_\ell(E_v,r) = 0 \quad (4.40)$$

One can also confirm the orthogonality condition between $R_\ell(E,r)$ and $\dot{R}_\ell(E,r)$ by differentiating Equation 4.8 with respect to energy:*

$$\int_0^a R_\ell(E,r)\dot{R}_\ell(E,r)r^2 dr = 0 \quad (4.41)$$

An insertion of Equation 4.38 into Equation 4.39 leads to

$$\omega_\ell\big(D(E)\big) = -\frac{R_\ell(E_v,a)}{\dot{R}_\ell(E_v,a)} \cdot \frac{D(E) - D_\ell(E_v)}{D(E) - \tilde{D}_\ell(E_v)} \quad (4.42)$$

where $D_\ell(E_v)$ and $\tilde{D}_\ell(E_v)$ are defined as

$$D_\ell(E_v) = a\frac{d}{dr}\ln R_\ell(E_v,r)\Big|_{r=a} \quad (4.43a)$$

and

$$\tilde{D}_\ell(E_v) = a\frac{d}{dr}\ln \dot{R}_\ell(E_v,r)\Big|_{r=a} \quad (4.43b)$$

---

* In both LMTO and LAPW methods, Equation 4.38 derived from a linear combination of the radial wave function and its derivative inside the MT sphere is used as basis functions. The wave function $\phi_\ell(E,r)$ thus constructed is confirmed from Equations 4.25 and 4.41 to be orthogonal to core electron wave function.

respectively*.

Now we evaluate the expectation value of energy when $\phi_\ell(D(E),r)$ is employed as a trial function. By expanding it into series with the use of Equation 4.8 under the assumption that $\omega_\ell(D(E))$ is small, we obtain

$$E(D) = \frac{\langle \phi_\ell(D) | H | \phi_\ell(D) \rangle_a}{\langle \phi_\ell(D) | \phi_\ell(D) \rangle_a} = \frac{E_v + \omega_\ell + N_\ell(E_v) E_v \omega_\ell^2}{1 + N_\ell(E_v) \omega_\ell^2}$$

$$= \left(E_v + \omega_\ell + N_\ell(E_v) E_v \omega_\ell^2 + \cdots \right)\left(1 - N_\ell(E_v) \omega_\ell^2 + \cdots \right) \quad (4.44)$$

$$= E_v + \omega_\ell - N_\ell(E_v) \omega_\ell^3 + O(\omega_\ell^4)$$

where

$$N_\ell(E_v) = \int_0^a \left[\dot{R}_\ell(E_v, r) \right]^2 r^2 dr$$

is called the norm. Moreover, insertion of Equations 4.39 and 4.43a,b into Equation 4.42 leads to

$$\omega_\ell(D) = -\frac{R(E_v)}{\dot{R}(E_v)} \cdot \frac{\left(R'(E)R(E_v) - R(E)R'(E_v) \right)}{R(E_v)R(E)}$$

$$\times \frac{R(E)\dot{R}(E_v)}{\left(R'(E)\dot{R}(E_v) - \dot{R}'(E_v)R(E) \right)} \quad (4.45)$$

$$= \frac{R'(E)R(E_v) - R(E)R'(E_v)}{\dot{R}'(E_v)R(E) - R'(E)\dot{R}(E_v)}$$

The two terms $R_\ell(E)$ and $R'_\ell(E)$ in Equation 4.45 are now expanded into series around a fixed energy E_v:

---

* A symbol ~ is placed over $D_\ell(E_v)$ to emphasize the logarithmic derivative of $\dot{R}_\ell(E_v,r)$, that is, the energy derivative of $R_\ell(E_v,r)$.

$$R_\ell(E,r) \approx R_\ell(E_v,r) + \dot{R}_\ell(E_v,r)\Delta + \frac{1}{2}\ddot{R}_\ell(E_v,r)\Delta^2$$

$$+\frac{1}{6}\dddot{R}_\ell(E_v,r)\Delta^3 + O(\Delta^4) \tag{4.46a}$$

$$R'_\ell(E,r) \approx R'_\ell(E_v,r) + \dot{R}'_\ell(E_v,r)\Delta + \frac{1}{2}\ddot{R}'_\ell(E_v,r)\Delta^2$$

$$+\frac{1}{6}\dddot{R}'_\ell(E_v,r)\Delta^3 + O(\Delta^4) \tag{4.46b}$$

where $\Delta = E - E_v$. Insertion of Equation 4.46a,b into Equation 4.45 results in*

$$
\begin{aligned}
\omega_\ell(E) &= \frac{\left(R\dot{R}' - R'\dot{R}\right)\Delta + \frac{1}{2}\left(R\ddot{R}' - R'\ddot{R}\right)\Delta^2 + \frac{1}{6}\left(R\dddot{R}' - R'\dddot{R}\right)\Delta^3 + O\left(\Delta^4\right)}{\left(R\dot{R}' - R'\dot{R}\right) + \frac{1}{2}\left(\ddot{R}\dot{R}' - \ddot{R}'\dot{R}\right)\Delta^2 + O\left(\Delta^3\right)} \\[2mm]
&= \frac{\left(-\dfrac{1}{a^2}\right)\Delta + \left(\dfrac{1}{6}\dfrac{3N_\ell}{a^2}\right)\Delta^3 + O\left(\Delta^4\right)}{\left(-\dfrac{1}{a^2}\right) + \left(\dfrac{1}{2}\dfrac{3N_\ell}{a^2}\right)\Delta^3 + O\left(\Delta^3\right)} \\[2mm]
&= \Delta\left[1 - \frac{1}{2}N_\ell\Delta^2 + O\left(\Delta^3\right)\right]\left[1 - \frac{3}{2}N_\ell\Delta^2 + O\left(\Delta^3\right)\right]^{-1} \\[2mm]
&= \Delta + \Delta^3 N_\ell(E_v) + O\left(\Delta^4\right)
\end{aligned}
\tag{4.47}
$$

Hence, Equation 4.38 is reduced to

$$
\begin{aligned}
\phi_\ell(E,r) &= R_\ell(E_v,r) + \dot{R}_\ell(E_v,r)\Delta + N_\ell\left(E_v\right)\dot{R}_\ell(E_v,r)\Delta^3 \\
&= R_\ell(E_v,r) + \dot{R}_\ell(E_v,r)\Delta + O\left(\Delta^3\right)
\end{aligned}
\tag{4.48}
$$

---

* Relations $R_\ell\dot{R}'_\ell - R'_\ell\dot{R}_\ell = -\dfrac{1}{a^2}$, $R_\ell\ddot{R}'_\ell - R'_\ell\ddot{R}_\ell = 0$, $R_\ell\dddot{R}'_\ell - R'_\ell\dddot{R}_\ell = \dfrac{3N_\ell}{a^2}$ and $\ddot{R}_\ell\dot{R}'_\ell - \ddot{R}'_\ell\dot{R}_\ell = \dfrac{3N_\ell}{a^2}$

were used [21].

It is important to realize that the term Δ^2 disappears in Equation 4.48 and, thus, the wave function can be calculated within the accuracy of Δ^2, provided that the radial wave function $R_\ell(E_\ell,r)$ is expanded into series around a fixed energy E_v and is approximated only up to the first-order. The energy eigenvalue is calculated by inserting Equation 4.47 into Equation 4.44:

$$
\begin{aligned}
E(D) &= E_v + \omega_\ell - N_\ell\left(E_v\right)\omega_\ell^3 + O\left(\omega_\ell^4\right) \\
&= E_v + \left(\Delta + N_\ell(E_v)\Delta^3\right) - N_\ell\left(E_v\right)\left(\Delta + N_\ell(E_v)\Delta^3\right)^3 + \cdots \\
&= E_v + \Delta + N_\ell(E_v)\Delta^3 - N_\ell\left(E_v\right)\Delta^3 + O\left(\Delta^4\right) \\
&= E_v + \Delta + O\left(\Delta^4\right)
\end{aligned}
\tag{4.49}
$$

indicating that the energy error $\Delta E_v = E(D) - E$ is proportional to $(E - E_v)^4$.

In summary, if we use a trial radial wave function $\phi_\ell(E,r)$ composed of a linear combination of the radial wave function $R_\ell(E_v,r)$ and its derivative $\dot{R}_\ell(E_v,r)$ inside the MT sphere, we can derive the energy eigenvalue and wave function with the accuracy of Δ^4 and Δ^2, respectively. The linearization method described above allows us to perform a fast-computation of the electronic structure without much sacrifice in accuracy. In reality, the error bar in determining the energy eigenvalue amounts to at most 1%, when the linearization method is applied for a typical metal having a valence band width of about 10 eV.

4.9 LMTO-ASA METHOD

The Linearized Muffin-Tin Orbital (LMTO) method refers to first-principles band calculation using the linearized MTO as basis functions [15,20]. The ASA (Atomic Sphere approximation) is combined with it to further enhance its computational efficiency. As discussed in Section 4.7, the MT sphere in the ASA is given by the atomic sphere with radius a having the same volume as that of the Wigner–Seitz cell. The resulting void and overlap regions are fairly small (see Figure 4.7). This allows us to choose an arbitrary value of κ or even $\kappa = 0$. This would not give rise to a serious error upon connecting the wave functions across the MT sphere, since the wavelength of the partial wave $2\pi/\kappa$ is much longer than the width of the intermediate region outside the MT sphere. Hence, the kinetic energy κ^2

in Equation 4.12 is forced to be zero.* The assumption $\kappa^2 = 0$ is tantamount to the disregard of the relation $E - \kappa^2 = V_{MTZ}$ and simplifies the LMTO method for solving the Schrödinger equation without seriously amplifying associated errors. Moreover, only s-, p-, and d-partial waves plus, in case of need, f-partial waves are retained and remaining higher-order partial waves are ignored.

The KKR-ASA equation 4.36 can be regarded as the $\kappa^2 = 0$ analog derived from the energy-dependent MTO. The LMTO-ASA equation is now derived by using the augmented, energy-*independent* MTO under the assumption of linearization [15]. First, the potential function in Equation 4.35 is linearized. An insertion of Equation 4.42 into Equation 4.38 leads to

$$\phi_\ell\big(D(E),r\big) = R_\ell(E_v,r) + \omega_\ell\big(D(E)\big)\dot{R}_\ell(E_v,r)$$

$$= R_\ell(E_v) - \frac{R_\ell(E_v)}{\dot{R}_\ell(E_v)} \cdot \frac{D(E) - D_\ell(E_v)}{D(E) - \tilde{D}_\ell(E_v)} \dot{R}_\ell(E_v) \qquad (4.50)$$

$$= R_\ell(E_v) \cdot \frac{D_\ell(E_v) - \tilde{D}_\ell(E_v)}{D(E) - \tilde{D}_\ell(E_v)}$$

Now we assume two $\omega_\ell(D)$ s at two different values of D_1 and D_2 and take their difference, into which Equation 4.50 is inserted:

$$\omega_\ell(D_1) - \omega_\ell(D_2) = -\frac{R_\ell}{\dot{R}_\ell} \cdot \left(\frac{D_1 - D_\ell}{D_1 - \tilde{D}_\ell} - \frac{D_2 - D_\ell}{D_2 - \tilde{D}_\ell} \right)$$

$$= -\frac{R_\ell}{\dot{R}_\ell} \cdot \left(\frac{D_\ell - \tilde{D}_\ell}{D_1 - \tilde{D}_\ell} \cdot \frac{D_1 - D_2}{D_2 - \tilde{D}_\ell} \right) \qquad (4.51)$$

$$= -\frac{\phi_\ell(D_1)\phi_\ell(D_2)}{R_\ell(E_v)\dot{R}_\ell(E_v)} \left(\frac{D_1 - D_2}{D_\ell - \tilde{D}_\ell} \right)$$

$$= -a\phi_\ell(D_1)\phi_\ell(D_2)(D_1 - D_2)$$

where the relation $(D_\ell - \tilde{D}_\ell)aR_\ell\dot{R}_\ell = 1$ is used [15]. In Equation 4.51, we set $D_1 = D_\ell(E)$, replace D_2 by D_1 in one case, retain D_2 in the other case, and take the ratio of the resulting two relations:

---

* Note that the wave number κ refers to that in the free electron model but not to that of the Bloch electron.

$$\frac{\omega_\ell(D_\ell(E))-\omega_\ell(D_1)}{\omega_\ell(D_\ell(E))-\omega_\ell(D_2)}=\frac{\phi_\ell(D_1)}{\phi_\ell(D_2)}\left(\frac{D_\ell(E)-D_1}{D_\ell(E)-D_2}\right) \tag{4.52}$$

The replacement of $D_1=-\ell-1$ and $D_2=\ell$ immediately leads to*

$$\frac{\omega_\ell(D_\ell(E))-\omega_\ell(-\ell-1)}{\omega_\ell(D_\ell(E))-\omega_\ell(\ell)}=\frac{\phi_\ell(-\ell-1)}{\phi_\ell(\ell)}\left(\frac{D_\ell(E)+\ell+1}{D_\ell(E)-\ell}\right) \tag{4.53}$$

The potential function $P_\ell(E)$ given by Equation 4.35 is now rewritten by inserting Equation 4.53:

$$P_\ell(E)=2(2\ell+1)\frac{D_\ell(E)+\ell+1}{D_\ell(E)-\ell}$$

$$=2(2\ell+1)\frac{\phi_\ell(\ell)}{\phi_\ell(-\ell-1)}\left(\frac{\omega_\ell(D_\ell(E))-\omega_\ell(-\ell-1)}{\omega_\ell(D_\ell(E))-\omega_\ell(\ell)}\right) \tag{4.54}$$

According to Equation 4.47, $\omega_\ell(D_\ell(E))=E-E_v$ holds in the accuracy of Δ^2. Thus, the potential function can be linearized in the accuracy of Δ^2 in the following form:

$$P_\ell(E)\approx 2(2\ell+1)\frac{\phi_\ell(\ell)}{\phi_\ell(-\ell-1)}\left(\frac{(E-E_v)-\omega_\ell(-\ell-1)}{(E-E_v)-\omega_\ell(\ell)}\right) \tag{4.55}$$

where $\phi_\ell(-\ell-1)$ and $\omega_\ell(\ell)$ are independent of energy, since energy enters only through $D(E)$, which is fixed as $D_1=-\ell-1$ and $D_2=\ell$. This is the linearization of the potential function in the LMTO-ASA method.

The matrix element in Equation 4.36 representing the KKR-ASA equation may be rewritten as

$$P_\ell(E)\delta_{\ell'\ell}\delta_{m'm}-S^k_{\ell'm',\ell m}\equiv\Lambda_{\ell'm',\ell m} \tag{4.56}$$

and the following relation is newly defined in relation to Equation 4.55:

---

* See footnote 7 on page 63 for the reason why we choose $D_1=-\ell-1$ and $D_2=\ell$.

$$\Delta_{\ell'm',\ell m} = \frac{\omega_\ell(\ell) - (E - E_v)}{\omega_\ell(\ell) - \omega_\ell(-\ell-1)} \delta_{\ell\ell'} \delta_{mm'} \tag{4.57}$$

Now the product of Equations 4.56 and 4.57 is calculated to be

$$\left[\Delta\Lambda\right]_{\ell m,\ell'm'} = \Omega^k_{\ell m,\ell'm'} - (E - E_v)\Pi^k_{\ell m,\ell'm'} \tag{4.58}$$

where the coefficients $\Omega^k_{\ell m,\ell'm'}$ and $\Pi^k_{\ell m,\ell'm'}$ are explicitly given as

$$\Omega^k_{\ell m,\ell'm'} = \sqrt{\frac{2}{a}} \cdot \frac{1}{\phi(-\ell-1)}$$

$$\times \left[\sqrt{\frac{2}{a}} \frac{\omega(-\ell-1)}{\phi(-\ell-1)} \delta_{\ell\ell'}\delta_{mm'} + \sqrt{\frac{a}{2}} \frac{\phi(-\ell-1)\omega(\ell)}{\omega(\ell)-\omega(-\ell-1)} S^k_{\ell m,\ell'm'}\right] \tag{4.59}$$

and

$$\Pi^k_{\ell m,\ell'm'} = \sqrt{\frac{2}{a}} \cdot \frac{1}{\phi(-\ell-1)}$$

$$\times \left[\sqrt{\frac{2}{a}} \frac{1}{\phi(-\ell-1)} \delta_{\ell\ell'}\delta_{mm'} + \sqrt{\frac{a}{2}} \frac{\phi(-\ell-1)}{\omega(\ell)-\omega(-\ell-1)} S^k_{\ell m,\ell'm'}\right] \tag{4.60}$$

where $\omega(-\ell-1) - \omega(\ell) = (2\ell+1)a\phi(-\ell-1)\phi(\ell)$ is used. Note that $\omega(\ell)$, $\omega(-\ell-1)$ and $\phi(-\ell-1)$ in Equations 4.59 and 4.60 are all energy independent.

The matrix element $\langle\chi_{\ell'm'}|H - E|\chi_{\ell m}\rangle$ arising from Equation 4.37 in the LMTO-ASA method is simplified by using Equations 4.59 and 4.60:

$$(H - EO)_{\ell m,\ell'm'} = \Pi^+_{\ell m,\ell'm'}\left\{\Omega_{\ell m,\ell'm'} - (E - E_v)\Pi_{\ell m,\ell'm'}\right\} \tag{4.61}$$

Since the quantity in the curly brackets in the right-hand side is nothing but Equation 4.58, we find that the LMTO-ASA equation is similar to the KKR-ASA equation 4.36 involving the term $\Lambda_{\ell'm',\ell m}$ and is explicitly written as:

$$\det\left[\Pi^{+}_{\ell m,\ell' m'}\left(\Omega_{\ell m,\ell' m'}-(E-E_{\mathrm{v}})\Pi_{\ell m,\ell' m'}\right)\right]=0 \qquad (4.62)$$

Note that the structure factor $S^{k}_{\ell m,\ell' m'}$ involved in Equations (4.59) and (4.60) is independent of energy. It is often called canonical in the sense that it depends neither on energy, on the MT-sphere radius, nor on the scale of the structure. Indeed, energy enters only in a linearized form in the second term of the matrix element in the secular determinant (4.62). This is called the LMTO-ASA equation. The characteristic feature of the ionic potential of a given element is found to enter through $\phi_{\ell}(\ell)$ and $\omega(\ell)$ involved in the potential function (4.55), which are solely determined from the radial wave function $R_{\ell}(E_{\mathrm{v}})$ and its derivative $\dot{R}_{\ell}(E_{\mathrm{v}})$ inside the MT sphere.

The canonical structure factor for a monatomic system is calculated from Equation 4.29 for a set of matrices distributed on a suitable grid specifying the wave vector **k** in the irreducible wedge of the Brillouin zone.[*] The canonical structure factor for a simple cubic lattice, for example, is alternatively expressed in the two-center notation of Slater and Koster [15,19]:

$$S_{s,s}=-2(a/R)$$

$$S_{p_x,p_x}=\left(3\left(\frac{x}{R}\right)^{2}-1\right)\times 6(a/R)^{3} \qquad (4.63)$$

$$S_{d_{xy},d_{3z^2-r^2}}=\left(-\sqrt{3}/2\right)5\left(7\left(\frac{z}{R}\right)^{2}-1\right)\left(\frac{x}{R}\right)\left(\frac{y}{R}\right)\times 10\left(\frac{a}{R}\right)^{5}$$

where R/a represents the atomic distance normalized with respect to the radius of an atomic sphere a and x/R, y/R and z/R are directional cosine. In the case where s-, p-, and d-partial waves are sufficient, we end up with the 9×9 matrix arising from a combination of nine different orbitals, that is, $(2\ell+1)$ ℓ-partial waves composed of a single s-, three p-, and five d-waves. It is explicitly written as follows:

---

[*] See the definition of the "irreducible wedge" for the fcc and bcc Brillouin zones in Section 5.5 and Figure 5.3.

$$
\begin{pmatrix}
S_{s,s} & S_{s,p_x} & S_{s,p_y} & S_{s,p_z} & S_{s,d_{xy}} & S_{s,d_{yz}} & S_{s,d_{zx}} & S_{s,d_{x^2-y^2}} & S_{s,d_{3z^2-r^2}} \\
S_{p_x,s} & S_{p_x,p_x} & S_{p_x,p_y} & S_{p_x,p_z} & S_{p_x,d_{xy}} & S_{p_x,d_{yz}} & S_{p_x,d_{zx}} & S_{p_x,d_{x^2-y^2}} & S_{p_x,d_{3z^2-r^2}} \\
S_{p_y,s} & S_{p_y,p_x} & S_{p_y,p_y} & S_{p_y,p_z} & S_{p_y,d_{xy}} & S_{p_y,d_{yz}} & S_{p_y,d_{zx}} & S_{p_y,d_{x^2-y^2}} & S_{p_y,d_{3z^2-r^2}} \\
S_{p_z,s} & S_{p_z,p_x} & S_{p_z,p_y} & S_{p_z,p_z} & S_{p_z,d_{xy}} & S_{p_z,d_{yz}} & S_{p_z,d_{zx}} & S_{p_z,d_{x^2-y^2}} & S_{p_z,d_{3z^2-r^2}} \\
S_{d_{xy},s} & S_{d_{xy},p_x} & S_{d_{xy},p_y} & S_{d_{xy},p_z} & S_{d_{xy},d_{xy}} & S_{d_{xy},d_{yz}} & S_{d_{xy},d_{zx}} & S_{d_{xy},d_{x^2-y^2}} & S_{d_{xy},d_{3z^2-r^2}} \\
S_{d_{yz},s} & S_{d_{yz},p_x} & S_{d_{yz},p_y} & S_{d_{yz},p_z} & S_{d_{yz},d_{xy}} & S_{d_{yz},d_{yz}} & S_{d_{yz},d_{zx}} & S_{d_{yz},d_{x^2-y^2}} & S_{d_{yz},d_{3z^2-r^2}} \\
S_{d_{zx},s} & S_{d_{zx},p_x} & S_{d_{zx},p_y} & S_{d_{zx},p_z} & S_{d_{zx},d_{xy}} & S_{d_{zx},d_{yz}} & S_{d_{zx},d_{zx}} & S_{d_{zx},d_{x^2-y^2}} & S_{d_{zx},d_{3z^2-r^2}} \\
S_{d_{x^2-y^2},s} & S_{d_{x^2-y^2},p_x} & S_{d_{x^2-y^2},p_y} & S_{d_{x^2-y^2},p_z} & S_{d_{x^2-y^2},d_{xy}} & S_{d_{x^2-y^2},d_{yz}} & S_{d_{x^2-y^2},d_{zx}} & S_{d_{x^2-y^2},d_{x^2-y^2}} & S_{d_{x^2-y^2},d_{3z^2-r^2}} \\
S_{d_{3z^2-r^2},s} & S_{d_{3z^2-r^2},p_x} & S_{d_{3z^2-r^2},p_y} & S_{d_{3z^2-r^2},p_z} & S_{d_{3z^2-r^2},d_{xy}} & S_{d_{3z^2-r^2},d_{yz}} & S_{d_{3z^2-r^2},d_{zx}} & S_{d_{3z^2-r^2},d_{x^2-y^2}} & S_{d_{3z^2-r^2},d_{3z^2-r^2}}
\end{pmatrix}
$$

$$(4.64)$$

The structure factor $S^{\mathbf{k}}_{\ell m,\ell'm'}$ defined by Equation 4.29 can be extended to a system containing more than two atoms in the unit cell:

$$
S^{\mathbf{k}}_{\mathbf{r}_s\ell m,\mathbf{r}_{s'}\ell'm'} = \left(\frac{a_{t'}}{a}\right)^{\ell'+\frac{1}{2}} \sum_{\mathbf{R}\neq 0} e^{i\mathbf{k}\cdot\mathbf{R}} S_{\mathbf{r}_{s'}\ell'm',(\mathbf{r}_s+\mathbf{R})\ell m} \left(\frac{a_t}{a}\right)^{\ell+\frac{1}{2}}
$$

$$
= \left(\frac{a_{t'}}{a}\right)^{\ell'+\frac{1}{2}} g_{\ell'm',\ell m} \sum_{\mathbf{R}\neq\mathbf{r}_{s'}-\mathbf{r}_s} e^{i\mathbf{k}\cdot\mathbf{R}} \left(\frac{a}{|\mathbf{R}-(\mathbf{r}_{s'}-\mathbf{r}_s)|}\right)^{\ell''+1} \quad (4.65)
$$

$$
\times \left[\sqrt{4\pi}\, i^{\ell''} Y_{\ell''m''}\left(\overline{\mathbf{R}-(\mathbf{r}_{s'}-\mathbf{r}_s)}\right)\right]^* \left(\frac{a_t}{a}\right)^{\ell+\frac{1}{2}}
$$

where $\mathbf{R}$ is the lattice vector, $\mathbf{r}_s$ is the coordinate of s-th atom in the unit cell, a_t is the radius of the atomic sphere of type t, a is an average radius of atomic spheres, $\ell'' = \ell'+\ell$, $m'' = m'-m$ and the coefficient $g_{\ell'm',\ell m}$ is defined in Equation 4.30 [15].

In the case of the CsCl-type binary alloy AB, we have four different atomic pairs A-A, A-B, B-A, and B-B, and the structure factor (4.65) is reduced to the form:

$$
\begin{pmatrix}
A-A & A-B \\
B-A & B-B
\end{pmatrix}
\quad (4.66)
$$

Let us consider, for example, the CuAu B2-compound. Now a matrix given by Equation 4.66 consists of 18 rows and 18 columns, since each atomic pair of Cu-Cu, Cu-Au, Au-Cu, and Au-Au forms a matrix of nine rows and nine columns. The electronic states $Cu-4s$, $Cu-4p$, $Cu-3d$, $Au-6s$, $Au-6p$ and $Au-5d$ are assigned to $m(n)=1$, $m(n)=2\sim4$, $m(n)=5\sim9$, $m(n)=10$, $m(n)=11\sim13$ and $m(n)=14\sim18$ for the m-th row and n-th column in the matrix element $S(m,n)$, respectively. Thus, $S(1,10)$ represents the matrix element between $Cu-4s$ and $Au-6s$, whereas $S(2,11)$ that between $Cu-4p$ and $Au-6p$ states.

The LMTO-wave function for the CuAu B2-compound is explicitly given by

$$\psi = \sum_R e^{i\mathbf{k}\cdot\mathbf{R}} \left[\sum_{lm} a_{Cu-lm}^{jk}\chi_{Cu-lm}\left(E, \mathbf{r}-\mathbf{R}-\mathbf{r}_{(000)}\right) + \sum_{l'm'} a_{Au-l'm'}^{jk}\chi_{Au-l'm'}\left(E, \mathbf{r}-\mathbf{R}-\mathbf{r}_{\left(\frac{111}{222}\right)}\right) \right] \quad (4.67)$$

where

$$\sum_{lm}\chi_{Cu-lm}$$

is the sum of MTOs associated with one Cu-4s, three Cu-4p, and five Cu-3d states and

$$\sum_{lm}\chi_{Au-lm}$$

that of MTOs associated with one Au-6s, three Au-6p, and five Au-5d states. Therefore, we have totally 18 different coefficients to be determined in Equation 4.67.

According to the variational principle, one varies ψ to make the energy functional stationary such that $\delta\langle\psi|H-E|\psi\rangle=0$, which has solutions whenever

$$\det\left\{\left\langle\chi_{l'l'm'r_{s'}}^{k}\left|H-E\right|\chi_{tlmr_s}^{k}\right\rangle\right\}=0 \quad (4.68)$$

where t stands for type t atom and $\mathbf{r}_s$ atom position in the unit cell. Equation 4.68 needs to be evaluated only over atoms in the unit cell [15].

4.10 EXTRACTION OF spd-d ORBITAL HYBRIDIZATIONS IN THE LMTO-ASA METHOD

In Chapter 1, an emphasis was laid on the historical importance of the gamma-brass as a material, which inspired Jones' curiosity in 1933 after he attended a lecture given by W. L. Bragg. It is also emphasized that gamma-brasses have played a special role in the interpretation for the Hume-Rothery electron concentration rule since the theory advanced by Mott and Jones in 1936. In the remaining chapters, we will systematically work on a series of gamma-brasses containing 52 atoms per unit cell and having space group of either $I\bar{4}3m$ or $P\bar{4}3m$.

To proceed with LMTO-ASA band calculations for Cu_5Zn_8 gamma-brass with space group $I\bar{4}3m$, we need to replace Equation 4.64 representing the matrix of the structure factor for a monatomic system by that given by Equation 4.65. As will be discussed in Chapter 6, Section 6.2, the atomic structure of Cu_5Zn_8 gamma-brass is described by arranging a 26-atom cluster to form a bcc lattice. Now the matrix of the structure factor $S^{\mathbf{k}}_{\mathbf{r}_s\ell m, \mathbf{r}'_s\ell'm'}$ with $\mathbf{r}_s$ and $\mathbf{r}'_s$ specifying Cu or Zn atom positions in the unit cell is composed of $26 \times 9 = 234$ rows and 234 columns, since we have 26 atoms in the cluster and each Cu or Zn atom in it has nine electronic states consisting of one 4s, three 4p and five 3d states.* The LMTO-wave function for Cu_5Zn_8 gamma-brass will be constructed in a way similar to Equation 4.67 for the CuAu B2-compound. Among $26 \times 9 = 234$ coefficients, we have 90 $a^{j\mathbf{k}}_{Cu-lm}$ and 144 $a^{j\mathbf{k}}_{Zn-lm}$ coefficients to be determined at a given energy and given state $\mathbf{k}$ for this gamma-brass.

Figure 4.8 shows energy dispersion relations along <411> direction calculated for Cu_5Zn_8 gamma-brass in the LMTO-ASA method [22]. The electronic states are bunched in the binding energies over $-2 \leq E \leq -4$ eV and $-7 \leq E \leq -8$ eV. They are identified as the Cu-3d and Zn-3d band, respectively. In contrast, electronic states are sparse in the neighborhood of the Fermi level. This leads to what we call a *pseudogap*. Its presence may be more clearly seen in the DOS shown in Figure 4.9 [6]. We see that the Fermi level is situated at an energy, where the Cu-3d band is almost terminated and its diminishing tail extends toward the bottom of the

---

* Spin states are assumed to be degenerate.

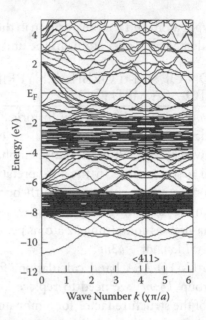

FIGURE 4.8 Energy dispersion relations along <411> direction for Cu_5Zn_8 gamma-brass calculated using the LMTO-ASA method. [From U. Mizutani, T. Takeuchi, and H. Sato, *Prog. Mat. Sci.* 49, No. 3–4 (2004) 227.]

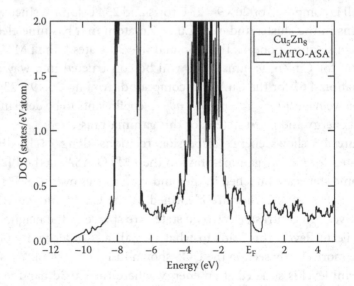

FIGURE 4.9 Total DOS for Cu_5Zn_8 gamma-brass calculated using the LMTO-ASA method. [From U. Mizutani, *The Science of Complex Alloy Phases* (edited by T.B. Massalski and P.E.A. Turchi), TMS (The Minerals, Metals & Materials Society, 2005) pp. 1–42.]

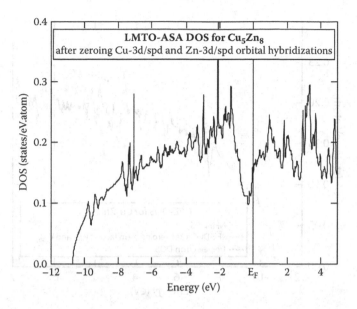

FIGURE 4.10   Total DOS for Cu_5Zn_8 gamma-brass obtained after zeroing all spd/ Cu-3d and all spd/Zn-3dorbitals in the LMTO-ASA method. [From U. Mizutani, *The Science of Complex Alloy Phases* (edited by T.B. Massalski and P.E.A. Turchi), TMS (The Minerals, Metals & Materials Society, 2005) pp. 1–42.]

pseudogap. From this, we perceive that the spd-d orbital hybridization can significantly affect the form of a pseudogap.

To realize the advantage of the LMTO-ASA method in exploring the role of orbital hybridizations in the formation of a pseudogap, we show in Figure 4.10 the total DOS of Cu_5Zn_8 gamma-brass calculated after intentionally eliminating product terms like $a_{Cu-3d}^{jk} \cdot a_{Zn-4p}^{jk}$ associated with all spd-orbitals hybridized with Cu-3d and Zn-3d orbitals. These terms appear upon explicit calculations of Equation 4.68, into which the LMTO-wave function like (4.67) is inserted [6]. Though both Cu-3d and Zn-3d bands are reduced to discrete levels, the pseudogap remains existent across the Fermi level. This strongly indicates that the pseudogap in this system is not caused by orbital hybridizations. However, the deletion of spd-d orbital hybridizations apparently shifts the pseudogap to the left relative to the Fermi level, or to higher binding energies. Indeed, the Fermi level in Figure 4.10 is now found at a rising slope of the pseudogap after passing its bottom [6]. Hence, we may conclude that, in the case of Cu_5Zn_8 gamma-brass, spd-d orbital hybridizations play a role in shifting the pseudogap structure relative to the Fermi level.

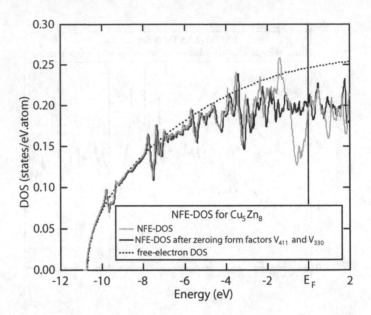

FIGURE 4.11 DOS for Cu_5Zn_8 gamma-brass calculated in the context of NFE band calculations (gray line). A black line refers to the DOS obtained after intentionally zeroing form factors associated with the set of {411} and {330} lattice planes. A dotted line represents the free electron parabolic band. [From U. Mizutani, *The Science of Complex Alloy Phases* (edited by T.B. Massalski and P.E.A. Turchi), TMS (The Minerals, Metals & Materials Society, 2005) pp. 1–42.]

Judging from the argument above, we can safely say that both the Cu-3d and Zn-3d bands have little to do with the formation of the pseudogap itself. Hence, it may well be justifiable to check the presence of a pseudogap by calculating the DOS in the Nearly Free Electron (NFE) model by ignoring the Cu-3d and Zn-3d states [6]. As shown in Figure 4.11, a pseudogap in this case is observed immediately below the Fermi level. However, it completely disappears, when the form factor (i.e., the Fourier component of the ionic potential associated with the {330} and {411} zone planes) is set to be zero (see black line in Figure 4.11).* This means that a pseudogap in

---

* The form factor in the NFE model is defined as

$$V_G = \frac{1}{N} \sum_{i=1}^{N} V_{i(\alpha)}(\mathbf{r}_i) \exp(-i\mathbf{G} \cdot \mathbf{r}_i)$$

where $V_{i(\alpha)}(\mathbf{r}_i)$ is the ionic potential due to an atomic species α at the position $\mathbf{r}_i$ in the unit cell. The sum is taken over N atoms in the unit cell. The energy gaps across the {411} and {330} zone planes are intentionally set to zero in the present case.

Cu_5Zn_8 gamma-brass originates from the FsBz interaction [6]. However, we *do* realize that the NFE model is still far removed from being realistic beyond the Mott and Jones free electron approach, because this model ignores spd-d orbital hybridizations associated with Cu-3d and Zn-3d states. Instead, we will show in the following sections that the FLAPW method is the most powerful in extracting the FsBz interaction, no matter how strong are spd-d orbital hybridizations. Nevertheless, NFE band calculations will be proved to be powerful in extracting the FsBz interactions for RT-type 1/1-1/1-1/1 approximants in place of FLAPW band calculations, since the latter needs much more computation time and memories, as will be described below (see also Chapter 9, Sections 9.2.3 and 9.3.3.).

In summary, the LMTO-ASA method has the following advantages: (1) it uses a minimal basis, which leads to high efficiency and makes calculations possible for crystals having large unit cells; (2) it is best suited for studying orbital hybridization effects, since atom-centered basis functions of well-defined angular momentum are used.*

4.11 APW METHOD

The principle of the augmented plane wave (APW) method prior to the introduction of its linearization needs to be discussed first [2,20]. We consider the two regions separated by the MT sphere in the Wigner–Seitz cell. As listed in Table 4.1, the radius a_{APW} of the MT sphere is the smallest among others, indicating that a wider intermediate region is assumed in the APW method. It should be noted that, in the APW method, the Bloch plane wave $\psi(\mathbf{r}) = \Omega^{-1/2} \exp[i(\mathbf{k}+\mathbf{G})\cdot\mathbf{r}]$ is superimposed over allowed reciprocal lattice vectors $\mathbf{G}$ to describe the motion of an electron outside the MT sphere, in sharp contrast with MTOs given by equation like (4.26) in the LMTO method.

We consider again a crystal consisting of a single atomic species. A MT sphere with radius a is placed at each atomic site. In the region $r \leq a$, a trial function is expressed as

$$\chi^{MT}_{\mathbf{k}+\mathbf{G}_n}(E,\mathbf{r}) = \sum_{lm} A^{lm}_{\mathbf{k}+\mathbf{G}_n} R_l(E,\mathbf{r}) Y_{lm}(\hat{\mathbf{r}}) \qquad (4.69)$$

---

* "A minimal basis sets" represent a basis set that describes only the most basic aspects of the orbitals.

TABLE 4.1 Radius of the MT Sphere for Several Elements Employed in LMTO-ASA and FLAPW Band Calculations

Element	Cu	Cr	Mo	W
Structure	fcc	bcc	bcc	bcc
Lattice constant (nm)	0.361	0.288	0.314	0.319
a_{ins} (nm)	$\dfrac{\sqrt{2}}{4}a = 0.35a = 0.127$	$\dfrac{\sqrt{3}}{4}a = 0.43a = 0.124$	$\dfrac{\sqrt{3}}{4}a = 0.43a = 0.136$	$\dfrac{\sqrt{3}}{4}a = 0.43a = 0.138$
a_{ASA} (nm)	$\sqrt[3]{\dfrac{3}{16\pi}}\,a = 0.39a = 0.141$	$\sqrt[3]{\dfrac{3}{8\pi}}\,a = 0.49a = 0.142$	$\sqrt[3]{\dfrac{3}{8\pi}}\,a = 0.49a = 0.154$	$\sqrt[3]{\dfrac{3}{8\pi}}\,a = 0.49a = 0.157$
a_{APW} (nm)	0.127[A] 0.116[B]	0.123	0.130	0.130

Note: a_{ins}: radius when the MT-sphere touches the plane of the Wigner–Seitz cell shortest to the origin. a_{ASA}: radius of the Wigner–Seitz sphere. a_{APW}: radius employed in the APW band calculations. Cu: (A) G.A. Burdick, Phys. Rev. 129 (1963) 138; (B) present work, see Section 5.3; Cr, Mo W: T.L. Loucks, Phys. Rev. 139 (1965) 223.

where $\mathbf{k}$ is the wave vector of the Bloch electron and $\mathbf{G}_n$ is an arbitrary reciprocal lattice vector in the present system. In the interstitial region $r > a$, a trial function is expressed in the form of a plane wave with the wave vector $\mathbf{k} + \mathbf{G}_n$:

$$\chi_{\mathbf{k}+\mathbf{G}_n}^{inter}(\mathbf{r}) = \Omega^{-1/2} \exp\left[i(\mathbf{k}+\mathbf{G}_n)\cdot\mathbf{r}\right] \tag{4.70}$$

Now Equation 4.70 is expanded into the spherical harmonics:

$$\chi_{\mathbf{k}+\mathbf{G}_n}^{inter}(\mathbf{r}) = \Omega^{-1/2} \exp\left[i(\mathbf{k}+\mathbf{G}_n)\cdot\mathbf{r}\right]$$
$$= 4\pi\Omega^{-1/2} \sum_{\ell=0}^{\infty}\sum_{m=-\ell}^{\ell} i^{\ell} j_{\ell}(|\mathbf{k}+\mathbf{G}_n|r) Y_{\ell m}^{*}(\widehat{\mathbf{k}+\mathbf{G}_n}) Y_{\ell m}(\hat{\mathbf{r}}) \tag{4.71}$$

By imposing the continuity condition of Equations 4.69 and 4.71 at the surface of the MT sphere, we can determine the coefficient $A_{\mathbf{k}+\mathbf{G}_n}^{\ell m}$ in Equation 4.69 as follows:

$$A_{\mathbf{k}+\mathbf{G}_n}^{\ell m} = 4\pi i^{\ell}\Omega^{-1/2} Y_{\ell m}^{*}(\widehat{\mathbf{k}+\mathbf{G}_n}) \frac{j_{\ell}(|\mathbf{k}+\mathbf{G}_n|a)}{R_{\ell}(E,a)} \tag{4.72}$$

In this way, we can construct the APW basis function, which is continuous across the MT sphere [2,20]. Note that the derivative is not continuous at the surface of the MT sphere.

The function having the coefficient given by Equation 4.72 is simply denoted as $\chi_{\mathbf{k}+\mathbf{G}_n}(E,\mathbf{r})$. This is called the APW orbital. In order to guarantee $\chi_{\mathbf{k}+\mathbf{G}_n}(E,\mathbf{r})$ to be the Bloch wave propagating throughout a crystal, we need to impose the Bloch condition:

$$\chi_{\mathbf{k}+\mathbf{G}_n}(E,\mathbf{r}+\mathbf{R}) = e^{i(\mathbf{k}+\mathbf{G}_n)\cdot\mathbf{R}}\chi_{\mathbf{k}+\mathbf{G}_n}(E,\mathbf{r}) \tag{4.73}$$

where $\mathbf{R}$ is the lattice vector satisfying the relation $e^{i\mathbf{G}_n\cdot\mathbf{R}} = 1$. The wave function to describe the motion of an electron in a crystal is now expressed by summing APW orbitals as basis functions over the allowed reciprocal lattice vectors $\mathbf{G}_n$:

$$\psi_{k+G_n}(E,r)=\sum_{G_n}C(k+G_n)\chi_{k+G_n}(E,r) \qquad (4.74)$$

The coefficient $C(k+G_n)$ in Equation 4.74 is determined as the solution of a set of linear equations by using the variational principle. Its secular determinantal equation is expressed as

$$\det\left[\left(\left|k+G_n\right|^2-E\right)\delta_{mn}+F_{mn}\right]=0 \qquad (4.75)$$

where the coefficient F_{mn} [2,20] is explicitly given as

$$F_{mn}=\frac{4\pi a^2}{\Omega}\left\{-\left[\left(k+G_m\right)\cdot\left(k+G_n\right)-E\right]\right\}\frac{j_\ell\left(\left|G_m-G_n\right|a\right)}{\left|G_m-G_n\right|}$$

$$+4\pi\sum_{\ell m}Y_{\ell m}\widehat{(k+G_m)}^{*}j_\ell\left(\left|k+G_m\right|a\right)\frac{R_\ell'(E,a)}{R_\ell(E,a)} \qquad (4.76)$$

$$j_\ell\left(\left|k+G_n\right|a\right)Y_{\ell m}\widehat{(k+G_n)}$$

The coefficient F_{mn} is similar to the Fourier component of the ionic potential $V_{mn}\equiv V(G_m-G_n)$, that is, the form factor in NFE band calculations (see Footnote * on page 84 in Section 4.10.). The secular determinantal equation 4.75 is derived from the condition so as to have the coefficient $C(k+G_n)$ in the secular equation physically meaningful. In contrast to NFE band calculations, the energy to be solved is involved in F_{mn} through the terms $R_\ell(E,a)$ and $R_\ell'(E,a)$. This is again the reason for the need of linearization in order to perform a fast, but efficient computation.

4.12 LAPW METHOD

In this section, we study the principle of LAPW method [2,20,21,23,24], which allows a fast computation by linearizing the APW method described in the preceding section. The determinant can be diagonalized, if energy-dependent logarithmic derivative of the radial wave function

$$\frac{R_\ell'(E,a)}{R_\ell(E,a)}$$

in F_{mn} is made energy independent. For this purpose, we assume the trial radial wave function in the same way as in Equation 4.38 [20]:

$$\phi_\ell(E,r) = R_\ell(E_v,r) + \omega_\ell(E)\dot{R}_\ell(E_v,r) \tag{4.77}$$

where E_v is a fixed energy for each partial wave ℓ. In the region $r \leq a$ in the Wigner–Seitz cell located at origin, the wave function is written as

$$\chi_{k+G_n}(\mathbf{r}) = \sum_{\ell m}\left[A^{\ell m}_{k+G_n} R_\ell(E_v,r) + B^{\ell m}_{k+G_n}\dot{R}_\ell(E_v,r)\right]Y_{\ell m}(\hat{\mathbf{r}}) \tag{4.78}$$

instead of Equation 4.69 [21]. It must be noted that the ratio

$$\frac{B^{\ell m}_{k+G_n}}{A^{\ell m}_{k+G_n}}$$

corresponds to $\omega(E)$ in Equation 4.77 but is no longer energy dependent. The energy E_v in Equation 4.78 is taken as energy at the center of gravity in the band for the partial wave ℓ. In the same way as in the APW method, the wave function of an electron with the wave vector $\mathbf{k}+\mathbf{G}_n$ outside the MT sphere is given by Equations 4.70 or 4.71 and is expanded into spherical harmonics as shown in Equation 4.71. However, there are two parameters $A^{\ell m}_{k+G_n}$ and $B^{\ell m}_{k+G_n}$ to be determined in Equation 4.78. Hence, we can make not only the wave functions (4.71) and (4.78) but also their derivatives to be continuous at the surface of the MT sphere. As a result, both coefficients $A^{\ell m}_{k+G_n}$ and $B^{\ell m}_{k+G_n}$ are explicitly determined as

$$A^{\ell m}_{k+G_n} = 4\pi a^2 \Omega^{-1/2} i^\ell\, Y^*_{\ell m}(\widehat{\mathbf{k}+\mathbf{G}_n})a_\ell \tag{4.79a}$$

$$a_\ell = j'_\ell(|\mathbf{k}+\mathbf{G}_n|a)\dot{R}_\ell(E_v,a) - j_\ell(|\mathbf{k}+\mathbf{G}_n|a)\dot{R}'_\ell(E_v,a) \tag{4.79b}$$

$$B^{\ell m}_{k+G_n} = 4\pi a^2 \Omega^{-1/2} i^\ell Y^*_{\ell m}(\widehat{\mathbf{k}+\mathbf{G}_n})b_\ell \tag{4.80a}$$

and

$$b_\ell = j_\ell(|\mathbf{k}+\mathbf{G}_n|a)R'_\ell(E_v,a) - j'_\ell(|\mathbf{k}+\mathbf{G}_n|a)R_\ell(E_v,a) \qquad (4.80b)$$

We have finally obtained the energy independent LAPW basis function, which is smoothly connected across the MT sphere. This is called the *LAPW orbital*.

The following relation is imposed on the LAPW orbital to satisfy the Bloch condition:

$$\chi_{\mathbf{k}+\mathbf{G}_n}(\mathbf{r}+\mathbf{R}) = e^{i(\mathbf{k}+\mathbf{G}_n)\cdot\mathbf{R}}\chi_{\mathbf{k}+\mathbf{G}_n}(\mathbf{r}) \qquad (4.81)$$

where $\mathbf{R}$ is the lattice vector and satisfies the relation $e^{i\mathbf{G}_n\cdot\mathbf{R}} = 1$ with the corresponding reciprocal lattice vector $\mathbf{G}_n$. In contrast to Equation 4.73, Equation 4.81 is independent of energy. Now the wave function of an electron moving through a crystal can be described by superimposing LAPW basis functions:

$$\psi_{\mathbf{k}+\mathbf{G}_n}(\mathbf{r}) = \sum_{\mathbf{G}_n} C(\mathbf{k}+\mathbf{G}_n)\chi_{\mathbf{k}+\mathbf{G}_n}(\mathbf{r}) \qquad (4.82)$$

where the coefficient $C(\mathbf{k}+\mathbf{G}_n)$ is determined as a solution of a set of linear equations by means of the variational principle. A secular equation can be derived from Equation 4.37 or

$$\sum_n \left(H_{mn} - E_j^k O_{mn} \right) a_n^{kj} = 0 \qquad (4.83)$$

where the coefficient H_{mn} and O_{mn} are explicitly written down as

$$H_{mn} = |\mathbf{k}+\mathbf{G}_m| \cdot |\mathbf{k}+\mathbf{G}_n| + \delta_{\mathbf{G}_m\mathbf{G}_n} - \frac{4\pi a^2}{\Omega}\frac{j_\ell(|\mathbf{G}_m-\mathbf{G}_n|a)}{|\mathbf{G}_m-\mathbf{G}_n|}$$

$$+ \frac{4\pi a^4}{\Omega}\sum_\ell (2\ell+1)P_\ell(E_v, s_{mn}^\ell + \gamma^\ell) \qquad (4.84a)$$

and

$$O_{mn} = \delta_{G_m G_n} - \frac{4\pi a^2}{\Omega} \frac{j_\ell\left(|G_m - G_n|a\right)}{|G_m - G_n|} + \frac{4\pi a^4}{\Omega} \sum_\ell (2\ell + 1) P_\ell s_{mn}^\ell \quad (4.84b)$$

where

$$s_{mn}^\ell = [j_\ell\left(|k - G_m|a\right)\dot{R}_\ell(a) - j_\ell\left(|k - G_n|a\right)\dot{R}_\ell'(a)] \quad (4.85a)$$

$$
\begin{aligned}
\gamma_{mn}^\ell = &\, \dot{R}_\ell(a)R_\ell'(a)\left[j_\ell'\left(|k - G_m|a\right) j_\ell\left(|k - G_n|a\right) \right. \\
&\left. + j_\ell\left(|k - G_m|a\right) j_\ell'\left(|k - G_n|a\right) \right] \\
&- \left[\dot{R}_\ell'(a)R_\ell'(a) j_\ell\left(|k - G_m|a\right) j_\ell\left(|k - G_n|a\right) \right. \\
&\left. + \dot{R}_\ell(a)R_\ell(a) j_\ell'\left(|k - G_m|a\right) j_\ell'\left(|k - G_n|a\right) \right]
\end{aligned}
\quad (4.85b)
$$

and

$$P_\ell \equiv P_\ell \left(\frac{(k - G_m) \cdot (k - G_n)}{|k - G_m| \cdot |k - G_n|} \right) \quad (4.86)$$

where P_ℓ is the Legendre polynomial. The energy-dependent logarithmic derivative of the radial wave function

$$j_\ell\left(|k + G_m|a\right) \frac{R_\ell'(a;E)}{R_\ell(a;E)} j_\ell\left(|k + G_n|a\right)$$

found in F_{mn} in Equation 4.76 in the APW representation is now replaced by

$$
\begin{aligned}
a &\left[\left(E_v - E\right)\left(A_{k+G_m}^\ell A_{k+G_m}^\ell + B_{k+G_m}^\ell B_{k+G_n}^\ell N_\ell\right) \right] \\
&+ \dot{R}_\ell R_\ell' \left\{ j_\ell'\left(|k + G_m|a\right) \cdot j_\ell\left(|k + G_n|a\right) + j_\ell\left(|k + G_m|a\right) \cdot j_\ell'\left(|k + G_n|a\right) \right\} \\
&- \left\{ \dot{R}_\ell'R_\ell' j_\ell\left(|k + G_m|a\right) \cdot j_\ell\left(|k + G_n|a\right) + \dot{R}_\ell R_\ell j_\ell'\left(|k + G_m|a\right) \cdot j_\ell'\left(|k + G_n|a\right) \right\}
\end{aligned}
$$

$$(4.87)$$

where N_ℓ is given by

$$N_\ell = \int_0^a \dot{R}_\ell^2 \left(E_v, r \right) r^2 dr$$

and is a constant called the *norm*, which already appeared in Equation 4.44. In this way, the linearization is completed and the secular determinantal equation can be efficiently solved. The restriction of spherical symmetry imposed on the MT potential can be lifted to perform the electronic structure calculation as precisely as possible. The full-potential method is introduced to cope with a potential of any arbitrary shape. This is called the *full-potential linearized augmented plane wave* (FLAPW) method [24,25], which is known as a tool capable of calculating the electronic structure with the highest accuracy among various all-electron first-principles band calculation methods.

Now we consider a crystal containing more than two atoms in the unit cell, including CMAs containing more than 50 atoms in the unit cell, in the framework of the LAPW method. The MT potential experienced by an electron at the position $\mathbf{r}$ in the Wigner–Seitz cell around any atom at the position vector $\mathbf{r}_s$ in the unit cell specified by the lattice vector $\mathbf{R}$ is expressed as

$$V_{MT}(\mathbf{r}) = \begin{cases} V(\mathbf{r} - \mathbf{R} - \mathbf{r}_s) & |\mathbf{r} - \mathbf{R} - \mathbf{r}_s| \leq a_s \\ 0 & |\mathbf{r} - \mathbf{R} - \mathbf{r}_s| > a_s \end{cases} \tag{4.88}$$

where $V(\mathbf{r} - \mathbf{R} - \mathbf{r}_s)$ represents the MT potential centered at the s-th atom in the unit cell and a_s is the radius of its MT sphere. Similarly to Equation 4.78 for a monatomic crystal, the wave function of an electron at the position $\mathbf{r}$ inside the MT sphere at the position $\mathbf{R} + \mathbf{r}_s$ is written as

$$\chi_{\mathbf{k}+\mathbf{G}_n}^{MT-\alpha}(\mathbf{r}) = \exp\left[i\left(\mathbf{k} + \mathbf{G}_n \right) \cdot \left(\mathbf{R} + \mathbf{r}_s \right) \right]$$

$$\times \sum_{\ell m} \left[A_{\mathbf{k}+\mathbf{G}_n}^{\ell m} R_\ell^\alpha \left(E_v, |\mathbf{r} - \mathbf{R} - \mathbf{r}_s| \right) + B_{\mathbf{k}+\mathbf{G}_n}^{\ell m} \dot{R}_\ell^\alpha \left(E_v, |\mathbf{r} - \mathbf{R} - \mathbf{r}_s| \right) \right] Y_{\ell m} \left(\overline{\mathbf{r} - \mathbf{R} - \mathbf{r}_s} \right) \tag{4.89}$$

where the subscript α specifies the atomic species at the s-th atom.* In the intermediate range outside the MT sphere, Equation 4.70 remains valid. By introducing the step function defined as

$$\theta\left(\left|\mathbf{r}-\mathbf{R}-\mathbf{r}_s\right|\right)=\begin{cases}0 & \left|\mathbf{r}-\mathbf{R}-\mathbf{r}_s\right|\leq a_s \\ 1 & \left|\mathbf{r}-\mathbf{R}-\mathbf{r}_s\right|>a_s\end{cases} \tag{4.90}$$

we can write down the LAPW basis function in the following form:

$$\chi_{\mathbf{k}+\mathbf{G}_n}(\mathbf{r})=\left[\sum_{s=1}^{N}\left\{1-\theta\left(\left|\mathbf{r}-\mathbf{R}-\mathbf{r}_s\right|\right)\right\}\chi_{\mathbf{k}+\mathbf{G}_n}^{MT-\alpha}(\mathbf{r})\right]$$

$$+\left[\prod_{s=1}^{N}\theta\left(\left|\mathbf{r}-\mathbf{R}-\mathbf{r}_s\right|\right)\right]\chi_{\mathbf{k}+\mathbf{G}_n}^{inter}(\mathbf{r}) \tag{4.91}$$

As is clear from Equation 4.91, we need to sum up contributions from all MT spheres for a crystal containing N atoms in the unit cell. As a final step, we need to require the Bloch condition on Equation 4.91 and construct the wave function of an electron propagating through a crystal by summing it over allowed reciprocal lattice vectors in the same way as that in Equations 4.74 and 4.82.

Let us consider the case of gamma-brass as an example again. Firstly, the summation over azimuthal quantum number ℓ and magnetic quantum number m is carried out up to about $\ell = 8$. There are 26 independent atoms for the gamma-brass with space group $I\bar{4}3m$. We construct Equation 4.91 by calculating the partial radial wave function $R_\ell(E_v,r)$ and its derivative $\dot{R}_\ell(E_v,r)$ in each MT sphere. One can clearly see how the construction

---

* In order to smoothly connect $\chi_{\mathbf{k}+\mathbf{G}_n}^{MT-\alpha}(\mathbf{r})$ centered at $\mathbf{R} + \mathbf{r}_s$ with the plane wave $\chi_{\mathbf{k}+\mathbf{G}_n}^{inter}(\mathbf{r})$ across the MT sphere, we can rewrite Equation 4.71 as

$$\chi_{\mathbf{k}+\mathbf{G}_n}^{inter}(\mathbf{r})=\Omega^{-1/2}\exp[i(\mathbf{k}+\mathbf{G}_n)\cdot\mathbf{r}]=\Omega^{-1/2}\exp[i(\mathbf{k}+\mathbf{G}_n)\cdot(\mathbf{R}+\mathbf{r}_s)]\cdot\exp[i(\mathbf{k}+\mathbf{G}_n)\cdot(\mathbf{r}-\mathbf{R}-\mathbf{r}_s)]$$

$$=\Omega^{-1/2}\exp[i(\mathbf{k}+\mathbf{G}_n)\cdot(\mathbf{R}+\mathbf{r}_s)]$$

$$\times 4\pi\sum_{\ell=0}^{\infty}\sum_{m=-\ell}^{\ell}i^\ell j_\ell\left(\left|\mathbf{k}+\mathbf{G}_n\right|\cdot\left|\mathbf{r}-\mathbf{R}-\mathbf{r}_s\right|\right)Y_{\ell m}^*\widehat{(\mathbf{k}+\mathbf{G}_n)}Y_{\ell m}\widehat{(\mathbf{r}-\mathbf{R}-\mathbf{r}_s)}$$

This is the reason why the phase shift $\exp[i(\mathbf{k}+\mathbf{G}_n)\cdot(\mathbf{R}+\mathbf{r}_s)]$ appears in Equation 4.89.

of the FLAPW basis function $\chi_{k+G_n}(\mathbf{r})$ becomes complex when dealing with CMAs. As mentioned above, the FLAPW basis function $\chi_{k+G_n}(\mathbf{r})$ must be summed over all allowed reciprocal lattice vectors, the number of which generally exceeds about 2500.* Obviously, the secular equation constructed from Equation 4.83 consists of the number of rows and columns equal to that of the reciprocal lattice vectors employed. In the case of the gamma-brass, one has to solve the secular determinantal equation having 2500×2500 up to 5000×5000 matrix. This is contrasted with the LMTO-ASA method, where the secular determinant consists of 239×239 matrix for Cu_5Zn_8 gamma-brass, as mentioned in Section 4.10. In addition, one needs to compute the canonical structure factor only once for a given structure in the LMTO-ASA.

The FLAPW method has to handle a much larger determinant to solve and to require a much larger capacity of memories and a longer computation time. This is a big disadvantage in comparison with the LMTO-ASA method upon applying to CMAs like 1/1-1/1-1/1 approximants containing more than one hundred atoms in the unit cell. This is the reason why the LMTO-ASA band calculations have been almost exclusively employed in the past to calculate the electronic structure of 1/1-1/1-1/1 approximants. However, some attempts have been recently made to perform FLAPW band calculations even for such CMAs. For example, Zijlstra and Bose [26] made FLAPW band calculations by approximating the structure of the Al-Pd-Mn quasicrystal by a model structure containing 65 atoms in the unit cell on the basis of the Quandt-Elser model [27]. More recently, Mizutani et al. [28] could determine the effective $\mathbf{e/a}$ value for transition metal elements Fe and Ru by performing FLAPW electronic structure calculations for $Al_{108}Cu_6TM_{24}Si_6$ (TM = Fe and Ru) approximants containing 144 atoms in the unit cell with space group $Pm\overline{3}$. The details will be discussed in Chapters 9 and 10.

Before ending Chapter 4, we discuss why the FLAPW method is best suited for extracting the FsBz interaction. Since we are interested in the mechanism for the formation of a pseudogap in the very vicinity of the Fermi level, we are in a position to examine if it is created by forming stationary waves as a result of interference of electron waves having the Fermi energy with a specific set of lattice planes. The FLAPW wave

---

* The number of reciprocal lattice vectors is generally taken up to about 50 times the number of atoms in the unit cell for a 3d-transition metal alloy system.

function (4.82) outside the MT sphere is constructed by superimposing the plane waves over reciprocal lattice vectors allowed to a given structure. The presence of a FsBz-induced pseudogap will be confirmed by extracting the dominant plane wave of the wave vector $\mathbf{k} + \mathbf{G}_n$ in the wave function (4.82) at the principal symmetry points like the points N in the case of the bcc Brillouin zone, where the stationary waves are formed. This can be done by plotting the square of the plane wave component outside the MT sphere as a function of the square of the allowed reciprocal lattice vector at the symmetry point chosen. This is indeed the execution of the Fourier spectrum analysis and will be introduced as the *FLAPW-Fourier method* in Chapter 7. This is obviously a standard approach in NFE band calculations. However, as emphasized in connection with Figure 4.11, the NFE model is not suitably applicable to a system involving the d-band in its valence band. In contrast, the FLAPW method chooses FLAPW-orbitals as basis functions and allows us to perform precise first-principles band calculations for any realistic crystals and to extract the FsBz interaction in the same way as in NFE band calculations.

The program package WIEN2k developed by P. Blaha, K. Schwarz, G. Madsen, D. Kvasnicka and J. Luitz, Institut für Materialchemie, Technische Universität Wien, Austria, is commercially available to perform electronic structure calculations of crystals using density functional theory (DFT) [29]. It is based on the FLAPW method plus local orbitals (lo) method, including relativistic effects. Both the FLAPW-package developed by Freeman's group [24,25] and the WIEN2k package were employed to perform first-principles FLAPW band calculations, as will be discussed in the remaining chapters.

REFERENCES

1. N.F. Mott and H. Jones, *The Theory of the Properties of Metals and Alloys* (Oxford University Press, England, 1936).
2. U. Mizutani, *Introduction to the Electron Theory of Metals* (Cambridge University Press, Cambridge, 2001).
3. C. Kittel, *Introduction to Solid State Physics* (Third edition, John Wiley & Sons, New York, 1967).
4. T. Fujiwara, *Phys. Rev. B* 40 (1989) 942.
5. U. Mizutani, *Prog. Mat. Sci.* 49 (2004) 227.
6. U. Mizutani, *The Science of Complex Alloy Phases* (edited by T.B. Massalski and P.E.A. Turchi), TMS (The Minerals, Metals & Materials Society, 2005) pp. 1–42.
7. G.T. Laissardiere, D. Nguyen-Manh, and D. Mayou, *Prog. Mat. Sci.* 50 (2005) 679.

8. P. Hohenberg and W. Kohn, *Phys. Rev.* 136 (1964) B864.

9. W. Kohn and L.J. Sham, *Phys. Rev.* 140 (1965) A1133.

10. M. Schlüter and L.J. Sham, *Physics Today* 35 (February 1982) 36.

11. J. Perdew, K. Burke, and M. Ernzerhof, *Phys. Rev. Lett.* 77 (1996) 3865.

12. E. Wigner and F. Seitz, *Phys. Rev.* 43 (1933) 804.

13. J.C. Slater, *Phys. Rev.* 51 (1937) 846.

14. U. Mizutani, *MATERIA* (in Japanese), 45, No.9 (2006) 677.

15. H.L. Skriver, *The LMTO Method*, Springer series in Solid-State Sciences 41 (Springer-Verlag, Berlin, 1984).

16. J. Korringa, *Physica*, 13 (1947) 392.

17. W. Kohn and N. Rostoker, *Phys. Rev.* 94 (1954) 1111.

18. J.S. Faulkner, *Prog. Mat. Sci.* 27 (1982) 1.

19. O.K. Andersen, W. Klose, and H. Nohl, *Phys. Rev.* B 17 (1978) 1209.

20. O.K. Andersen, *Phys. Rev.* B12 (1975) 3060.

21. D.D. Koelling and G.O. Arbman, *J. Phys. F: Metal Phys.* 5 (1975) 2041.

22. U. Mizutani, T. Takeuchi, and H. Sato, *Prog. Mat. Sci.* 49, Nos. 3–4 (2004) 227.

23. D.J. Singh, *Planewaves, Pseudopotentials and the LAPW Method* (Kluwer Academic Publishers, Boston, 1994).

24. E. Wimmer, H. Krakauer, M. Weinert, and A.J. Freeman, *Phys. Rev.* B 24 (1981) 864.

25. M. Weinert, E. Wimmer, and A.J. Freeman, *Phys. Rev.* B 26 (1982) 4571.

26. E.S. Zijlstra and S.K. Bose, *Phil. Mag.* 86, Nos. 6–8 (2006) 717.

27. A. Quandt and V. Elser, *Phys. Rev.* B 61 (2000) 9336.

28. U. Mizutani, R. Asahi, T. Takeuchi, H. Sato, O.Y. Kontsevoi, and A.J. Freeman, *Z. Kristallogr.* 224 (2009) 17.

29. http://www.wien2k.at/; P. Blaha, K. Schwarz, P. Sorantin and S.B. Trickey, *Comput. Phys. Commun.* 59 (1990) 399.

CHAPTER 5

Hume-Rothery Electron Concentration Rule Concerning the α/β Phase Transformation in Cu-X (X = Zn, Ga, Ge, etc.) Alloy Systems

5.1 STABLE FCC-Cu VERSUS METASTABLE BCC-Cu

In Chapter 3, we introduced the model of Jones (I) [1], which was put forward in 1937 to interpret the Hume-Rothery electron concentration rule concerning the α/β phase transformation in the Cu-Zn alloy system. However, we had to point out various difficulties in his theory based on the NFE model. The best way to overcome difficulties in the model of Jones (I) will be to perform first-principles electronic structure calculations (see Chapter 4) for fcc- and bcc-Cu as accurately as possible and to compare the respective valence-band structure energies as a function of **VEC** within the context of a rigid-band model. Though the present monograph is dedicated to the structurally complex metallic alloys (CMAs), we will attempt in this Chapter 5 to perform the FLAPW band calculations for fcc- and bcc-Cu in order to make clear to what extent the Hume-Rothery electron concentration rule concerning the α/β phase transformation can be explained within the rigid-band model.

Paxton et al. [2] made such attempts in 1997 by performing LMTO-ASA band calculations for fcc- and bcc-Cu within the rigid-band model. They employed the frozen potential approximation such that a self-consistent potential generated for fcc-Cu is simply transferred to bcc-Cu. This is quite convenient, since, to the first order in the potential difference, the valence-band structure energy difference ΔU_v between fcc- and bcc-Cu is equal to the total-energy difference ΔU_{total} [3]. But their approach is not fully self-consistent. They further assumed that "all calculations were done at the measured volume of β-CuZn." This is apparently meant to use the volume per atom V_a of the CuZn B2-compound for both fcc- and bcc-Cu under the assumption of $V_a^{bcc} = V_a^{fcc}$.*

In Chapter 5, we determine the volume per atom for both the fcc- and bcc-Cu by minimizing the respective total-energies in FLAPW band calculations with respect to their lattice constants. As will be shown below, V_a^{bcc} is found to be slightly larger than V_a^{fcc} beyond the accuracy of the resolution. Using the optimized atomic structure of fcc- and bcc-Cu thus obtained, we calculate their DOSs and the valence-band structure energy difference ΔU_v and discuss the α/β phase transformation using their **VEC** dependences within the rigid-band model.

5.2 FIRST-PRINCIPLES BAND CALCULATIONS FOR FCC- AND BCC-Cu IN LITERATURE

Prior to the discussion on the α/β phase transformation in the Cu-Zn system, we first review the first-principles band calculations performed on both fcc- and bcc-Cu in the past two decades. Typical ground-state properties for fcc- and bcc-Cu are summarized in Table 5.1 [5–9]. The cohesive energy difference between fcc- and bcc-Cu, which is defined as $\Delta\varepsilon_{coh} = (\varepsilon_{coh}^{fcc} - \varepsilon_{coh}^{bcc})$, is always positive in favor of the fcc-phase relative to the bcc-phase and is distributed over the range from 0.02 to 0.04 eV/atom or 2 to 4 kJ/mol.† This is in a perfect agreement with the experimental

---

* According to literature [4], the measured lattice constant is 0.29539 nm for the CuZn B2-compound at the composition 47.66 at.% Zn. Its volume per atom is calculated to be 0.01289 nm³, which is 9.8% larger than the literature value of 0.01174 nm³ for fcc-Cu [4].

† The relation $\Delta\varepsilon_{coh} = -\Delta U_{total}$ holds, where $\Delta\varepsilon_{coh} = \varepsilon_{coh}^{fcc} - \varepsilon_{coh}^{bcc}$ and $\Delta U_{total} = U_{total}^{fcc} - U_{total}^{bcc}$. As shown in Figure 2.1, ε_{coh} is defined relative to the ionization energy and represents a quantity positive in sign, whereas U_{total} is defined relative to an infinity and represents a quantity negative in sign.

evidence that fcc-Cu exists as a stable phase in the equilibrium phase diagram. The lattice constant and, hence, the volume per atom can be determined from a minimum in the volume dependence of the total-energy. As listed in Table 5.1, the difference in the calculated volume per atom between fcc- and bcc-Cu is quite small and is scattered over the range of +0.55% to −0.49%, depending on the data in literature [5–9]. Thus, attention is directed to the need of determining the most accurately the volume per atom for both fcc- and bcc-Cu.

5.3 FLAPW ELECTRONIC STRUCTURE CALCULATIONS FOR FCC- AND BCC-Cu

The electronic structure and the total-energy for fcc- and bcc-Cu are calculated, using the commercially available WIEN2k program package based on the FLAPW method [10] and the FLAPW-program package developed by Freeman et al. [11–13]. In both cases, the exchange and correlation energy functional of the electrons is calculated within the generalized gradient approximation coupled with Perdew–Burke–Ernzerhof hybrid scheme (GGA-PBE) [14]. The generation of the muffin-tin (MT) potential is self-consistently made for both fcc- and bcc-Cu. Preliminary FLAPW band calculations are performed by varying the radius of the MT sphere, r_{MT}, over the range 0.1164 to 0.128 nm for both fcc- and bcc-Cu. It turned out that the effect of r_{MT} on the DOS and ΔU_V was very small. In the present calculations using both the WIEN2k and Freeman-program packages, the value of r_{MT} is fixed to be 0.1164 nm (=2.2 a.u.) for both fcc- and bcc-Cu.

At first, the total-energy U_{total} for both fcc- and bcc-Cu is computed, using the WIEN2k package, by varying the respective lattice constants. Note that the value of r_{MT} = 0.1164 nm is small enough to avoid any overlap of the neighboring MT spheres upon volume contraction in both phases. As shown in Figure 5.1, U_{total} takes its minimum at the lattice constants of 0.36301 and 0.28864 nm for fcc- and bcc-Cu, respectively. The same conclusion is reached, using the Freeman-program package. The volume per atom for bcc-Cu becomes 0.54% larger than that for fcc-Cu, that is, $V_a^{bcc} = 1.0054 V_a^{fcc}$. This is taken as a clear indication of the breakdown of the condition $V_a^{bcc} = V_a^{fcc}$ in the self-consistent calculations. The value of U_{total} for fcc-Cu turns out to be 0.038 eV/atom or 3.7 kJ/mol lower than that of bcc-Cu. All these ground-state properties are quite consistent with calculated ones reported in literature [5–9], as summarized in Table 5.1.

TABLE 5.1 Ground-State Properties of fcc-Cu and bcc-Cu in Literature

	Lattice Constant a [nm]		Volume per Atom V_a [(nm)^3]		Cohesive Energy ε_{coh} [eV/atom]		$\Delta\varepsilon_{coh}$ (=$\varepsilon_{coh}^{fcc} - \varepsilon_{coh}^{bcc}$) [eV/atom]	$(V_{fcc} - V_{bcc})/V_{fcc}$ [%]	Ref
	fcc-Cu	bcc-Cu	fcc-Cu	bcc-Cu	fcc-Cu	bcc-Cu			
PP-GO-LDA	0.362	0.287	0.01186	0.01182	3.83	3.81	0.02	+0.33	5
LAPW-NR-LDA	0.361	0.286	0.01176	0.01169	4.14	4.12	0.02	+0.55	6
PP-LDA	0.361	0.287	0.01176	0.01182	4.37	4.33	0.0372	−0.49	7
VASP-GGA	0.364	0.289	0.01207	0.01212	3.763	3.727	0.036	−0.41	8
FLAPW-GGA	0.362	0.290	0.01186	0.01219	3.76	3.72	0.04	−0.02	9

Source: J.R. Chelikowsky and M.Y. Chou, *Phys. Rev.* B38 (1988) 7966; Z.W. Lu, S.-H. Wei, and A. Zunger, *Phys. Rev.* B 41 (1990) 2699; S. Jeong, *Phys. Rev.* B 53 (1996) 13973; C. Domain and C.S. Becquart, *Phys. Rev.* B 65 (2001) 024103; Z. Tang, M. Hasegawa, Y. Nagai, and M. Saito, *Phys. Rev.* B 65 (2002) 195108.

Note: PP-GO: pseudopotentials with local orbital basis consisting of Gaussians; LDA: local density approximation; NR: non-relativistic; PP-LDA: pseudopotential plane-wave basis; VASP-GGA: Vienna ab-initio simulation package with general-ized gradient approximations

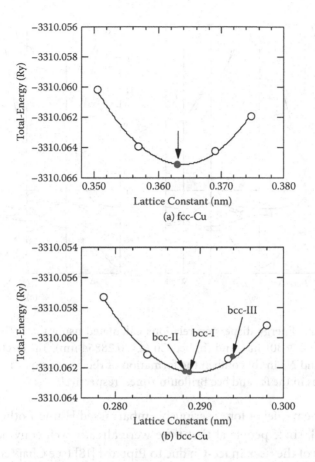

FIGURE 5.1 Total-energy calculated using the WIEN2k, as a function of the lattice constant for fcc- and bcc-Cu. The radius of the MT sphere r_{MT} is fixed to be 0.1164 nm. Data points marked with solid circle refer to fcc-Cu, bcc-I, -II, and -III, for which the DOS and valence-band structure energies were calculated.

The energy dispersion relations for the optimized structures of fcc- and bcc-Cu are calculated using the WIEN2k and are depicted in Figure 5.2a,b, respectively. The results for fcc-Cu are in good agreement with previous results [2,5,9,15,16]. As repeatedly mentioned in Chapter 1, Section 1.2 and Chapter 3, Section 3.2, the Fermi surface of fcc-Cu is characterized by the neck centered at the principal symmetry points L of its Brillouin zone. The corresponding electronic state is marked as L_2' in Figure 5.2a. Its location below the Fermi level can be taken as a theoretical confirmation of the presence of the neck across the {111} zone planes of the fcc Brillouin zone. As a matter of fact, the assumption of the Fermi surface with a diminishing

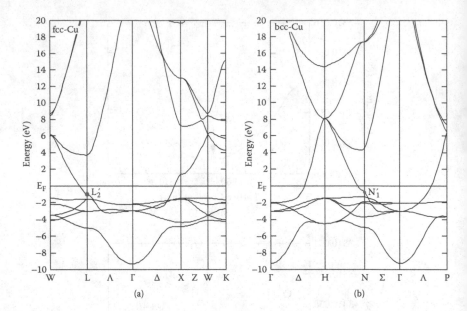

FIGURE 5.2 Energy dispersion relations calculated using the WIEN2k for (a) fcc-Cu (a = 0.36301 nm) and (b) bcc-Cu (a = 0.28864 nm). The electronic states L_2' in (a) and N_1' in (b) confirm the formation of the neck across {111} and {110} zone planes in the fcc and bcc Brillouin zones, respectively.

neck in the model of Jones seriously embarrassed Hume-Rothery in early 1960s [17], since people at that time were already well convinced by the discovery of the neck in fcc-Cu due to Pippard [18] (see Chapter 1, Section 1.2 and Chapter 3, Section 3.2).

As shown in Figure 5.2b, the electronic state marked as N_1' in bcc-Cu is also located below the Fermi level. This confirms the presence of the neck against the {110} zone planes in agreement with previous reports [2,5,9]. Thus, the van Hove singularity due to the contact of the Fermi surface with the {110} zone planes should appear below the Fermi level in the DOS. This is again at variance with the model of Jones [1], in which it is located far above the Fermi level (see Chapter 3, Figure 3.3a).

5.4 TOTAL-ENERGY AND VALENCE-BAND STRUCTURE ENERGY

According to the Kohn-Sham formulation of the Density-Functional Theory (DFT) [19] (see Chapter 4, Section 4.3), the total-energy of a system is expressed as

$$U_{\text{total}} = \sum_i \varepsilon_i - \left\{ \begin{array}{l} \dfrac{1}{2} \iint \dfrac{n(\mathbf{r})n(\mathbf{r}')}{|\mathbf{r}-\mathbf{r}'|} d\mathbf{r} d\mathbf{r}' \\[12pt] - \displaystyle\int n(\mathbf{r}) \left[\varepsilon_{XC}\big(n(\mathbf{r})\big) - \mu_{XC}\big(n(\mathbf{r})\big) \right] d\mathbf{r} \end{array} \right\} \tag{5.1}$$

where the first term represents the one-electron energy due to both core and valence electrons, the second term known as the Hartree energy represents the contribution due to the electron–nuclei interaction plus average electron-electron interaction, and the third term represents the exchange and correlation energy. As mentioned by Cohen et al. [20], the Kohn–Sham formulation allows the eigenvalues to be shifted by an arbitrary constant V_0 called the "muffin-tin zero." In self-consistent total-energy calculations, the value of this constant completely cancels, as long as we discuss a quantity given by a difference in the total-energy between the two competing phases.

Fcc-Cu with the lattice constant of 0.36301 nm is taken as a reference. Instead, we consider the bcc-Cu structures having the following three different lattice constants (see Figure 5.1): (I) 0.28864 nm derived from a minimum in the total-energy curve, (II) 0.28812 nm derived by inserting 0.36301 nm into the relation $V_a^{bcc} = V_a^{fcc}$, that is, assuming the same volume per atom for both phases and (III) 0.29388 nm, which is intentionally increased by 1.8% relative to bcc-I. Note that a difference in the lattice constant between bcc-I and –II is merely 0.18%.

The total-energy is calculated for these four structures, using the WIEN2k package. The total-energy difference ΔU_{total} $(= U_{\text{total}}^{fcc} - U_{\text{total}}^{bcc})$ between fcc- and bcc-Cu is summarized in Table 5.2. The value of ΔU_{total} = −0.0384 eV/atom is found, regardless of whether bcc-I or bcc-II is employed, and is well consistent with the data in Table 5.1. This means that no discernible difference arises in the total-energy difference ΔU_{total}, regardless of whether the bcc-I or the bcc-II is employed. However, the value of ΔU_{total} is increased to −0.05207 eV/atom, when bcc-III is employed.

Equation 5.1 may be rearranged into the following form:

$$U_{\text{total}} = U_v + U_c + \left\{ U_{pot.\,inside\,MT} + U_{pot.\,outside\,MT} \right\} \tag{5.2}$$

TABLE 5.2   Total-Energies for fcc-Cu and Three bcc Structures I, II, and III, Calculated Using the WIEN2k Package

	fcc-Cu	bcc-I	bcc-II	bcc-III
Lattice constant a [nm]	0.36301	0.28864	0.28812	0.29388
Volume per atom V_a [(nm)^3]	0.011959	0.012024	0.011959	0.012691
Radius of the MT sphere r_{MT} [nm]	0.1164	0.1164	0.1164	0.1164
Total-energy difference $\Delta U_{total} = U_{total}^{fcc} - U_{total}^{bcc}$ [eV/atom]	—	−0.03842 (−3.71 kJ/mol)	−0.03844 (−3.71 kJ/mol)	−0.05207 (−5.02 kJ/mol)
Binding energy at the bottom of the valence band E_o [eV]	−9.33834	−9.26705	−9.30773	−8.86663

where the first term in Equation 5.1 is decomposed into the kinetic energy of valence electrons, U_v, and the binding energy of core electrons, U_c, and the potential energy in the curly bracket in Equation 5.1 is divided into those from the regions inside and outside the MT sphere. Each contribution in Equation 5.2 for the fcc- and bcc-Cu is evaluated, using the Freeman-program package [12,13], as listed in Table 5.3.

According to Table 5.3, the valence-band structure energy U_v looks small compared with U_c and $U_{pot.outside\,MT}$. This is not true, since the energy scale used in the calculation of the total-energy is arbitrary due to the presence of the muffin-tin zero V_0. According to the free electron model, it is given by $U_v = 3E_F/5$, which roughly amounts to about +4 eV/atom or +400 kJ/mole for fcc-Cu and is almost one-half in magnitude as large as the potential energy caused by the remaining terms in Equation 5.2. More important is that only U_v in Equation 5.2 acts against stabilization, since it represents the repulsive energy of valence electrons caused by the Pauli exclusion principle. In other words, nature would always try to reduce U_v as much as possible through structural transformation for a given alloy.

As emphasized above, the total-energy difference ΔU_{total} is crucially important in the discussion of relative stability of the two competing phases. An energy difference ΔU due to each contribution in Equation 5.2 is listed in a separate column in Table 5.3 in the units of eV/atom.

TABLE 5.3   Contributions to the Total-Energy in Equation 5.2 for fcc- and bcc-Cu Calculated Using the FLAPW Freeman-Package

	fcc-Cu	bcc-Cu (bcc-I)	$\Delta U = U^{fcc} - U^{bcc}$	[eV/atom]
Lattice constant a [nm]	0.36301	0.28864	ΔU_{total}	−0.035813
Volume per atom V_a [(nm)^3]	0.011959	0.012024	ΔU_v	+0.5585849
U_{total} [Ry]	−3310.07765	−3310.07502	ΔU_c	+0.9356014
U_v [Ry]	4.42024	4.37918	$\Delta U_{pot.inside\,MT}$	+0.02720
U_c [Ry]	−1895.97947	−1896.04824	$\Delta U_{pot.outside\,MT}$	−1.557201
$U_{pot.inside\,MT}$ [Ry]	−0.20409	−0.20609		
$U_{pot.outside\,MT}$ [Ry]	−1418.31433	−1418.19988		

Surprisingly, all the contributions except for the potential energy outside the MT sphere act in favor of bcc-Cu. Indeed, this last term, the potential energy outside the MT sphere, is responsible for the stabilization of fcc-Cu.

Ideally speaking, one should evaluate accurately each term in Equation 5.2 for every Cu-Zn alloy, say at 10 at.%Zn interval across both α- and β-phase regions and show if the total-energy difference ΔU_{total} changes its sign at the α/β phase boundary. However, FLAPW band calculations become an extremely formidable task, as soon as we start to work on the α-phase Cu-Zn alloys. This is because chemical disorder due to the addition of Zn into the Cu matrix hampers a well-defined unit cell. As a possible compromise, one may use either the virtual-crystal approximation (VCA) or the super-cell approximation (see Footnote * on page 44 in Chapter 3, Section 3.2). A more practical and conventional way to circumvent this difficulty is to employ a rigid-band model. When applied to the Cu-Zn alloy system, the electronic structures of fcc- and bcc-Cu are assumed to be the same as those of any Zn concentrations, and the position of the Fermi level is simply determined by the amount of electrons accommodated in the respective valence bands. In addition to the rigid-band model above, we assume the **VEC** dependence of ΔU_{total} to originate solely from that of ΔU_v in Equation 5.2. Now the discussion on the Hume-Rothery electron concentration rule regarding the α/β-phase transformation will be made in an attempt to interpret the **VEC** dependence of ΔU_v in terms of the FsBz interactions involved in fcc- and bcc-phases.

In the framework of the rigid-band model discussed above, the **VEC** dependence of $\Delta U_v = (U_v^{fcc} - U_v^{bcc})$ can be easily pursued in the same way as Jones did [1]. But a basic difference from Jones exists in the use of **VEC** instead of **e/a** as an electron concentration parameter in first-principles band calculations. The value of **VEC** is calculated by integrating the relevant DOS from the bottom of the valence band, E_0, up to an arbitrary energy E:

$$\mathbf{VEC}(E) = \int_{E_0}^{E} D(E')dE' \qquad (5.3)$$

and the valence-band structure energy U_v is likewise expressed as

$$U_v(E) = \int_{E_0}^{E} D(E')\left(E' - E_0\right)dE' \qquad (5.4)$$

where $D(E)$ is the DOS of the valence band. Note that Equation 5.3 is the same as Equation 3.8 except for the use of **VEC** in place of **e/a**, since the Cu-3d band is now fully taken into account. The value of ΔU_v can be easily evaluated, once the two DOSs for the competing phases are calculated (see Equation 3.5).

In the remainder of this chapter, we try to determine the **VEC** dependence of ΔU_v as accurately as possible in the framework of the rigid-band model. For this purpose, the DOSs for both fcc- and bcc-Cu must be calculated with the least statistical errors to locate the extremely small van Hove singularities due to the FsBz interactions.

5.5 DOS FOR FCC- AND BCC-Cu

Now we are ready to discuss the DOS for fcc-Cu and bcc-I, -II and –III. Energy bands remain unchanged under operations of some particular rotations and permutations in the Brillouin zone. Thus, the energy information we wish to obtain can be reduced to an "irreducible wedge" containing only 1/48th of the fcc and bcc Brillouin zones. This is illustrated in Figure 5.3. To minimize a statistical scatter in counting the DOS and to reliably extract small van Hove singularities, we partitioned the irreducible wedge of the respective Brillouin zones into 125000 or $50 \times 50 \times 50$ segments.

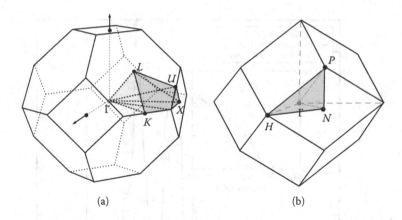

FIGURE 5.3   Irreducible wedges of the Brillouin zones of (a) fcc and (b) bcc lattices.

The DOSs for fcc-Cu, bcc-I, and bcc-III are shown in Figure 5.4. The DOS for bcc-II is not shown, since it is almost superimposed onto that for bcc-I on the scale chosen. In all three cases, a large DOS is observed over −5 to −1.5 eV in the binding energy. This obviously represents the Cu 3d-band. A deep valley in energies over −2.8 to −2.4 eV in bcc-I and -III is due to the splitting of the 3d-band into bonding and antibonding subbands. This is a distinctive feature of bcc transition metals. The valley is much shallower in fcc-Cu, resulting in a characteristic difference between the two DOSs. It is also important to mention that the d-band width for bcc-III is much narrower than that for bcc-I. This is naturally understood as the fact that the volume per atom in bcc-III is 5.5% larger than that in bcc-I: the larger the volume per atom, the weaker are the orbital hybridizations, and the narrower is the width of the valence band.

As shown in the insert to Figure 5.4, there exists a noticeable difference in the binding energy at the bottom of the valence band, E_0, depending on the structure involved. Its value is listed in Table 5.2. Rather than E_0, we take more specifically a difference in E_0 between fcc- and bcc-Cu, which is defined as $\Delta E_0 = E_0^{fcc} - E_0^{bcc}$. Figure 5.5 shows that ΔE_0 falls onto a straight line as a function of ΔV_a, which is defined as a difference in the volume per atom between fcc- and bcc-Cu, that is, $V_a^{fcc} - V_a^{bcc}$. Now a difference between bcc-I and -II sharply emerges. It is interesting to note that the value of ΔE_0 is the smallest but is still finite even for bcc-II, where the condition $V_a^{bcc} = V_a^{fcc}$ is imposed. We will show later that ΔV_a strongly affects the absolute value of ΔU_v and even its sign through ΔE_0, signaling the importance of a proper choice of ΔV_a.

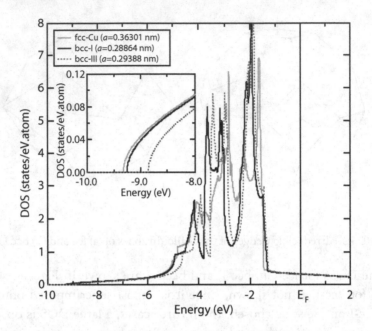

FIGURE 5.4   DOSs calculated using the WIEN2k for fcc-Cu, bcc-I, and bcc-III. An insert shows the blow-up of the DOS near the bottom of the valence band.

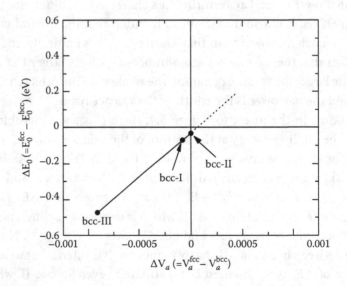

FIGURE 5.5 $\Delta E_0\ (= E_0^{fcc} - E_0^{bcc})$ as a function of $\Delta V_a\ (= V_a^{fcc} - V_a^{bcc})$ for three bcc structures, bcc-I, -II, and -III. E_0 refers to the binding energy at the bottom of the valence band.

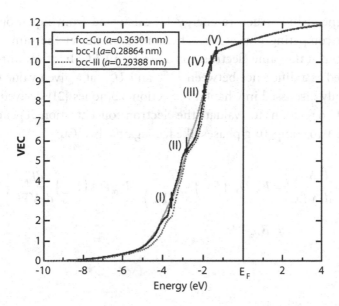

FIGURE 5.6 **VEC** as a function of the binding energy for fcc-Cu, bcc-I and bcc-III, calculated using the WIEN2k. All three curves meet at **VEC** = 11.0 at the Fermi level. Crossings between the fcc-Cu and bcc-I curves within the Cu-3d bands are marked with symbols (I) to (V).

One can easily calculate the electron concentration **VEC** by integrating the relevant DOS over energies using equation (5.3). The resulting **VEC** is plotted in Figure 5.6 as a function of the binding energy for fcc-Cu, bcc-I, and -III. We see that the value of the **VEC** reaches 11 electrons per atom at the Fermi level for all three cases, being consistent with the fact that one 4s and ten 3d electrons are accommodated in the valence band for pure Cu. It is noted that the **VEC** curve for bcc-III runs consistently below that for fcc-Cu and never crosses the fcc curve over the range from the bottom of the valence band up to the Fermi level, while the curves for bcc-I and -II (the latter not shown in Figure 5.6) cross the fcc curve several times in this energy range. Five crossings between the fcc-Cu and bcc-I curves are confirmed and are marked with symbols (I) to (V) below the top of the Cu-3d band. As will be discussed in Section 5.6, these crossings play a key role in identifying the extrema in the ΔU_v-**VEC** curve.

5.6 RELATIVE STABILITY OF THE α- AND β-PHASES IN Cu-X (X = Zn, Ga, Ge, ETC.) ALLOY SYSTEMS

Now we concentrate on studies of the **VEC** dependence of ΔU_v between fcc-Cu and bcc-I. An important remark should be exercised at this stage:

the valence-band structure energy U_v calculated from Equation 5.4 for the two competing phases must be compared not at the same binding energy but at the same electron concentration **VEC**. In other words, ΔU_v is defined as a difference between U_v^{fcc} and U_v^{bcc} at a given value of **VEC**. As already discussed in Chapter 3, Section 3.3, Jones [21] derived the following two relations to evaluate the electron concentration dependence of ΔU_v for two competing phases like fcc- against bcc-Cu:

$$\frac{d(\Delta U_V)}{d(\textbf{VEC})} = E_{fcc} D_{fcc}(E_{fcc}) \frac{dE_{fcc}}{d(\textbf{VEC})} - E_{bcc} D_{bcc}(E_{bcc}) \frac{dE_{bcc}}{d(\textbf{VEC})} \quad (5.5)$$

$$= E_{fcc} - E_{bcc}$$

and

$$\frac{d^2(\Delta U_V)}{d(\textbf{VEC})^2} = \frac{dE_{fcc}}{d(\textbf{VEC})} - \frac{dE_{bcc}}{d(\textbf{VEC})} = \frac{1}{D_{fcc}(E_{fcc})} - \frac{1}{D_{bcc}(E_{bcc})} \quad (5.6)$$

where $D_{fcc}(E_{fcc})$ and $D_{bcc}(E_{bcc})$ represent the DOSs at maximum energies E_{fcc} and E_{bcc} obtained by accommodating a given amount of **VEC** for fcc- and bcc-Cu, respectively. Note that the electron concentration parameter e/a in Equations 3.8 and 3.9 is replaced by the **VEC** in first-principles band calculations. It is now important to remind that upper limits E_{fcc} and E_{bcc} of the integral in Equation 5.3 for the two phases would be generally different to yield the same **VEC**, because the DOSs involved are different. This is particularly evident, when the upper limit is inside the Cu-3d band. However, there exist *critical energies* at which the condition $E_{fcc} = E_{bcc}$ holds and, hence, $(\textbf{VEC})_{fcc} = (\textbf{VEC})_{bcc}$ as well. This is read off from the energy at which the two **VEC** versus energy curves cross each other (see crossings marked with symbols (I) to (V) between the fcc-Cu and bcc-I in Figure 5.6).

The **VEC** dependence of ΔU_v is easily calculated by inserting the DOSs for fcc- and bcc-Cu into Equations 5.3 and 5.4 and taking the difference between U_v^{fcc} and U_v^{bcc} at the same **VEC** value. Figures 5.7a,b are constructed so as to help readers identify the maxima and minima on the resulting ΔU_v versus **VEC** curve with respect to the critical energies (I) to (V) shown by vertical dotted lines inside the Cu-3d bands in the respective

DOSs duplicated from Figure 5.4. We found that the maxima and minima in ΔU_v appear exactly at crossings marked with symbols (I) to (V) in the two **VEC** curves shown in Figure 5.6. This is taken as a demonstration for the existence of the one-to-one correspondence between the condition $E_{fcc} = E_{bcc}$ caused by singularities in the DOS and extrema in the ΔU_v-**VEC** curve in accordance with Equations 5.5 and 5.6.

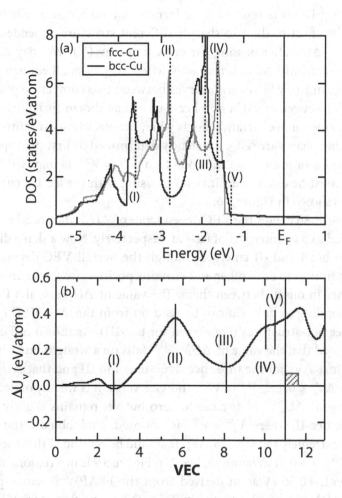

FIGURE 5.7 (a) DOS for fcc- and bcc-Cu (bcc-I). The condition $E_{fcc} = E_{bcc}$ and, hence, $(\mathbf{VEC})_{fcc} = (\mathbf{VEC})_{bcc}$ holds at energies marked with dotted lines (I) to (V). (b) **VEC** dependence of ΔU_v for fcc- and bcc-Cu. Vertical lines (I) to (V) are located at **VEC**, where the condition $E_{fcc} = E_{bcc}$ in Equation 5.5 holds within the Cu-3d band. A shaded rectangle in (b) refers to the α-phase region in the Cu-Zn alloy system. The extrema from (I) to (V) in (b) appear at **VEC**=2.74, 5.65, 8.09, 10.04, and 10.50, respectively.

Figure 5.8a shows the DOS near the edge of the Cu-3d band for the bcc-I and -II Cu in comparison with that for fcc-Cu. It is found that the DOSs for bcc-I and -II are almost superimposed onto each other and that the critical energy (V) satisfying the condition $E_{fcc} = E_{bcc}$ falls at the very top of the Cu-3d band in both cases. This indicates that the condition $E_{fcc} = E_{bcc}$ is well satisfied, once the Cu-3d band is completely filled. Each time when the condition $E_{fcc} = E_{bcc}$ is met, the history behind it in the competing DOSs is reset to zero. Hence, one no longer needs to worry about the effect of the two sharply different structure-dependent Cu-3d DOSs on ΔU_v after passing the critical energy (V). We may call it the "golden rule," which holds true only when ΔV_a is small enough to allow crossings in the VEC-energy curves between two competing phases like (I) to (V) between fcc-Cu and bcc-I or -II, as shown in Figure 5.6. The "golden rule" above certainly holds when the electronic structures of two competing phases are self-consistently determined by first-principles band calculations or even when the condition $V_a^{bcc} = V_a^{fcc}$ is imposed. Instead, bcc-III must be discarded, since no crossing with the fcc-Cu curve takes place, as shown in Figure 5.6.

Figures 5.8b,c show the VEC dependence of ΔU_v, when bcc-I and -II are employed as a counterpart of fcc-Cu, respectively. Now a sharp difference between bcc-I and -II emerges. Though the overall VEC dependence of ΔU_v is similar to each other and remains positive, we found an apparent difference in offsets between them. The value of ΔU_v, say, at VEC = 11.0 corresponding to pure Cu can be read off from the ΔU_v-VEC curve for the three bcc-structures (not shown for bcc-III). As shown in Figure 5.9, we revealed that the value of $\Delta U_v^{VEC=11.0}$ falls on a straight line when plotted against ΔV_a for the three bcc structures I to III and that $\Delta U_v^{VEC=11.0} > 0$ when $\Delta V_a < 0$, i.e., $V_a^{fcc} < V_a^{bcc}$ and vice versa. It is worthy of noting that the value of $\Delta U_v^{VEC=11.0}$ is close to zero but yet remains slightly positive even for bcc-II, where $V_a^{fcc} = V_a^{bcc}$ is imposed. Undoubtedly, the value of ΔU_v is extremely sensitive to ΔV_a. It should be reminded that the value of $\Delta U_v^{VEC=11.0} = +0.31$ eV/atom for bcc-I in Figure 5.9 is in a reasonable agreement with +0.56 eV/atom derived from the FLAPW Freeman-program package [11–13], which is listed in Table 5.3, while that of +0.10 eV/atom for bcc-II may be too small.

Now we are ready to compare the electronic structures above the critical energy (V) between fcc-Cu and bcc-I (hereafter simply referred to as bcc-Cu) and to study how the FsBz interaction or the van Hove singularity is reflected on the VEC dependence on ΔU_v. The DOSs for fcc- and

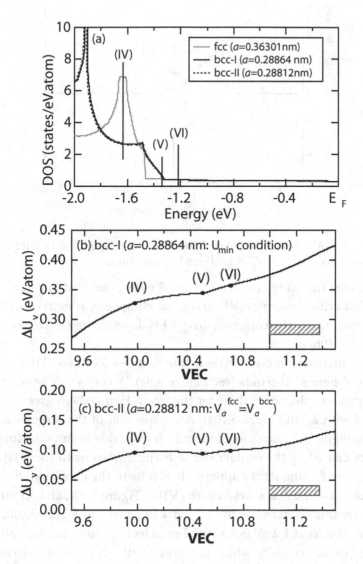

FIGURE 5.8 (a) DOS near the edge of the Cu-3d band for fcc-Cu, bcc-I, and bcc-II. (b) **VEC** dependence of $\Delta U_V = (U_v^{fcc} - U_v^{bcc-I})$, (c) **VEC** dependence of $\Delta U_V = (U_v^{fcc} - U_v^{bcc-II})$. Symbols (IV) and (V) in (a) are located at energies, where the condition $E_{fcc} = E_{bcc}$ in Equation 5.5 holds. A shaded rectangle indicates the α-phase region in the Cu-Zn alloy system.

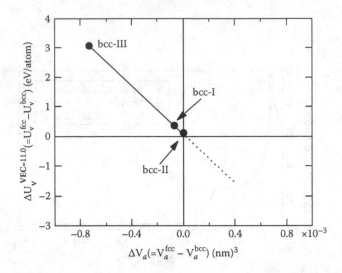

FIGURE 5.9 $\Delta U_v (= U_v^{fcc} - U_v^{bcc})$ at **VEC** = 11.0 corresponding to pure Cu as a function of $\Delta V_a (= V_a^{fcc} - V_a^{bcc})$ for three bcc structures.

bcc-Cu over the energy range –2 up to +7 eV are depicted in Figure 5.10a, where the critical energies (VI) to (IX) newly appear, as marked with vertical dotted lines. The corresponding **VEC** dependence of ΔU_v is shown in Figure 5.10b.

Sharp increases in both DOSs below about –1.5 eV signal the onset of the two different 3d-bands (see Figure 5.8a). A very weak anomaly can be located on a declining slope for the fcc-DOS at -1.39 eV (see an arrow marked with L_2' in Figure 5.10a). An inspection of the dispersion relations in Figure 5.2 allows us to identify it to be due to the van Hove singularity caused by the contact of the Fermi surface with the {111} zone planes. The L_2' singularity apparently results in the condition $E_{fcc} = E_{bcc}$ at about E = –1.2 eV, as marked with (VI) in Figure 5.10a, and, in turn, an extremely small maximum in ΔU_v at **VEC** = 10.7 in Figure 5.10b. Note that this occurs at 1.2 eV below the Fermi level of pure Cu. A small cusp at –0.63 eV in bcc-DOS, which is marked with N_1', is easily ascribed to the contact of the Fermi surface with the {110} zone planes in bcc-Cu. The N_1' singularity obviously gives rise to a tiny minimum in ΔU_v at **VEC** = 11.0 corresponding to pure Cu, as marked with (VII).

The cusp at +1.47 eV in fcc-DOS can be easily designated as the contact of the Fermi surface with the symmetry points X or the center of the {200} zone planes of the fcc Brillouin zone and is denoted as X_4'. This apparently results in a broad maximum in ΔU_v at **VEC** = 11.65, contributing to the

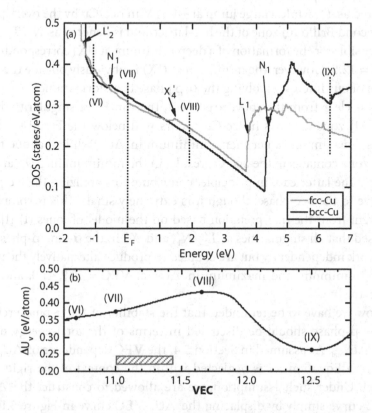

FIGURE 5.10 (a) DOS and (b) **VEC** dependence of ΔU_V in the range above the 3d-band for fcc- and bcc-Cu (bcc-I). Note that the scale on the ordinate is only one-twentieth as small as those in Figures 5.7a and 5.8a. The dotted vertical lines (VI) to (IX) in (a) refer to energies, at which the condition $E_{fcc} = E_{bcc}$ holds in equation (5.5). The corresponding extrema in (b) are marked using the same symbols as in (a). The extrema from (VI) to (VIII) in (b) appear at **VEC** = 10.7, 11.0, 11.65, and 12.50, respectively. A shaded rectangle indicates the α-phase region in the Cu-Zn alloy system. Symbols L_2' and X_4' refer to the van Hove singularities due to the contact of the Fermi surface with {111} and {002} zone planes in the fcc-Cu, respectively, while N_1' due to contact of the Fermi surface to the {110} zone planes in the bcc-Cu. Two large jumps marked as L_1 and N_1 reflect the overlap of electrons into the second Brillouin zone in both fcc- and bcc-Cu, respectively.

stabilization of the β-phase, as marked with (VIII). Finally, a large jump in the DOS at +3.74 eV in fcc-Cu is caused by the overlap of electrons across the {111} zone planes into the second Brillouin zone of the fcc lattice and is denoted as L_1, while a large jump at +4.3 eV in bcc-Cu by the overlap into the second Brillouin zone of the bcc-lattice and is denoted as N_1. They are responsible for the formation of a deep minimum at (IX) corresponding to **VEC** = 12.5. However, the minimum at (IX) is obviously outside the **VEC** range of our interest involving the α/β phase transformation.

As is clear from the above argument, the van Hove singularity due to the {111} zone contacts in fcc-Cu occurs well below the Fermi level and yields a maximum rather than a minimum in ΔU_v, while that due to the {110} zone contacts in the bcc-Cu results in the minimum in ΔU_v at **VEC** = 11.0. The latter can, in principle, contribute to stabilizing the fcc-phase relative to the bcc-phase, though it is extremely small. This is completely different from the interpretation based on the model of Jones (I) [1]. It is stressed that all singularities at L_2', N_1', and X_4' in the α- and β-phases do not work independently but are coupled to produce alternatively the maximum, minimum, and maximum in ΔU_v at **VEC** = 10.7, 11.0, and 11.6, respectively.

Now we have to be reminded that the stability of the α-phase relative to the β-phase should be discussed in terms of the total-energy difference ΔU_{total}. As assumed in Section 5.4, the **VEC** dependence of ΔU_c and ΔU_{pot} in Equation 5.2 is neglected within the context of the rigid-band model. Under such assumption, we are allowed to construct the ΔU_{total} -**VEC** curve simply by displacing the ΔU_v-**VEC** curve in Figure 5.10b as a whole downward so as to meet the condition ΔU_{total} = 0.036 eV/atom at **VEC** = 11.0 (see Table 5.3). In addition, thanks to the "golden rule" discussed above, the electron concentration parameter **VEC** may be replaced by **e/a** using the relation **e / a = VEC − 10**.* The ΔU_{total} -**e/a** curve thus obtained is shown in Figure 5.11.

The alternatively appearing minima and maxima are caused by the van Hove singularities characteristic of fcc- and bcc-Cu and are found to

---

* The relation **e/a = VEC** − 10 holds, only when a partner element to Cu gives rise to its d-band in the valence band. Included are Zn, Ga, Ge, etc. This simple relation no longer holds when a partner element is free from d-electrons. For example, in the case of the α-phase $Cu_{80}Al_{20}$ alloy, the **VEC** at the Fermi level amounts to 0.8 * 11 + 0.2 * 3 = 9.4, which is less than 10. To discuss the α/β-phase transformation more universally by including those like Cu-Al alloy system, the electron concentration parameter **e/a** should be employed. Its extraction from the FLAPW band calculations will be discussed in Chapter 7.

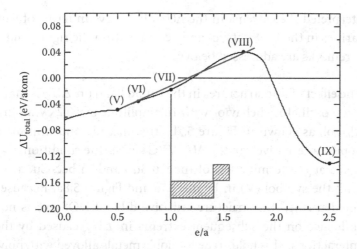

FIGURE 5.11 e/a dependence of the total-energy difference ΔU_{total} between fcc- and bcc-Cu (bcc-I). The extrema from (VI) to (VIII) caused by the van Hove singularities in the respective DOSs are located at e/a = 0.7, 1.0, 1.65, and 2.50, respectively. Red and blue rectangles indicate the α- and β-phase regions in the Cu-Zn alloy system, respectively. A red line is drawn to pass through the center of the oscillations due to van Hove singularities. Note that the depths of the mini-mum at (VII) and maximum at (VIII) relative to the line are only 0.008 and 0.02 eV/atom or 0.8 and 2 kJ/mol, respectively.

oscillate around a straight line with a positive slope, as drawn in Figure 5.11. Let us assume the background contribution drawn with the straight line to represent the VEC dependence of terms other than ΔU_v in equation (5.2). Now we can say that, relative to the straight line, the stabilization of the α-phase is driven by the van Hove singularity leading to the minimum at (VII) over the range up to about e/a = 1.2–1.3 and that of the β-phase by the van Hove singularity leading to the maximum at (VIII) in the vicinity of e/a = 1.7. But it is true that the effect of van Hove singularities on ΔU_{total} is merely of the order of 0.01 to 0.02 eV/atom or 1 to 2 kJ/mol. The present analysis based on the rigid-band model is applicable not only to the Cu-Zn alloy system but also to Cu-based alloys like the Cu-X (X = Zn, Ga, Ge, etc.), since the VEC or e/a is only a critical parameter involved.

5.7 SUMMARY

We discussed the Hume-Rothery electron concentration rule regarding the α/β phase transformation in Cu-X (X = Zn, Ga, Ge, etc.) alloy systems by performing self-consistent FLAPW band calculations for fcc- and bcc-Cu and interpreted the extrema in the ΔU_v-**VEC** curve in terms of van Hove singularities in the DOSs of fcc- and bcc-Cu within the rigid-band model. Five remarks are addressed below:

1. The effect of fine structures in the Cu-3d band on ΔU_v is manifested as an oscillatory behavior with an amplitude of ±0.2 eV/atom or ±20 kJ/mol, as shown in Figure 5.7b. The maxima and minima always appear alternatively in the ΔU_v-**VEC** curve. The condition $E_{fcc} = E_{bcc}$ holds at the termination of the Cu-3d-band in bcc-Cu, as marked with the symbol (V) in Figure 5.7a and Figure 5.8a. Because of the "golden rule," the structure-sensitive d-band DOS exerts no direct influence on the subsequent extrema in ΔU_v caused by the FsBz interactions. This holds true for noble metals alloyed with polyvalent elements.

2. An offset in ΔU_v is found to depend sensitively on ΔV_a between fcc- and bcc-Cu. A choice of the fcc- and bcc-structures to satisfy $V_a^{bcc} = V_a^{fcc}$ leads to the smallest offsets of ΔE_0 and ΔU_v. But we consider it to be artificial and, instead, should determine both structures self-consistently by minimizing the total-energy with respect to the lattice constant (see Tables 5.2 and 5.3). The contributions other than ΔU_v in Equation 5.2 amount to approximately 0.4 eV/atom or 40 kJ/mol, which is quite large. The evaluation of these contributions is left for future work.

3. We naturally doubt whether the rigid-band model based on fcc- and bcc-Cu is extendable to **VEC** = 11.5, where the CuZn B2-compound exists as a stable phase in the phase diagram. The electronic structure of the CuZn B2-compound can be also rigorously calculated from first-principles FLAPW method, since it is again free from any chemical disorder [22] (see Chapter 10, Section 10.3). However, we cannot construct the fcc-structure free from any chemical disorder at the same **VEC** as its counterpart. In this sense, the choice of fcc- and bcc-Cu would be the best at the moment to study the α/β phase transformation from first-principles band calculations.

4. In principle, we have to deal with disordered alloys through solid solution ranges of the α- and β-phases. As mentioned in Section 5.2, either the virtual crystal approximation (VCA) or a super-cell method may be better used rather than the rigid-band model. This

will allow us to evaluate the **VEC** dependence of all the terms in Equation 5.2 and to demonstrate that the **VEC** dependence of ΔU_v would determine an essential feature in the ΔU_{total} -**VEC** curve.

5. The van Hove singularities in both fcc- and bcc-Cu are so small that their effect on ΔU_v is at most 0.01 to 0.02 eV/atom or 1 to 2 kJ/mol at the extrema (VII) and (VIII) in Figure 5.11. Because of this delicate argument involved, we may still need some reservation to say that Figure 5.11 well explains the **VEC** dependent α/β-phase transformation in noble metal alloys. However, we do not need to take this so pessimistically in the case of structurally complex metallic alloys (CMAs), since ΔU_v becomes much larger than that revealed in structurally simple metals like pure Cu treated in this chapter. As described in Chapter 2, Section 2.3, ΔU_v should become 20–60 kJ/mol or 0.2–0.6 eV/atom in systems, where a deep pseudogap of 500–1500 meV wide is formed across the Fermi level (see both Chapter 2, Section 2.3 and Chapter 7, Section 7.2).

REFERENCES

1. H. Jones, *Proc. Phys. Soc.* (1937) 49, 250.
2. A.T. Paxton, M. Methfessel, and D.G. Pettifor, *Proc. Roy. Soc. Lond.* A 453 (1997) 1493.
3. D.G. Pettifor, *Commun. Phys.* 1 (1977) 141.
4. P. Villars, *Pearson's Handbook* (ASM, Materials Park, OH, 1997).
5. J.R. Chelikowsky and M.Y. Chou, *Phys. Rev.* B38 (1988) 7966.
6. Z.W. Lu, S.-H. Wei, and A. Zunger, *Phys. Rev.* B 41 (1990) 2699.
7. S. Jeong, *Phys. Rev.* B 53 (1996) 13973.
8. C. Domain and C.S. Becquart, *Phys. Rev.* B 65 (2001) 024103.
9. Z. Tang, M. Hasegawa, Y. Nagai, and M. Saito, *Phys. Rev.* B 65 (2002) 195108.
10. http://www.wien2k.at/.
11. E. Wimmer, H. Krakauer, M. Weinert, and A.J. Freeman, *Phys. Rev.* B 24 (1981) 864.
12. M. Weinert, E. Wimmer, and A.J. Freeman, *Phys. Rev.* B 26 (1982) 4571.
13. H.J.F. Jansen and A.J. Freeman, *Phys. Rev.* B 30 (1984) 561.
14. J. Perdew, K. Burke, and M. Ernzerhof, *Phys. Rev. Lett.* 77 (1996) 3865.
15. B. Segall, *Phys. Rev.* 125 (1962) 109.
16. G.A. Burdick, *Phys. Rev.* 129 (1963) 138.
17. W. Hume-Rothery, *J. Inst. Met.* 9 (1961–62) 42.
18. A.B. Pippard, *Phil. Trans. R. Soc. Lond.* A250 (1957) 325.
19. W. Kohn and L.J. Sham, *Phys. Rev.* 140 (1965) A1133.
20. R.E. Cohen, M.J. Mehl, and D.A. Papaconstantopoulos, *Phys. Rev.* B 50 (1994) 14694.
21. H. Jones, *J. de Phys. et le radium*, Paris, 23 (1962) 637.

22. U. Mizutani, R. Asahi, T. Takeuchi, H. Sato, O.Y. Kontsevoi, and A.J. Freeman, *Z. Kristallogr.* 224 (2009) 17.

CHAPTER 6

Structure of Structurally Complex Metallic Alloys

6.1 WHAT ARE STRUCTURALLY COMPLEX ALLOYS?

The structurally complex metallic alloys (CMAs) are a class of metallic compounds characterized by the possession of giant unit cells ranging from some tens up to thousands of atoms with well-defined atom clusters. Many of them possess a solid solution range so that the composition can be varied within a single-phase field while some are stable only as line compounds. There generally exist chemical disorder and vacancies in the structure. It has been gradually established that physical properties of CMAs exhibit unique features different from those of normal metallic alloys possessing small unit cells like fcc, bcc, hcp, and so on.

As examples of CMAs, we point to three of the most complicated intermetallic phases, β-Al_3Mg_2, Cu_3Cd_4, and $NaCd_2$, all their structures having been solved by Samson in 1960s. The β-Al_3Mg_2 compound contains 1168 atoms in the unit cell with clusters characterized by the Friauf polyhedra with space group $Fd\bar{3}m$ [1,2]. According to the phase diagram [3], a solid solution range exists over 37.5 to 40.0 at .%Mg. In the case of Cu_3Cd_4, there are 1124 atoms in the unit cell with the lattice constant of 2.5871 nm. Its space group was deduced to be $F\bar{4}3m$. Dominant coordination shells in Cu_3Cd_4 are composed of Friauf polyhedra and icosahedra [4]. The $NaCd_2$ with space group $Fd\bar{3}m$ contains 1120 to 1190 atoms in the unit cell with the lattice constant $a = 3.056$ nm [5].

As mentioned in Chapter 1, Sections 1.2 and 1.3, quasicrystals and their approximants have been regarded as being typical of CMAs. They can be described in terms of six- or five-dimensional hyper-cubic lattice in the framework of the so-called cut-and-projection method. In the case of an icosahedral quasicrystal, an appropriate cut of a periodic density distribution $\rho_6(\mathbf{r})$ in the six-dimensional space R_6 by the three-dimensional physical space $R_{3//}$ can generate a set of atom positions, provided that $\rho_6(\mathbf{r})$ is flat without any thickness in $R_{3//}$ and, hence, is completely embedded in the three-dimensional perpendicular space $R_{3\perp}$ [6,7]. These three-dimensional objects in the space $R_{3\perp}$ are called the *atomic surfaces* (AS).

To see the situation in a visual way, a square lattice is chosen as a hyper-cubic lattice in the two-dimensional space R_2 and the straight line segment with a length Δ is taken as the object AS, which is embedded in the space $R_{1\perp}$ and is perpendicular to the physical space $R_{1//}$, as shown in Figure 6.1. The position of $R_{1//}$ in the space R_2 is fixed by assigning its angle θ relative to the square lattice in R_2. The one-dimensional quasi-lattice or the *Fibonacci chain** is obtained, if θ is set equal to $\tan^{-1}(1/\tau)$, where τ is the golden mean given by

$$\tau = \frac{\sqrt{5}+1}{2} \approx 1.6180339887\cdots$$

Similarly, icosahedral quasi-lattice is generated in $R_{3//}$, if $R_{3//}$ is fixed by the angle θ relative to R_6. Likewise, approximant lattices of different orders can be constructed by rationalizing the angle so as to conform with any Fibonacci ratio, 1/0, 1/1, 2/1, 3/2, 5/3, 8/5, ..., τ.

A quasicrystal is recognized as a solid achieved by setting an angle to $\tan^{-1}(1/\tau)$, whereas an approximant by substituting the rationalized Fibonacci ratio for $1/\tau$. A cubic approximant can be obtained by assigning the same rationalized Fibonacci ratios for $\tan\theta_i$, where θ_i ($i = x, y$ and z) is an angle between the three principal x-, y-, and z-axes in the $R_{3//}$ space

---

* A Fibonacci chain can be generated by algorithm such that a segment of length L on a straight line is subdivided into a long segment L′ and short segment S′, inflate L′ = L and S′ = S, replace L by L + S and S by L, and so on. In its first generation, the set of LS is created from the 0th generation of the parent L. In the second, the third, fourth ... generations, sets of LSL, LSLLS, LSLLSLSL, ... appear. The number ratio of L over S is 1/0, 1/1, 2/1, 3/2, 5/3, 8/5, ... and eventually converges to an irrational number τ = 1.618 ..., the golden mean τ. This is called the Fibonacci chain. The n-th order approximants can be constructed by terminating the operation at the n-th generation.

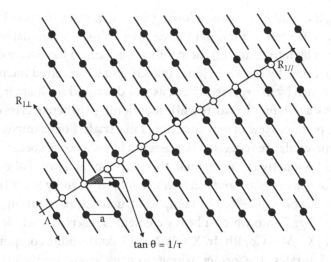

$$\tan \theta = 1/\tau$$

FIGURE 6.1 Construction of one-dimensional quasicrystal or Fibonacci lattice by cut-and-projection from a two-dimensional square lattice. An angle θ relative to the two-dimensional square lattice in R_2 is chosen to be $\tan^{-1}(1/\tau)$, where $\tau = (1 + \sqrt{5}/2)$.

relative to those in the six-dimensional space R_6. The structure of the 1/1-1/1-1/1 approximant is achieved by fixing $\tan\theta_x = \tan\theta_y = \tan\theta_z = 1/1$. For example, the $Al_xMg_{39.5}Zn_{60.5-x}$ ($20.5 \leq x \leq 50.5$) compounds known as the Bergman phase are identified as the 1/1-1/1-1/1 approximant containing 160 atoms in its unit cell with space group $Im\overline{3}$ [8]. Similarly, $Al_{15}Mg_{43}Zn_{42}$ compound is identified as the 2/1-2/1-2/1 approximant with space group $Pa3$ and contains as large as 672–692 atoms in its cubic unit cell with the lattice constant of 2.291–2.303 nm [9].[*] The approximants having orders higher than 2/1-2/1-2/1 are too complex to perform first-principles band calculations. Instead, 1/1-1/1-1/1 approximants can be handled and have been chosen to comprehensively understand the stability mechanism of the CMAs.

Included in the category of CMAs are a series of gamma-brasses containing 52 atoms in the unit cell. The number of atoms in the unit cell is much smaller than those mentioned above. Nevertheless, the advantage of working with gamma-brasses as CMAs essentially stems from the following reasons: (1) the number of atoms in the unit cell is still large enough to produce a sizable pseudogap at the Fermi level but is yet small enough to perform efficiently FLAPW band calculations, (2) a large number of combinations of elements in the periodic table give rise to stable gamma-brasses in more

---

[*] See more details in Footnote on pg. 274, Chapter 10.

than 20 binary alloy systems, allowing us to carry out systematic studies, (3) many reliable structural data have already been accumulated and are available in literature, and (4) the stability mechanism has been worked out for the first time by Mott and Jones in 1936 and has received intense attention since then by many researchers, as was discussed in Chapter 1.

At this stage, it may be worthwhile mentioning a different class of CMAs from the point of view of new functional materials. For example, thermoelectric materials are requested to possess a large Seebeck coefficient, large electrical conductivity, and small thermal conductivity. Following the concept of *phonon glass and electron crystal* proposed by Slack [10], more attention has been directed to complex structures, particularly, possessing "open cages" into which a heavy element is inserted. Included are the unfilled MX_3 (M = Co, Rh, Ir, X = P, As, Sb) skutterudite compounds and Si or Ge clathrates. The former belongs to space group $Im\bar{3}$ and is characterized by the possession of two cages in its unit cell containing 32 atoms. The rare earth element R can be inserted as a guest atom into the cage to form filled skutterudite compounds RM_4X_{12} like $CeFe_4Sb_{12}$, in which the rattling of the rare earth atom Ce will produce significant phonon scattering and reduce significantly the thermal conductivity [11].

Si clathrates were first synthesized by encaging alkaline elements like Na, K, Rb, and Cs into the three-dimensional Si-sp³ network in 1960s. In 1998, Nolas et al. [12] synthesized the $Sr_8Ga_{16}Ge_{30}$ clathrate as candidate for thermoelectric applications and achieved the dimensionless figure of merit ZT = 0.25 at 300 K.* More recently, even the superconductivity was observed at 8.0 K in Ba_8Si_{46}, which was synthesized under high pressures of about 3 GPa and temperatures at about 800°C [13]. The intermetallic compound consists of 46 Si atoms and 8 Ba atoms in the cubic unit cell of a = 1.0328 nm with space group $Pm\bar{3}n$ and, hence, can be included into the family of CMAs [14].

As described above, we will select both a series of gamma-brasses and several 1/1-1/1-1/1 approximants as working substances to explore the stabilization mechanism of CMAs, which indeed constitutes the present main subjects in the remaining chapters of this book. Both gamma-brasses and 1/1-1/1-1/1 approximants will be reviewed from the point of view of the atomic structure in this chapter.

---

* ZT is defined as $TS^2/\kappa\rho$, where T is the absolute temperature, S is the Seebeck coefficient, κ is the thermal conductivity, and ρ is the electrical resistivity. Solids possessing ZT higher than unity have been searched as good thermoelectric materials.

6.2 GAMMA-BRASSES

As has been briefly mentioned in Chapter 1, gamma-brasses can be basically described in terms of the 26-atom cluster [15]. The cluster is made up of four shells: four atoms are positioned on four vertices of the inner tetrahedron (IT), four atoms on those of the outer tetrahedron (OT), six atoms on those of the octahedron (OH), and twelve atoms on those of the cubo-octahedron (CO), as illustrated in Figure 6.2 [16]. Here, the cubo-octahedron represents a polyhedron formed by meeting edges of the cube at the middle of octahedral faces.

Gamma-brasses are divided into four families, depending on the space group. The most abundant forms a bcc lattice with two identical 26-atom clusters with space group $I\bar{4}3m$. They are hereafter called the I-cell gamma-brasses. The second largest family forms a CsCl-type structure composed of two different 26-atom clusters. Because the space group is $P\bar{4}3m$, they are referred to as the P-cell gamma-brasses. Thus, both I- and P-cell gamma-brasses normally contain 52 atoms in its cubic unit cell. As

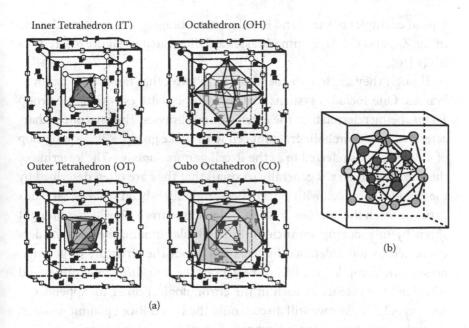

FIGURE 6.2 (a) Four shell structures consisting of inner tetrahedron (IT), outer tetrahedron (OT), octahedron (OH) and cubo-octahedron (CO) for the gamma-brass. (b) 26-atom cluster consisting of four atoms on IT, four atoms on OT, six atoms on OH, and twelve atoms on CO. [From Mizutani, *MATERIA* (in Japanese), 45 No.11 (2006) 803.]

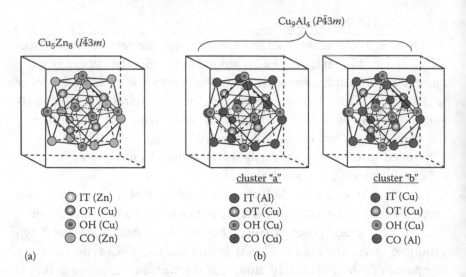

FIGURE 6.3 The 26-atom cluster in (a) Cu_5Zn_8 and (b) Cu_9Al_4 gamma-brasses with space groups $I\bar{4}3m$ and $P\bar{4}3m$, respectively. [From Mizutani, *MATERIA* (in Japanese), 45 No.11 (2006) 803.]

typical examples of the I- and P-cell gamma-brasses, the 26-atom clusters in Cu_5Zn_8 and Cu_9Al_4 gamma-brasses are illustrated in Figure 6.3, respectively [16].

Though they are less frequent, we have two other families of gamma-brasses. One includes systems that form super-lattices with space group $F\bar{4}3m$, being referred to as the F-cell gamma-brasses. The final one is characterized by a rhombohedral unit cell with space group $R3m$, a subgroup of $P\bar{4}3m$, and is referred to as the R-cell gamma-brasses. The lowering of the cubic symmetry is generally so small that they are often described by the pseudo-cubic cell with a distortion angle slightly off from 90°. It tends to occur when the number of atoms is decreased from 52, which is brought about by introducing vacancies into particular atomic sites. For readers interested in more detailed information about the structure of gamma-brasses, we compiled the literature data in the four different families and described the essence of each in the chronological order in Appendix 2, Section A2.2. Below we will discuss only the I- and P-cell gamma-brasses, since they were employed in first-principles band-calculations.

As listed in Table 6.1, we found I- and P-cell gamma-brasses in 24 binary alloy systems [3,17]. In most cases, the gamma-brass phase has a finite solid solution range. Since first-principle band calculations can be performed only for an ordered phase without involving any chemical and/

TABLE 6.1 Gamma-Brasses with Space Groups $I\bar{4}3m$ and $P\bar{4}3m$ in Three Different Groups in Binary Alloy Systems

Group	Gamma-Brass	e/a	Solid Solution Range (%)	Group	Gamma-Brass	e/a	Solid Solution Range (%)	Group	Gamma-Brass	e/a	Solid Solution Range (%)
I	Cu_5Zn_8(I)	21/13	57<Zn<70	I	Cu_9In_4(P)	21/13	27.7<In<31.3	II	Ni_2Be_{11} (I)	?	11.5<Ni<?
	Ag_5Cd_8 (I)	21/13	57<Cd<62.5		Ag_9In_4(P)	21/13	31.1<In<33.6		Ni_2Cd_{11} (I)	?	12<Ni<19.5
	Ag_5Zn_8 (I)	21/13	58<Zn<64.7		Au_9In_4 (P)	21/13	28.8<In<31.4		Mn_2Zn_{11} (I)	?	15.2<Mn<23
	Cu_5Cd_8 (I)	21/13	52.2<Cd<66	II	Ni_2Zn_{11} (I)	?	15<Ni<30		Pt_2Zn_{11} (?)	?	19<Pt<23
	Au_5Cd_8 (I)	21/13	61.2<Cd<67.6		Pd_2Zn_{11} (I)	?	14.5<Pd<24		Al_8V_5 (I)	?	Al_8V_5
	Au_5Zn_8 (I)	21/13	62.5<Zn<76		Fe_2Zn_{11} (I)	?	17<Fe<31		Mn_3In (P)	?	Mn_3In
	Cu_9Al_4 (P)	21/13	31.5<Al<37		Co_2Zn_{11} (I)	?	14.6<Co<31	III	Ag_5Li_8 (I)	?	63.5<Li<76
	Cu_9Ga_4 (P)	21/13	29.5<Ga<40		Ir_2Zn_{11} (I)	?	15.3<Ir<15.7		$Li_{10}Pb_3$?	?	22.5<Pb<23.5

Note: (I): space group $I\bar{4}3m$; (P): space group $P\bar{4}3m$, Solid solution range was taken from [2].

or geometrical disorder, it is important to fill a specific atomic species into each shell in the 26-atom cluster without vacancies. Thus, the stoichiometric composition of the ideal gamma-brass should fall in one of A_2B_{11}, A_4B_9, A_5B_8, A_7B_6, and $A_{10}B_3$ with space group $I\bar{4}3m$ and a combination of these two with space group $P\bar{4}3m$. However, only the formula A_5B_8 has been experimentally found for I-cell gamma-brasses, while the formula A_4B_9 has been found for P-cell gamma-brasses.* Such stoichiometric compounds are found within a solid solution range. A maximum solid solution range and the stoichiometric compound in 24 gamma-brass alloy systems are summarized in Table 6.1. Among them, the composition for Mn_3In cannot be realized, unless chemical disorder is introduced over shells in the 26-cluster (see Appendix 2, Section A2.2.2.10).

We have classified the I- and P-cell gamma-brasses into three groups in terms of a combination of constituent elements in the periodic table, as listed in Table 6.1. Group I consists of a combination of monovalent noble metal and polyvalent element, whose valency is well defined. The validity of the Hume-Rothery electron concentration rule of $e/a = 21/13$ has been claimed for group I. As a matter of fact, people in the 1920s had already been aware that the prototype Cu_5Zn_8 and Cu_9Al_4 gamma-brasses are stabilized at $e/a = 21/13$ in spite of different solute concentrations [18–20]. We will elucidate the FsBz-induced stabilization mechanism in Cu_5Zn_8 and Cu_9Al_4 by performing first-principles FLAPW band calculations in Chapter 7.

The group II includes totally eleven gamma-brasses, which consist of 3d-transition metal elements like TM = V, Mn, Fe, Co, and Ni and either divalent elements like Be, Zn, and Cd or trivalent elements like Al and In. Since the e/a value for the TM element is not a priori known, it is not clear whether the gamma-brass in the group II is stabilized at $e/a = 21/13$ or not. This has been regarded for a long time as one of the most challenging themes in the electron theory of metals. In Chapter 8, we will show that the d-states-mediated-splitting or the d-states-mediated-FsBz interaction plays a key role in the stabilization of group (II) gamma-brasses and will also discuss the determination of the value of e/a for the TM element involved.

Group III includes gamma-brasses consisting of a combination of two non-transition metal elements, for which a use of their nominal valencies

---

* Gamma-brasses with the composition TM_2Zn_{11} (TM=Fe, Co, Ni, Pd, etc.) are marginal but are still within a solid solution range. The gamma-brasses with this composition generally involve chemical disorder as well as vacancies so that the number of atoms per unit cell is generally lower than 52 (see Chapter 8).

does not yield the characteristic value of 21/13. The most notable example in this group is the $Ag_{100-x}Li_x$ (x = 63.5-76) gamma-brass, a combination of two monovalent elements leading to e/a = 1.0, regardless of the Li concentration. Hence, it has been wondered if Ag_5Li_8 gamma-brass obeys the Hume-Rothery electron concentration rule or not. We will discuss its stability mechanism in Chapter 8 in relation to the Ag-4d-states-mediated-splitting. $Li_{10}Pb_3$ had been also considered as another example in group III, since one can easily estimate its average e/a to be 22/13, if valencies of Pb and Li are assumed to be four and unity, respectively. However, our recent experimental studies confirmed that the $Li_{10}Pb_3$ gamma-brass does not exist in the phase diagram (see Appendix 2, Section A2.1).

6.3 1/1-1/1-1/1 APPROXIMANTS

Icosahedral quasicrystals and their approximants can be classified with respect to the atomic clusters building up its structure and the structure type. Three different atomic cluster types are known to exist in literature: the first described is the rhombic triacontahedron (abbreviated as RT) containing 44 atoms in the cluster, the second is the Mackay icosahedron (abbreviated as MI) containing 54 atoms in the cluster, and the third is the Tsai-type cluster. The first one is denoted as the RT-type or the Frank Kasper-type cluster, since the Al-Mg-Zn 1/1-1/1-1/1 approximant mentioned above has been also referred to as the Frank–Kasper compound [21]. The RT-type cluster is found in systems like Al-Mg-X (X = Zn, Ag, Cu, and Pd) and Al-Li-Cu quasicrystals and their approximants [22].* The MI-type cluster involves TM element as one of the major constituent elements. The MI-type cluster is found in systems composed of Al and TM elements like in Al-Mn, Al-Cu-Fe, Al-Cu-Ru, and Al-Pd-Re quasicrystals and their approximants. In 2000, Tsai and his colleagues [23] discovered quasicrystals in the Cd-Yb alloy system and pointed out the possession of atomic clusters, which are different from the RT- and the MI-type clusters, as described below.

---

* Following the pioneering work by Bergman et al. [21], the $Mg_{32}(Al, Zn)_{49}$ compound has been often referred to as possessing 162 atoms per unit cell, though the occupancy at the center of the cluster is less than unity (i.e., 0.8 in their Table 1). Mizutani et al. [8] claimed from their powder diffraction Rietveld analysis for a series of $Al_xMg_{39.5}Zn_{60.5-x}$ 1/1-1/1-1/1 approximants (20.5 ≤ x ≤ 50.5) that the fractional occupancies at the center of the cluster is at most 0.1 at 20.5 at.%Al and is decreased to zero when Al concentration exceeds 30 at.%. Hence, the total number of atoms in the unit cell is better described as possessing 160 atoms per unit cell.

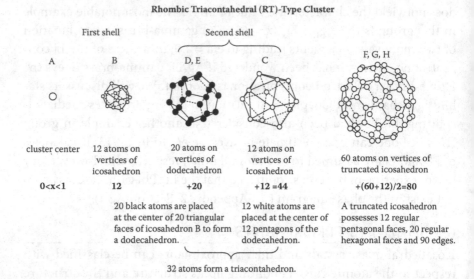

Rhombic Triacontahedral (RT)-Type Cluster

	First shell	Second shell		F, G, H
A	B	D, E	C	
cluster center	12 atoms on vertices of icosahedron	20 atoms on vertices of dodecahedron	12 atoms on vertices of icosahedron	60 atoms on vertices of truncated icosahedron
0<x<1	12	+20	+12 =44	+(60+12)/2=80
		20 black atoms are placed at the center of 20 triangular faces of icosahedron B to form a dodecahedron.	12 white atoms are placed at the center of 12 pentagons of the dodecahedron.	A truncated icosahedron possesses 12 regular pentagonal faces, 20 regular hexagonal faces and 90 edges.

32 atoms form a triacontahedron.

FIGURE 6.4 Successive shell structures of atoms in the atomic cluster of the rhombic triacontahedral (RT)-type quasicrystals and their approximants. solid circles: largest atom like Mg and Li in Al-Mg-Zn and Al-Li-Cu, respectively. The assembly consisting of the first and second shells is called the RT-cluster. Atoms on the sites F, G, H are shared by two neighboring Wigner–Seitz cells.

The successive shell structure of the RT-, MI-, and Tsai-type atomic clusters is illustrated in Figures 6.4, 6.5, and 6.6, respectively. In both RT- and MI-type clusters, 12 atoms are located around a given lattice site to form an icosahedron and constitutes the first shell. The center of the icosahedron thus formed is fully or partially filled with an atom or more frequently completely vacant. In the Tsai-type atomic cluster, the first shell is a tetrahedron composed of three or four Cd or Zn atoms at its vertices around the cluster center, as shown in Figure 6.6. It is noteworthy that the presence of this unique tetrahedral unit internally breaks the icosahedral symmetry [23].

There exist twenty triangular faces on an icosahedron. In the case of the RT-type cluster, a dodecahedron (D and E in Figure 6.4) is formed by locating twenty larger atoms like Mg at the center of the triangular face of the icosahedron "B" on the first shell. The center of 12 pentagonal faces on the resulting dodecahedron is then filled with smaller atoms Al or X = Zn, Cu, Ag, and Pd in the case of the Al-Mg-X system. In total, 32 atoms constitute the rhombic triacontahedron as the second shell. The RT-type atomic cluster is, therefore, composed of 44 atoms, including 12 atoms in the first

Mackay Icosahedral (MI)-Type Cluster

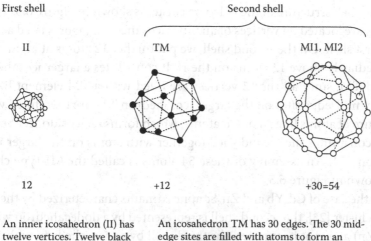

First shell	Second shell	
II	TM	MI1, MI2
12	+12	+30≈54

An inner icosahedron (II) has twelve vertices. Twelve black atoms are located immediately above them to form a larger icosahedron TM.	An icosahedron TM has 30 edges. The 30 mid-edge sites are filled with atoms to form an icosidodecahedron. An icosidodecahedron has 30 identical vertices, which are divided into six sites MI1 and twenty-four sites MI2.

FIGURE 6.5 Successive shell structures of atoms in the atomic cluster of the Mackay Icosahedral (MI)-type quasicrystals and their approximants. Solid circles: the TM atom like Fe.

Tsai-Type Atomic Cluster

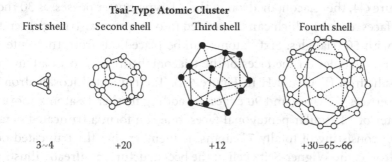

First shell	Second shell	Third shell	Fourth shell
3~4	+20	+12	+30=65~66

FIGURE 6.6 Successive shell structures of atoms in the atomic cluster of the Tsai-type quasicrystals and their approximants. Open circles: Cd or Zn, solid circles: Yb or Sc.

shell and 32 atoms in the second shell. This is a characteristic feature of the atomic cluster found in both the RT-type quasicrystals and approximants.

The shell structure of the MI-type cluster is shown in Figure 6.5. Twelve atoms are located on vertices of an inner icosahedron (abbreviated as II) as the first shell. On the second shell, we put further 12 atoms at atomic sites immediately above 12 atoms on the II. It constitutes a larger icosahedron called TM, since all the 12 vertices are filled with a TM element like Fe. Thirty mid-edge sites on the larger icosahedron TM are then filled with a mixture of atoms like Al or Cu atoms, which forms an icosidodecahedron. This constitutes the second shell together with atoms on the larger icosahedron TM. An assembly of these 54 atoms is called the MI-type cluster, as shown in Figure 6.5.

In the case of Cd_6Yb and Zn_6Sc approximants characterized by the Tsai-type cluster [23], the second shell is represented by a dodecahedron with 20 Cd (Zn) atoms at its vertices, the third shell by an icosahedron with 12 Yb (Sc) atoms, and the fourth shell is a Cd (Zn) icosidodecahedron obtained by placing 30 Cd (Zn) atoms on the mid-edges of the Yb (Sc) icosahedron. In total, the icosahedral cluster consists of 65–66 atoms, as shown in Figure 6.6. In an icosahedral quasicrystal, the atomic cluster shown in Figure 6.4 to 6.6 is distributed so as to satisfy an overall icosahedral symmetry on a quasi-periodic lattice. The structure of both RT- and MI-type clusters is characterized by the m35 symmetry but a slight distortion is always observed in 1/1-1/1-1/1 approximants. The RT-, MI- and Tsai-type atomic clusters are located on the body-center and corner of a cubic lattice to make a bcc packing in the 1/1-1/1-1/1 approximant, as described below.

The formation of a cubic unit cell may be explained by using the RT-type 1/1-1/1-1/1 approximant. As shown in the right-hand corner of Figure 6.4, the triacontahedron in the RT-type cluster possesses 30 rhombic faces, each of which can be divided into two regular triangles, ending up with 60 triangles. Sixty atoms can be placed out from the centers of the 60 triangles on the faces of the triacontahedron to form a truncated icosahedron (F, G and H in Figure 6.4). The truncated icosahedron has 60 vertices, 32 faces, and 90 edges. By adding further 12 atoms above the center of 12 regular pentagonal faces, one can form a truncated octahedron consisting of totally 72 atoms. Remember that the truncated octahedron is the Wigner-Seitz cell of the bcc-structure, as already illustrated in Figure 4.4b. It can fill the space without any overlap and/or void, when packed together in a body-centered cubic lattice, and ensure the possession of a periodic lattice.

Figure 6.7 illustrates how atoms are grown into clusters and clusters into unit cells for both RT- and MI-type 1/1-1/1-1/1 approximants [22]. For example, the $Al_{40.5}Mg_{39.5}Zn_{20}$ 1/1-1/1-1/1 approximant contains 160 atoms in the unit cell with the lattice constant of 1.4443 nm [8]. Remember that 72 atoms on the truncated octahedron are shared with their neighboring ones so that 36 atoms together with 44 atoms in the RT-cluster belong

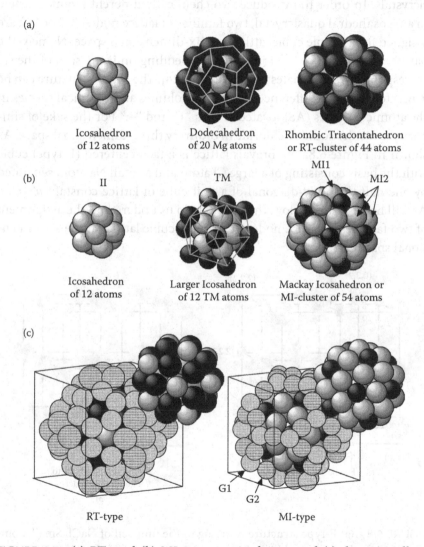

(a)

Icosahedron of 12 atoms   Dodecahedron of 20 Mg atoms   Rhombic Triacontahedron or RT-cluster of 44 atoms

(b)

II   TM   MI1   MI2

Icosahedron of 12 atoms   Larger Icosahedron of 12 TM atoms   Mackay Icosahedron or MI-cluster of 54 atoms

(c)

G1
G2

RT-type   MI-type

FIGURE 6.7 (a) RT- and (b) MI-type atomic clusters and (c) the unit cells of the RT- and MI-type 1/1-1/1-1/1 approximants [from U. Mizutani et al., *J. Phys.: Condens. Matter*, 14 (2002) R767–R788.]. See symbols II, TM, MI1, and MI2 in Figure 6.5. G1 and G2 refer to glue atoms (G3 and G4 are not shown).

to each truncated octahedron. Therefore, we can say that each truncated octahedron contains 80 atoms and forms a bcc lattice in the RT-type 1/1-1/1-1/1 approximant.

The icosahedral quasicrystals and their approximants must be also distinguished from types of a quasi-lattice involved. As shown below, we need to consider two different quasi-lattices of P- and F-types in real quasicrystals. In order to introduce two chemically different atomic clusters in an icosahedral quasicrystal, two families of lattice nodes, "+" or "−", are assigned for a simple cubic lattice in a six-dimensional space. Namely, the parity of either "+" or "−" is assigned, depending on if the sum of the six corresponding coordinates is even or odd [24]. The super-structure can be generated by small differences in shapes, volumes and chemical species in the atomic surfaces (AS) located at sites "+" and "−." For the sake of simplicity, consider sodium-chloride in ordinary three-dimensional space. As shown in Figure 6.8a, its Bravais lattice is a face-centered (F-type) cubic with the basis consisting of a large Cl atom and a small Na atom separated by one-half the body-diagonal of a unit cube of lattice constant $a' = 2a$. As will be described below, this would help us understand the assignment of two families of lattice nodes in a simple cubic lattice in the six-dimensional space.

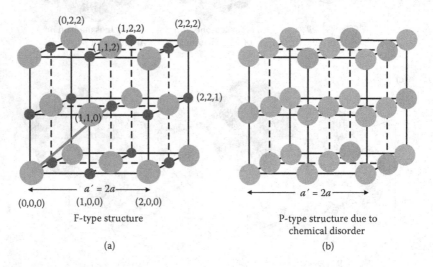

F-type structure

(a)

P-type structure due to chemical disorder

(b)

FIGURE 6.8 (a) F-type structure referring to the unit cell of NaCl. Small atoms are positioned with an odd lattice node where the sum of three coordinates is odd, while large atoms have an even node where the sum of the coordinates is even. (b) P-type structure. Small and large atoms are not differentiated by x-ray diffraction measurements, when they are randomly distributed.

All indices in reflection lines observed for NaCl must be either even or odd, being characteristic of an fcc lattice with the lattice constant a'. However, the structure is reduced to a simple cubic lattice (P-type) of lattice constant a, provided that the two atoms happen to possess equal numbers of electrons like K^+ and Cl^- in KCl or two different atomic species cannot be differentiated because of their random distributions over the lattice. This is illustrated in Figure 6.8b. Now the crystal looks to x-rays as if it were a monatomic simple cubic lattice of lattice constant a. In other words, only even integers occur in the reflection indices when indexed with respect to a cubic lattice of lattice constant a'. They are called fundamental reflections. In the case of NaCl (F-type), all reflections of the fcc lattice of lattice constant a' must be present, including weak lines with odd reflection indices, which are called superlattice reflections. In the case of the six-dimensional Bravais lattice discussed above, only fundamental reflections are observed in P-type quasicrystals, whereas superlattice reflections are additionally observed in F-type quasicrystals, which signifies the presence of two different atomic clusters arranged quasi-periodically.

The argument above is extended to the generation of the two types of 1/1-1/1-1/1 approximants from the fcc lattice (F-type) in the six-dimensional space. The P-type 1/1-1/1-1/1 approximant is generated by assigning only the even lattice node with the lattice constant a' in the six-dimensional space. The F-type 1/1-1/1-1/1 approximant is likewise generated by assigning both even and odd lattice nodes with the lattice constant a'. The latter gives rise to two different atomic clusters at the center and the corner of the cubic lattice in the three-dimensional physical space.

The 1/1-1/1-1/1 approximants are classified with respect to the type of lattice in the six-dimensional space and the type of the atomic cluster in the three-dimensional space. Typical examples are listed in Table 6.2 [8, 25–32]. They belong to space group of either $Im\overline{3}$ or $Pm\overline{3}$. Space group of $Im\overline{3}$ is assigned, if all diffraction lines are indexed using the Miller indices, in which the sum $h + k + l$ is even. Space group $Pm\overline{3}$ is assigned, if extra diffraction lines appear, which are indexed with the sum of the Miller indices odd.

Gamma-brasses are simply differentiated with respect to the space group. However, we learned that 1/1-1/1-1/1 approximants must be differentiated with respect to not only the space group, that is, I- or P-cell, but also the lattice-type in the six-dimensional space. Let us consider why such complexity exists in the approximants. For example, one may easily understand that F-type approximants like $Al_{54}Cu_{25.5}Fe_{12.5}Si_8$ and

TABLE 6.2 Classification of 1/1-1/1-1/1 Approximants in Terms of the Type of Lattice and the Type of Atomic Cluster

Type of Lattice	Type of Atomic Cluster	1/1-1/1-1/1 Approximant	Space Group	Features	Ref. No.
P	RT	$Al_xMg_{40}X_{60-x}$ (X = Zn, Cu, Ag and Pd)	$Im\bar{3}$	Two identical clusters; chemical disorder in Al-X	8, 25
P	MI	$Al_{75}(Mn_{1-x}Fe_x)_{17}Si_8$ ($0.32 \leq x \leq 0.72$)	$Im\bar{3}$	Two identical clusters; chemical disorder in Mn-Fe and Al-Si	26
P	MI	$Al_{68}Cu_7(Fe_{1-x}Ru_x)_{17}Si_8$ ($x = 0 \leq x \leq 1$)	$Im\bar{3}$	Two identical clusters; chemical disorder in Fe-Ru and Al-Si	30
P	MI	$Al_{75}(Mn_{1-x}Fe_x)_{17}Si_8$ ($0 \leq x \leq 0.29$)	$Pm\bar{3}$	Two identical clusters; chemical disorder in Al-Si and Mn-Fe	26-28
F	MI	$Al_{54}Cu_{25.5}Fe_{12.5}Si_8$ $Al_{54}Cu_{25.5}Ru_{12.5}Si_8$	$Pm\bar{3}$	Two different clusters; chemical and structural disorder	29, 30
P	Tsai	Cd_6M (M = Nd, Sm, Gd, Dy Yb, Ca, Y)	$Im\bar{3}$	Two identical clusters; innermost disordered Cd tetrahedra	31
P	Tsai	Zn_6Sc	$Im\bar{3}$	Two identical clusters; innermost disordered Zn tetrahedra	32

$Al_{54}Cu_{25.5}Ru_{12.5}Si_8$ listed in Table 6.2 ought to be specified by space group $Pm\overline{3}$, since two different atomic clusters are located at the center and corner of the bcc-lattice [29,30]. However, in the case of $Al_{75}(Mn_{1-x}Fe_x)_{17}Si_8$ ($0 \le x \le 0.29$), including well-studied $Al_{75}Mn_{17}Si_8$ approximant, the P-type lattice is assigned, since only a single atomic cluster is involved [26–28]. However, space group is deduced to be $Pm\overline{3}$, since the diffraction lines, in which the sum of the Miller indices is odd, are observed. This is because the two Wigner–Seitz cells in the unit cell are no longer regarded as being identical as a result of different arrangements of "glue" atoms connecting neighboring identical atomic clusters.* The presence of so-called glue atoms connecting two neighboring atom clusters in the 1/1-1/1-1/1 approximants makes their classification more complex than the gamma-brasses free from glue atoms.

In contrast to the perfectly ordered Cu_5Zn_8 and Cu_9Al_4 gamma-brasses, all cubic approximants listed in Table 6.2 involve various degrees of chemical and geometrical disorder. Only the geometrical disorder exists in the first shell in the case of both Cd_6M (M = Yb, Ca) [31] and Zn_6Sc [32] (see Figure 6.6). In first-principles band calculations, such disorder must be avoided. Hence, we have to construct a model structure free from any chemical and geometrical disorder with the least sacrifice from the observed structure.

REFERENCES

1. S. Samson, *Acta Crystallogr.* 19 (1965) 401.
2. M. Feuerbacher, C. Thomas, J.P.A. Makongo, S. Hoffmann, W. Carrillo-Cabrera, R. Cardoso, Y. Grin, G. Kreiner, J.-M. Joubert, T. Schenk, J. Gastaldi, H. Nguyen-Thi, N. Mangelinck-Noël, B. Billia, P. Donnadieu, A. Czyrska-Filemonowicz, A. Zielinska-Lipiec, B. Dubiel, T. Weber, P. Schaub, G. Krauss, V. Gramlich, J. Christensen, S. Lidin, D. Fredrickson, M. Mihalkovic, W. Sikora, J. Malinowski, S. Brühne, T. Proffen, W. Assmus, M. de Boissieu, F. Bley, J.-L. Chemin, J. Schreuer, and W. Steurer, *Z. Kristallogr.* 222 (2007) 259.
3. H. Okamoto, *Phase Diagrams for Binary Alloys* (ASM International, OH, 2000).
4. S. Samson, *Acta Crystallogr.* 23 (1967) 586.
5. S. Samson, *Nature* 195 (1962) 259.
6. C. Janot, *Quasicrystals*, 2nd edition (Clarendon Press, Oxford, 1994).
7. J.M. Dubois, *Useful Quasicrystals* (World Scientific, Singapore, 2005).

---

* Glue atoms are those located on the truncated octahedron in the case of the RT-type approximant (see Figure 6.7a). They are shared with neighboring ones and should be distinguished from atoms in the atomic cluster.

8. U. Mizutani, W. Iwakami, T. Takeuchi, M. Sakata and M. Takata, *Phil. Mag. Lett.* 76 (1997) 349.

9. T. Takeuchi and U. Mizutani, *Phys. Rev. B* 52 (1995) 9300; K. Sugiyama, W. Sun and K. Hiraga, *J. Alloys and Compounds* 342 (2002) 139; Q. Lin and J.D. Corbett, *Proc. Natl. Acad. Sci. U.S.A.*, 2006 September 12; 103(37): 13589; G. Kreiner, presented at Quasicrystal Meeting, Stuttgart, Germany, January 30, 2009.

10. G.A. Slack, *CRC Handbook of Thermoelectrics*, edited by D.M. Rowe (CRC, Boca Raton, Florida, 1995), p. 407.

11. D.T. Morelli and G.P. Meisner, *J. Appl. Phys.* 77 (1995) 3777.

12. G.S. Nolas, J.L. Cohn, G.A. Slack, and S.B. Schujman, *Appl. Phys. Letters* 73 (1998) 178.

13. S. Yamanaka, E. Enishi, H. Fukuoka, and M. Yasukawa, *Inorg. Chem.* 39 (1) (2000) 56.

14. A. Kitano, K. Moriguchi, M. Yonemura, S. Munetoh, A. Shintani, H. Fukuoka, S. Yamanaka, E. Nishibori, M. Takata, and M. Sakata, *Phys. Rev. B* 64 (2001) 045206.

15. A.J. Bradley and P. Jones, *J. Inst. Met.* 51 (1933) 131.

16. U. Mizutani, *MATERIA* (in Japanese), 45 No. 11 (2006) 803.

17. P. Villars, *Pearson's Handbook* (ASM International, Materials Park, OH, 1997).

18. A. Westgren and G. Phragmén, *Z. anorg. Chemie* 175 (1928) 80.

19. A. Westgren and G. Phragmén, *Metallwirtschaft* 7 (1928) 700.

20. A. Westgren and G. Phragmén, *Trans. Farad. Soc.* 25 (1929) 379.

21. G. Bergman, J.L.T. Waugh, and L. Pauling, *Acta Crystallogr.* 10 (1957) 254.

22. U. Mizutani, T. Takeuchi, and H. Sato, *J. Phys.: Condens. Matter* 14 (2002) R767–R788.

23. A.P. Tsai, J.Q. Guo, E. Abe, H. Takakura, and T.J. Sato, *Nature* 408 (2000) 537.

24. M. Boudard, M. de Boissieu, C. Janot, G. Herger, C. Beeli, H.-U. Nissen, H. Vincent, R. Ibberson, M. Audier, and J.M. Dubois, *J. Phys.: Condens. Matter* 4 (1992) 10149.

25. T. Takeuchi, T. Mizuno, E. Banno, and U. Mizutani, *Mater. Sci. Eng. B* 294–296 (2000) 522.

26. T. Takeuchi, T. Onogi, E. Banno, and U. Mizutani, *Mater. Trans.* 42 (2001) 933.

27. M. Cooper and K. Robinson, *Acta Cryst.* 20 (1966) 614.

28. K. Sugiyama, N. Kaji, and K. Hiraga, *Acta Cryst. C* 54 (1998) 445.

29. H. Yamada, T. Takeuchi, U. Mizutani, and N. Tanaka, *Mat. Res. Soc. Symp. Proc.* vol. 553 (1999 Materials Research Society) Edited by J.M. Dubois, P.A. Thiel, A.P. Tsai and K. Urban, pp. 117–122.

30. T. Takeuchi and U. Mizutani, *J. Alloys Compounds* 342 (2002) 416.

31. C.P. Gomez and S. Lidin, *Phys. Rev. B* 68 (2003) 024203.

32. Q. Lin and J.D. Corbett, *Inorganic Chem.* 43 (2004) 1912.

CHAPTER 7

Stabilization Mechanism of Gamma-Brasses Characterized by a FsBz-Induced Pseudogap

7.1 STABILIZATION MECHANISM DRIVEN BY A FsBz-INDUCED PSEUDOGAP

We have estimated in Chapter 2, Section 2.3 how deep and wide a pseudogap at the Fermi level must be to stabilize structurally complex metallic alloys (CMAs), and concluded that the experimentally observed pseudogap in many quasicrystals and approximants is indeed deep and large enough. It is also noted in Chapter 4 that first-principles band calculations can be performed successfully even for crystals containing more than 100 atoms in the unit cell, once their atom positions are defined without any ambiguities, such as chemical disorder or fractional site occupancy.*

Both I- and P-cell gamma-brasses in binary alloy systems may be classified into three groups, as discussed in Chapter 6, Section 6.2. Among them, group (I) gamma-brasses refer to those based on the noble metals Cu, Ag, or Au alloyed with the polyvalent elements such as Al, Zn, and so on (see Table 6.1). They are, we believe, the best suited to study the role of

---

* A fractional site occupancy means that given crystallographic sites are partially occupied by atoms, the remaining being left vacant.

the FsBz interaction and its impact on the stability of the CMAs. As will be shown below, the electronic structure of group (I) gamma-brasses is characterized by a FsBz-induced pseudogap at the Fermi level. In the present chapter, we consider Cu_5Zn_8 and Cu_9Al_4 gamma-brasses as the representative alloys of group (I), and demonstrate that a pseudogap is indeed induced by the FsBz interaction. We will also explain why the particular electron concentration e/a = 21/13 plays a predominant and key role in their stabilization.

In the latter half of this chapter, we address ourselves to another fundamental issue in group (I) gamma-brasses. All gamma-brasses in this group are known to exhibit a finite solid solution range. We consider it important to elucidate why the phase stability is maintained even at off-stoichiometric compositions. In principle, any alloy with an off-stoichiometric composition would not remain stable at absolute zero, since the configurational entropy remains finite. This is a consequence of the "third law of thermodynamics." Hence, a handling of off-stoichiometric alloys requires arguments about phase stability at finite temperatures. Here the role of vacancies has been suggested to be crucially important for many years. In order to deepen understanding of phase stability of off-stoichiometric group (I) gamma-brasses, we consider it important to collect, as a first step, reliable experimental data on the solute concentration dependence of vacancies in the unit cell and then to discuss its effect on stabilization mechanism in terms of the FsBz interactions rigorously evaluated for stoichiometric compounds like Cu_5Zn_8 and Cu_9Al_4. Such subjects will be discussed in Sections 7.6 and 7.7.

7.2 FLAPW BAND CALCULATIONS FOR Cu_5Zn_8 AND Cu_9Al_4 GAMMA-BRASSES

Both Cu_5Zn_8 and Cu_9Al_4 gamma-brasses contain 52 atoms in the cubic unit cell with lattice constants of 0.884 and 0.8675 nm, respectively. The 26-atom cluster forms a bcc lattice with space group $I\bar{4}3m$ in the former [1–4], whereas the CsCl structure with space group $P\bar{4}3m$ in the latter [1,5–8]. Hence, both are typical of I- and P-cell gamma-brasses. The powder x-ray diffraction spectra for alloys close to Cu_5Zn_8 and Cu_9Al_4 gamma-brass compositions are shown in Figure A2.1 and A2.3, respectively, in Appendix 2, Section A2.1. The diffraction lines (210), (221), and (300) underlined in Figure A2.3 are present in Cu_9Al_4 and indexed with the Miller indices, the sum of which is odd. They are observed only in P-cell gamma-brasses. As illustrated in Figure 6.3, chemical disorder is absent and occupancy is unity in sites IT, OT, OH, and CO in the 26-atom clusters [1–8], allowing us to rigorously perform first-principles band calculations for these two stoichiometric compounds.

In the FLAPW method discussed in Chapter 4, the motion of d-electrons having a strongly localized tendency is rigorously described in terms of the product of the radial wave function and spherical harmonics inside the MT sphere. On the other hand, the motion of electrons outside the MT sphere is described in terms of the superposition of plane waves over reciprocal lattice vectors allowed for a given lattice. Therefore, we can directly extract the FsBz interactions from the wave function outside the MT sphere, regardless of whether or not the transition metal TM is involved as a constituent element.

As emphasized in Chapter 4, the FLAPW method [9–11] treats all electrons and has no shape approximations for the potential and charge density. The exchange-correlation energies are treated within the local density approximation using the Hedin–Lundqvist parameterization of the exchange-correlation potential [12]. The core states are calculated fully relativistically and updated at each iteration, whereas the valence and semi-core states are treated semi-relativistically, that is, spin-orbit coupling is neglected.

Cut-offs of the plane-wave basis, 217 eV, and of the potential representation, 1360 eV, and an expansion in terms of spherical harmonics with $l = 8$ inside the MT sphere, were used for the calculations. The resulting numbers of plane waves were about 2500 and 5000 for Cu_5Zn_8 and Cu_9Al_4, respectively. Summations over the Brillouin zone to calculate self-consistent charge densities were made, using 10 and 4 special k-points [13] in the irreducible wedge for Cu_5Zn_8 and Cu_9Al_4, respectively (see Chapter 5, Figure 5.3). Convergence was judged to be sufficient, when an average root-mean-square difference between the input and output charge densities was reduced to the value less than 5×10^{-5} e/(a.u.)^3. We used 200 and 90 sampling k-points in the irreducible wedge and the linear tetrahedron scheme [14,15] for calculating the DOS for both Cu_5Zn_8 and Cu_9Al_4, respectively.

The DOSs calculated by using the FLAPW method for Cu_5Zn_8 and Cu_9Al_4 are shown in Figure 7.1a,b, respectively [8]. A large DOS in energies centered at −7.5 eV in Cu_5Zn_8 is due to the Zn-3d band, whereas that over the range from −2.5 to −4 eV is due to the Cu-3d band. A pseudogap clearly exists across the Fermi level, as shown in the insert. Its width is about 1.2 eV and H/H_0 is about 0.5 (see Chapter 2, Figure 2.6b in Section 2.3). A pseudogap is also seen across the Fermi level in Cu_9Al_4 and its

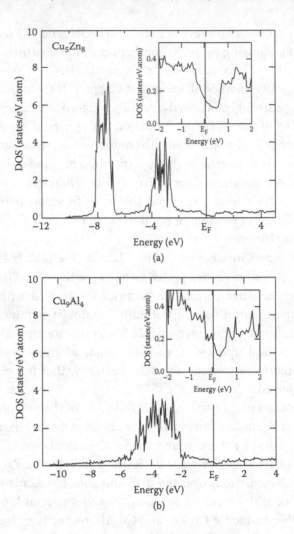

FIGURE 7.1 DOS derived from FLAPW band calculations for (a) Cu_5Zn_8 and (b) Cu_9Al_4 gamma-brasses. A pseudogap exists across the Fermi level in both cases. Inserts show the blow-up of the DOS across the Fermi level. [From R. Asahi et al., *Phys. Rev. B* 71 (2005) 165103.]

width is about 1.0 eV. The Cu-3d band of about 4 eV in width is wider than that in Cu_5Zn_8, reflecting a higher Cu concentration in Cu_9Al_4. Note that the Fermi level in both Cu_5Zn_8 and Cu_9Al_4 gamma-brasses is determined by filling

$$\frac{11 \times 5 + 12 \times 8}{13} = \frac{151}{13} = 11.615 \quad \text{and} \quad \frac{11 \times 9 + 3 \times 4}{13} = \frac{111}{13} = 8.538$$

electrons per atom into the respective DOSs. As noted in the Introduction, these numbers account for the number of valence electrons per atom, **VEC**, accommodated in the total DOS and they differ from the electron per atom ratio, **e/a**, employed in the Hume-Rothery electron concentration rule. As is evident from the argument above, entirely different values of **VEC** are assigned to Cu_5Zn_8 and Cu_9Al_4 gamma-brasses, even though they are considered to have the same **e/a** value of 21/13. Thus, the Hume-Rothery electron concentration rule is meaningful only when the **e/a** is used as a measure of electron concentration. We will learn more about its physical implication in the following sections.

7.3 EXTRACTION OF THE FsBz INTERACTION AT THE SYMMETRY POINTS N OF THE BRILLOUIN ZONE

The FLAPW one-electron wave function is expressed as

$$\psi_i(\mathbf{r},\mathbf{k}) = \sum_{G_n} C^i_{\mathbf{k}+G_n} \chi_{\mathbf{k}+G_n}(\mathbf{r})$$

(7.1)

where **k** is an arbitrary Bloch wave vector inside the reduced Brillouin zone, **G** is the allowed reciprocal lattice vector in a given system like the bcc lattice in the present case and i is the band index. As discussed in Chapter 4, Sections 4.11 and 4.12, the FLAPW basis function is given in the form:

$$\chi_{\mathbf{k}+G}(\mathbf{r}) =$$

$$\begin{cases} \Omega^{-1/2} e^{i(\mathbf{k}+\mathbf{G})\cdot\mathbf{r}} & r \in \text{outside MT-sphere} \\ \sum_{\ell m} \left[A^\alpha_{\ell m}(\mathbf{k}+\mathbf{G}) R_\ell(E^\alpha_v, r_\alpha) + B^\alpha_{\ell m}(\mathbf{k}+\mathbf{G}) \dot{R}_\ell(E^\alpha_v, r_\alpha) \right] Y_{\ell m}(\hat{r}_\alpha) & r \in \text{inside MT-sphere} \end{cases}$$

(7.2)

where $R_\ell(E^\alpha_v, r_\alpha)$ and $\dot{R}_\ell(E^\alpha_v, r_\alpha)$ represent the radial wave function and its energy derivative at a selected energy E^α_v, respectively, $Y_{\ell m}(\hat{r}_\alpha)$ is the spherical harmonics and the coefficients $A^\alpha_{\ell m}(\mathbf{k}+\mathbf{G})$ and $B^\alpha_{\ell m}(\mathbf{k}+\mathbf{G})$ are determined for the radial wave function and plane wave and their energy derivatives to be continuous across the MT sphere specified by the atomic species α.

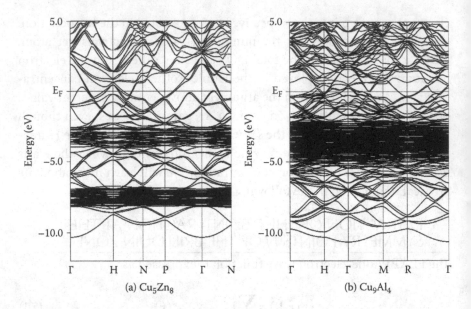

FIGURE 7.2 Energy dispersion relations derived from FLAPW band calculations for (a) Cu_5Zn_8 and (b) Cu_9Al_4 gamma-brasses. [From R. Asahi et al., *Phys. Rev.* B 71 (2005) 165103.]

Once the self-consistent wave function given by Equation 7.1 is calculated throughout a crystal, the coefficient C^i_{k+G} thus derived represents the Fourier component of the plane wave in the intermediate region between neighboring MT spheres (see Chapter 4, Figure 4.6.). By plotting the product of C^i_{k+G} and its complex conjugate as a function of the square of the reciprocal lattice vector, we can immediately find the plane wave component dominating the wave function in the intermediate region. If we carry out this procedure for the wave function at the edge of a pseudogap at the principal symmetry points of the bcc Brillouin zone, we can easily judge which reciprocal lattice vectors or the sets of the lattice planes are responsible for the formation of stationary waves and, thereby, opening a pseudogap. We hereafter call them *critical reciprocal lattice vectors* or the *sets of critical lattice planes*.* This is nothing but the process of extracting the FsBz interaction.

The energy dispersion relations for both Cu_5Zn_8 and Cu_9Al_4 gamma-brasses are shown in Figures 7.2a,b, respectively [8]. It is clear that the band gap opens across the Fermi level at the symmetry points N in I-cell

---

* We call in the present monograph the reciprocal lattice vector to be *critical*, when the relevant FsBz interaction is strong enough to produce a pseudogap at the Fermi level.

Cu$_5$Zn$_8$ (M in the P-cell Cu$_9$Al$_4$). Both the symmetry points N and M correspond to the center of the {110} zone planes of the bcc Brillouin zone in the reduced zone scheme but they also refer to the center of the {330} and {411} zone planes for gamma-brass in the extended zone scheme.* For instance, we focus on the wave function at the lower and higher edges of the pseudogap, that is, at energies of −0.54 (−0.41) and +0.58 (+0.56) eV relative to the Fermi level E$_F$ for Cu$_5$Zn$_8$ (Cu$_9$Al$_4$), respectively, with the band index i at the symmetry points N (or M). The square of its coefficient C_{k+G}^i in Equation 7.1 is taken and summed up over the equivalent reciprocal lattice vectors like (411), (141), (114) for {411}.

The resulting

$$\sum |C_{k+G}^i|^2$$

is plotted on the semi-logarithmic scale against the square of the Miller indices,

$$\sum h^2$$

in Figures 7.3a,b for Cu$_5$Zn$_8$ and in Figures 7.4a,b for Cu$_9$Al$_4$, respectively. It is clear that the sum of the squared Fourier coefficients is extremely large at

$$\sum h^2 = 18$$

or $|G|^2 = 18$ in the units of

---

* The center of the {110} plane of the bcc Brillouin zone is called the symmetry point N. It is known that all the zones specified by the set of Miller indices {hkl}, in which two of them are odd, pass through the points N upon reduction to the first zone. The Cu$_9$Al$_4$ crystallizes into the CsCl structure so that zone planes having the set of Miller indices, in which the sum of three integers are odd, also appear (see x-ray diffraction spectrum shown in Figure A2.3). For example, a cube bounded by six {100} zone planes forms the first Brillouin zone and accordingly, the rhombic dodecahedron becomes the second Brillouin zone. The center of the {110} planes coincides with the center of the edge of the {100} zone planes. This is referred to as the symmetry points M (see p. 237 in [16]).

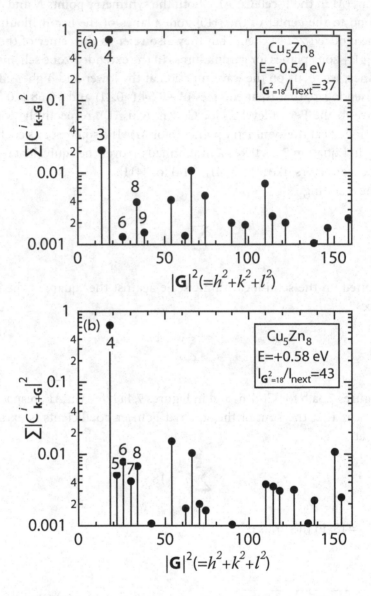

FIGURE 7.3 Sum of the square of Fourier coefficients over equivalent zone planes for the wave function at the symmetry points N at energies of -0.54 and $+0.58$ eV immediately below and above the Fermi level, respectively, as a function of the sum of the squared Miller indices or $|\mathbf{G}|^2$ for Cu_5Zn_8 gamma-brass [from R. Asahi et al., *Phys. Rev. B* 71 (2005) 165103]. Major zone planes are numbered as follows. 3: {321}, 4: {330}+{411}, 5: {332}, 6: {510}+{431}, 7: {521}, 8: {530}+{433}, 9: {611}+{532}.

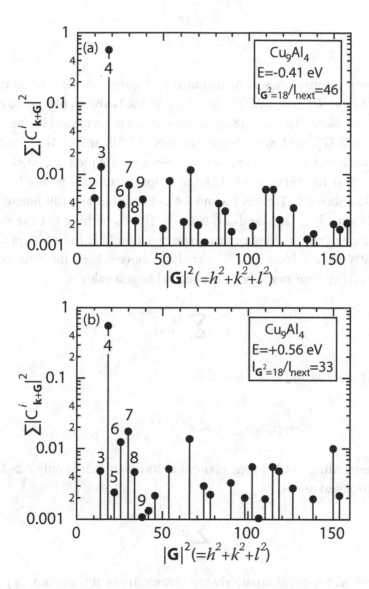

FIGURE 7.4   Sum of the square of Fourier coefficients over equivalent zone planes for the wave function at the symmetry points M at energies of -0.41 and $+0.56$ eV immediately below and above the Fermi level, respectively, as a function of the sum of the squared Miller indices or $|G|^2$ for Cu_9Al_4 gamma-brass [from R. Asahi et al., *Phys. Rev.* B 71 (2005) 165103]. Major zone planes are numbered as follows. 2: {310}, 3: {321}, 4: {330}+{411}, 5: {332}, 6: {510}+{431}, 7: {521}, 8: {530}+{433}, 9: {611}+{532}.

$$\left(\frac{2\pi}{a}\right)^2$$

for both gamma-brasses. As included in Figures 7.3 and 7.4, the ratio of intensity at $|\mathbf{G}|^2 = 18$ over the next largest one is always higher than 30 in all cases. Note that $|\mathbf{G}|^2 = 18$ specifies the sets of reciprocal lattice vectors $\mathbf{G}_{330}$ and $\mathbf{G}_{411}$ and, accordingly, the sets of {330} and {411} lattice planes in both I- and P-cell gamma-brasses. Since we are focusing on electronic states near the Fermi level at the principal symmetry points N (or M), the data shown in Figures 7.3 and 7.4 can be taken as a fulfillment of the condition $2k_F \cong |\mathbf{G}_{330}|$ and $2k_F \cong |\mathbf{G}_{411}|$. This is nothing but the demonstration of the matching condition in equation (4.1) from first-principles FLAPW band calculations. The analysis above is hereafter referred to as the *FLAPW-Fourier* method. The second largest value of

$$\sum \left|C_{\mathbf{k+G}}^i\right|^2$$

in Figures 7.3a and 7.4a is found at

$$\sum h^2 = 14$$

corresponding to {321} zone planes but its magnitude is only 2.2~2.7% of the strongest one at

$$\sum h^2 = 18$$

Therefore, we can alternatively say that electrons at the symmetry points N (or M) at energy very close to the Fermi level exclusively interact with the set of {330} and {411} lattice planes. Hereafter, both $\mathbf{G}_{330}$ and $\mathbf{G}_{411}$ are altogether referred to as $\mathbf{G}_0$.

We are now ready to discuss how a pseudogap is formed at the Fermi level, when the matching condition $2(\mathbf{k+G}) = \mathbf{G}_0$ is fulfilled. The Fourier coefficient $C_{\mathbf{k+G}}^i$ satisfying the relation $2(\mathbf{k+G}) = \mathbf{G}_0$ for Cu_5Zn_8 is summarized in Table 7.1 [8]. It is seen that the value of $C_{\mathbf{G}_0/2}^i$ in modes 1, 2,

TABLE 7.1 Largest Fourier Coefficient of the FLAPW Wave Function at the Symmetry Points N at Energies $E = E_1$ to E_4 Near the Fermi Level for Cu_5Zn_8 Gamma-Brass[a]

Mode	k at Point N	G	$k + G$ $(= G_0/2)$	$\|C^i_{k+G}\|$			
				E_1 (eV) -1.37	E_2 (eV) -0.62	E_3 (eV) $+0.58$	E_4 (eV) $+0.68$
1	(1/2)(1,1,0)	(1/2)(3,−2,−1)	(1/2)(4,−1,−1)	0.42	0.41	0.39	0.41
1'	(1/2)(1,1,0)	(1/2)(−5,0,1)	(1/2)(−4,1,1)	0.42	0.41	0.39	0.41
2	(1/2)(1,1,0)	(1/2)(3,0,1)	(1/2)(4,1,1)	0.42	0.41	0.39	0.41
2'	(1/2)(1,1,0)	(1/2)(−5,−2,−1)	(1/2)(−4,−1,−1)	0.42	0.41	0.39	0.41
3	(1/2)(1,1,0)	(1/2)(−1,2,3)	(1/2)(0,3,3)	0.02	0.17	0.16	0.01
3'	(1/2)(1,1,0)	(1/2)(−1,−4,−3)	(1/2)(0,−3,−3)	0.02	0.17	0.16	0.01

Source: R. Asahi, H. Sato, T. Takeuchi, and U. Mizutani, *Phys. Rev. B* 71 (2005) 165103.

[a] Positive and negative signs in energy represent the unoccupied and occupied states relative to the Fermi level, respectively.

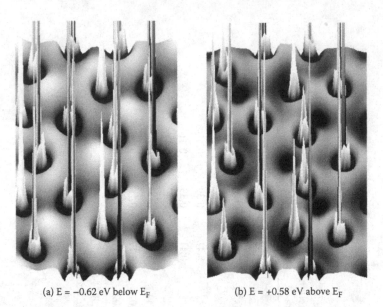

(a) E = −0.62 eV below E_F (b) E = +0.58 eV above E_F

FIGURE 7.5 Electron charge density distribution over (110) plane obtained by superimposing two waves with modes 1 and 1′ at (a) $E_2 = -0.62$ eV (occupied state) and (b) $E_3 = +0.58$ eV (unoccupied state) for Cu_5Zn_8 gamma-brass. A sharp cusp at each nucleus represents the atomic wave function inside the MT sphere. An intermediate region between the neighboring ions is the brightest corresponding to the highest charge density and the darkest corresponding to the lowest charge density in (a) and (b), respectively. This confirms the formation of stationary waves. [From R. Asahi et al., *Phys. Rev.* B 71 (2005) 165103.]

and 3 coincides with that of $C^i_{-G_0/2}$ in its counter modes 1′, 2′, and 3′, regardless of the choice of G_0. It is now obvious that the two plane waves running in opposite directions with coefficients $C^i_{G_0/2}$ and $C^i_{-G_0/2}$ having equal magnitudes dominate in the wave function given by Equation 7.1 at the symmetry points N (or M). This leads to the formation of either $\cos(\mathbf{k}\cdot\mathbf{r})$- or $\sin(\mathbf{k}\cdot\mathbf{r})$-type stationary waves. For example, we calculated $|\psi_i(\mathbf{r},\mathbf{k})|^2$, using the waves with modes 1 and 1′ listed in Table 7.1. As shown in Figures 7.5a,b, the electron density for the occupied state is enhanced in between the two nuclei, whereas that for the unoccupied state is suppressed [8]. The same conclusion is drawn for Cu_9Al_4. This is the confirmation of the formation of stationary waves.

We may add a few more words to explain why stationary waves thus formed yield a pseudogap at the Fermi level. As discussed in Chapter 4, Section 4.2, electron waves start interfering with a periodic lattice and form either a cosine- or sine-type stationary wave, when the wave number of the nearly free electron wave increases and reaches one-half the reciprocal lattice vector along direction perpendicular to the set of lattice planes like {411} and {330} in the case of gamma-brasses (see Chapter 4, Figure 4.2). The energy of the cosine-type stationary wave is lowered, whereas that of the sine-type one is increased relative to that of the free electron. This obviously gives rise to an energy gap along this direction as a result of the formation of bonding and antibonding states caused by the FsBz interaction (see Chapter 4, Figure 4.3).

As can be seen from Table 7.1, there exist several energy eigen-values $E_1, E_2 \cdots$ at the symmetry points N near the Fermi level. If attention is paid to directions away from the symmetry points N in Figure 7.2, one can immediately find that the dispersion curves cross the Fermi level several times. As the consequence, small finite states remain at the Fermi level. The DOS is then characterized by a trough across the Fermi level having finite states even at its bottom, as is seen in Figure 7.1. This is already named a pseudogap. In summary, we proved that a pseudogap originates from the formation of stationary waves at the symmetry points N (or M) as a result of the FsBz interaction involving the set of {330} and {411} lattice planes in both Cu_5Zn_8 and Cu_9Al_4. This is what we shall refer to as a *FsBz-induced pseudogap*. Now we can say that $|\mathbf{G}|^2 = 18$ or the set of {330} and {411} lattice planes must be *critical* to give rise to a sizable pseudogap across the Fermi level in gamma-brasses in group (I).

As noted in Chapter 4, Section 4.1, we refer to "the Hume-Rothery stabilization mechanism" as an electronic stabilization condition whereby an alloy is stabilized because a FsBz-induced pseudogap is formed across the Fermi level. In the next section, we will study further why the parameter **e/a** serves as a critical role, when a pseudogap is induced by the FsBz interaction.

7.4 THE HUME-ROTHERY STABILIZATION MECHANISM FOR Cu_5Zn_8 AND Cu_9Al_4 GAMMA-BRASSES

In terms of simple chemical valence considerations, the value of **e/a** is rather self-evident in alloys involving the noble metals and the polyvalent elements in the periodic table and is simply given as an averaged number of valencies per number of constituent element atoms in the unit cell. But we consider it necessary to verify the validity of this conventional approach. In particular, a clear-cut determination of **e/a** is critically important for alloys containing TM elements located to the left of the noble metals. The prescription we will describe below is based on the FLAPW–Fourier method, which will allow us to determine the **e/a** value in a theoretically sound way, regardless of whether a given alloy contains the TM element as a major constituent element [17].

First of all, the irreducible wedge corresponding to 1/48th of the Brillouin zone of the bcc lattice, which was already shown in Figure 5.3b, is partitioned into 200 elements. The FLAPW wave function is calculated within an energy interval E and $E+\Delta E$ at the wave vector $\mathbf{k}_i$ corresponding to the center of each element. The electronic state $2(\mathbf{k}_i+\mathbf{G})$ having the largest Fourier coefficient in the FLAPW wave function outside the MT sphere is extracted in the same manner as in Figures 7.3 and 7.4. This is done for 200 elements and its average value at energy E is calculated:

$$\left|2(\mathbf{k}+\mathbf{G})\right|_E \equiv 2\sum_{i=1}^{200}\omega_i\left|\mathbf{k}_i+\mathbf{G}\right|_E \tag{7.3}$$

where ω_i represents the number of degeneracies. The energy eigen-values are not distributed at equal intervals. Hence, the value of $\left|2(\mathbf{k}_i+\mathbf{G})\right|_E$ at a given energy is determined by interpolating values calculated at neighboring energy eigen-values. The resulting $\left|2(\mathbf{k}+\mathbf{G})\right|^2 - E$ relation yields an isotropic and single-branch dispersion relation in the extended zone scheme. We consider this to represent the motion of itinerant electrons outside the MT sphere. In this way, we can extract the energy dispersion relation for electrons of a highly itinerant character. An "effective Fermi sphere" can be constructed in the reciprocal space from the $\left|2(\mathbf{k}+\mathbf{G})\right|^2 - E$ dispersion relation. Indeed, the value of $\left|2(\mathbf{k}+\mathbf{G})\right|^2$ at the Fermi level gives rise to the square of the Fermi diameter, from which the number of electrons, $(\mathbf{e/a})_{\text{total}}$, contained in the Fermi sphere can be deduced. For the sake of convenience, the method may be hereafter called the *Hume-Rothery plot*.

Some caution is needed to evaluate the Fermi diameter and $(\mathbf{e/a})_{\text{total}}$ from a Hume-Rothery plot. Since we deal with an average quantity of $\left|2(\mathbf{k}_i+\mathbf{G})\right|_E$ over 200 elements in the irreducible wedge at a given energy E, its variance must be small enough to make it physically meaningful. The variance $\sigma^2(E)$ is defined as

$$\sigma^2(E)=\left[2\left\{k_G(E)+\sigma_G\left(E\right)\right\}\right]^2-\left\{2k_G(E)\right\}^2$$

$$=8k_G(E)\sigma_G\left(E\right)+4\sigma_G\left(E\right)^2 \tag{7.4}$$

where $\sigma_G(E)$ is called the standard deviation defined as

$$\sigma_G(E) \equiv \sqrt{\sum_i \omega_i \left(\left| \mathbf{k}_i + \mathbf{G} \right|_E - k_G(E) \right)^2} \qquad (7.5)$$

Figure 7.7 shows the energy dependence of $\left| 2(\mathbf{k}+\mathbf{G}) \right|^2$ and its variance for Cu_5Zn_8 gamma-brass [17]. The variance $\sigma^2(E)$ is extremely large over energy ranges, where the Zn-3d and Cu-3d bands are located. Otherwise, the variance is so small that the dispersion relation is justified to be reliable and meaningful. We can draw a straight line passing through the data points in the vicinity of the Fermi level and near the bottom of the valence band, as indicated in Figure 7.6a. The intercept at the Fermi level is read off as 18.47. This is nothing but the square of the Fermi diameter of an "effective Fermi sphere" and is found to be in excellent agreement with $\left| \mathbf{G}_0 \right|^2 = 18$ deduced from the FLAPW-Fourier method above. *This is indeed a theoretical confirmation of the validity of the matching condition (4.1)*. Similarly, the data for Cu_9Al_4 gamma-brass are plotted in Figures 7.7a,b [17]. The square of the Fermi diameter is again found to be 18.45 from the intercept at the Fermi level. In this way, we can prove from first-principles band calculations that both Cu_5Zn_8 and Cu_9Al_4 gamma-brasses are associated with a Fermi sphere with the same diameter $2k_F$. In other words, the Hume-Rothery plot enables us to extract a hidden electron concentration parameter **e/a** from the electronic structure calculated using FLAPW first-principles band calculations, in which **VEC** appears as an explicit electron concentration parameter.

The total number of electrons per atom $(e/a)_{total}$ is calculated by inserting the Fermi radius k_F thus obtained into the relation

$$\left(e/a \right)_{total} = \frac{8\pi k_F^3}{3N}$$

where the number of atoms per unit cell, N, is equal to 52 and k_F is in units of

$$\frac{2\pi}{a}$$

As listed in Table 7.2, the value of $(e/a)_{total}$ is very close to 21/13 (=1.615) for both Cu_5Zn_8 and Cu_9Al_4. This is a theoretical proof of the Hume-

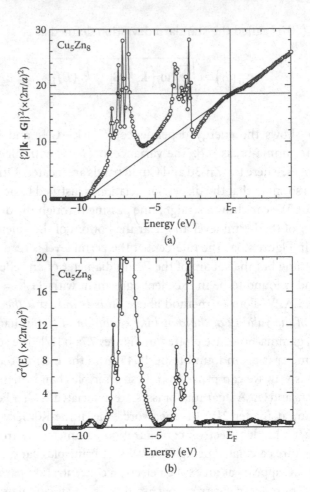

FIGURE 7.6 (a) Hume-Rothery plot calculated using equation (7.3). A straight line passing the data points, where the variance is small, is drawn as a guide. (b) Energy dependence of the variance defined by Equation 7.4 for Cu_5Zn_8 gamma-brass. [From R. Asahi et al., *Phys. Rev.* B 72 (2005) 125102.]

Rothery electron concentration rule for both Cu_5Zn_8 and Cu_9Al_4 gamma-brasses. The valency of the noble metal Cu, $(e/a)_{Cu}$, is easily calculated, if valencies of partner elements Zn and Al are assigned to be two and three, respectively. As listed in Table 7.2, the value is found to be close to unity for Cu consistent with the fact that Cu is monovalent in the metallic state.

Before ending this section, we should remark why the parameter **e/a** is more important than **VEC** in the formation of a FsBz-induced pseudogap. An essence in the FLAPW-Fourier analysis is to extract the Fermi diameter from the momentum distribution of electrons outside the MT sphere,

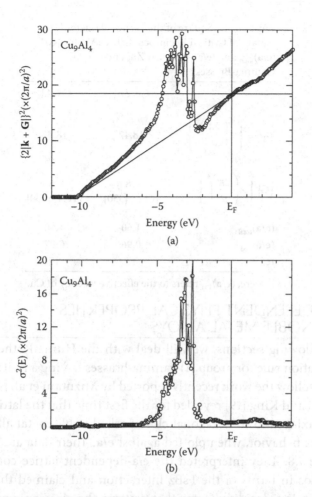

FIGURE 7.7 (a) Hume-Rothery plot calculated using equation (7.3). A straight line passing the data points, where the variance is small, is drawn as a guide. (b) Energy dependence of the variance defined by Equation 7.4 for Cu_9Al_4 gamma-brass. [From R. Asahi et al., *Phys. Rev.* B 72 (2005) 125102.]

while skillfully circumventing the contribution from d-electrons, which are mostly localized inside the MT sphere. Using such itinerant electrons, we have dealt with a FsBz-induced pseudogap caused by the interference effect with the set of the lattice planes, that is, {330} and {411} lattice planes for the case of gamma-brasses. As a result, the stability range is directly scaled with respect to e/a. This clearly demonstrates that the parameter e/a must be employed as a crucial electron concentration parameter in dealing with the Hume-Rothery stabilization mechanism or the Hume-Rothery electron concentration rule.

TABLE 7.2 Square of the Fermi Diameter $(2k_F)^2$, Square of Critical Reciprocal Lattice Vector $|G|^2$, $(e/a)_{total}$ and $(e/a)_{TM}$ for Cu_5Zn_8 and Cu_9Al_4 Gamma-Brasses[a]

	Cu_5Zn_8	Cu_9Al_4
$(2k_F)^2\left(\times\left(\dfrac{2\pi}{a}\right)^2\right)$	18.47	18.45
$|G|^2\left(\times\left(\dfrac{2\pi}{a}\right)^2\right)$	18 $\{411\}, \{330\}$	18 $\{411\}, \{330\}$
$(e/a)_{total}$	1.60	1.60
$(e/a)_{TM}$	0.96	0.97
ref	17	17

[a] Here, $(e/a)_{TM}$ refers to the effective valency of Cu.

7.5 e/a DEPENDENT PHYSICAL PROPERTIES OF NOBLE METAL ALLOYS

In the following sections, we will deal with the Hume-Rothery electron concentration rule for group (I) gamma-brasses having a finite solid solution and follow the work recently reported by Mizutani et al. [18]. In 1960, Massalski and King [19] revealed for the first time that the lattice constant and the axial ratio in hexagonal close packed noble metal alloys exhibit systematic behavior, when plotted against e/a. Their data are reproduced in Figure 7.8. They interpreted the e/a-dependent lattice constants and axial ratios in terms of the FsBz interaction and claimed that the FsBz interaction is a deciding factor that governs the electronic properties of noble metal alloys. Massalski further attempted to confirm the proposal above by studying physical quantities more directly relevant to the electronic structure than the lattice constant. As shown in Figure 7.9, the magnetic susceptibility due to itinerant electrons obtained after subtracting the ionic contribution was found to exhibit universal e/a dependences for various Hume-Rothery phases. In particular, the fact that the data for a series of gamma-brass alloys fall on a rapidly declining universal curve and are strongly negative in sign, was taken as the evidence for the manifestation of the FsBz interaction associated with $\{411\}$ and $\{330\}$ Brillouin zones [18].

Massalski and Mizutani (1978) [20] revealed, as shown in Figure 7.10, that the electronic specific heat coefficient in a series of phases α, β, γ, ε,

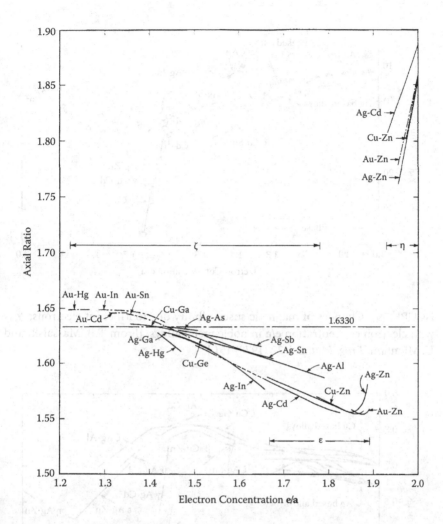

FIGURE 7.8   Changes of axial ratio with electron concentration **e/a** in hexagonal close-packed noble metal alloys. [From T.B. Massalski and U. Mizutani, *Prog. Mat. Sci.* 22 (1978) 151.]

and η of noble metal alloys falls on a universal curve for each phase, when plotted against **e/a**, and interpreted the results in terms of the respective FsBz interactions. In good agreement with the data for magnetic suscepti-bilities, the electronic specific heat coefficient in gamma-brass alloys was also found to decrease sharply with increasing **e/a**. They extended the eight-cone model developed by Ziman for pure Cu [21] to gamma-brasses and explained its rapidly declining **e/a** dependence by approximating its Fermi surface in the context of the "36-cone model." Based on the model of Jones

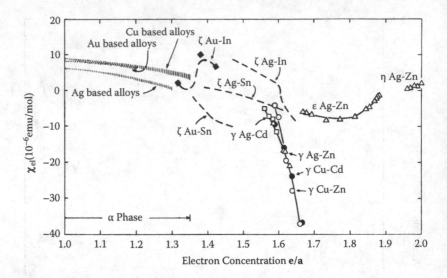

FIGURE 7.9 Changes of magnetic susceptibility due to itinerant electrons, χ_{el}, with electron concentration e/a in noble metal alloys. [From T.B. Massalski and U. Mizutani, *Prog. Mat. Sci.* 22 (1978) 151.]

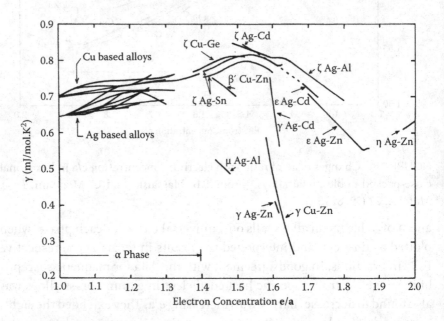

FIGURE 7.10 Changes of electronic specific heat coefficient with electron concentration e/a in noble metal alloys. [From T.B. Massalski and U. Mizutani, *Prog. Mat. Sci.* 22 (1978) 151.]

(II) discussed in relation to Figure 3.4, they insisted that the composition range, where gamma-brass is stabilized, should be located on a sharply declining slope of its DOS. In this way, the electronic energy should be most profoundly affected when the Fermi level falls inside the pseudogap, rather than when it is on its top point "A" in Figure 3.1. It is interesting to note that their interpretation was made without using the word *pseudogap* at that time, though this is nothing but the pseudogap concept. As is clear from the arguments in this section, one is led to conclude that the Hume-Rothery stabilization mechanism holds true throughout all solid solution ranges of the Hume-Rothery phases in noble metal alloys. We will discuss this subject further in the following sections, by taking into consideration the data on the concentration dependence of vacancies in the unit cell in a solid solution range for a series of group (I) gamma-brasses [18].

7.6 Cu-Zn AND Cu-Cd GAMMA-BRASSES IN SOLID SOLUTION RANGES

As listed in Table 6.1, most gamma-brasses possess a rather wide solubility range. For example, Cu_5Zn_8 and Cu_9Al_4 compounds discussed in Sections 7.2 to 7.4 are formed at particularly stoichiometric compositions within the respective solid solution ranges [22]. Our objective in this section is to study if the Hume-Rothery stabilization mechanism works throughout the gamma-brass solid solution range in noble metal alloys. However, as emphasized repeatedly, first-principles band calculations are powerful only for ordered alloys, free from any chemical disorder or fractional site occupancies. Hence, under such limited condition, we are obliged to discuss the stabilization mechanism by assuming the DOS constructed for perfectly ordered Cu_5Zn_8 and Cu_9Al_4 alloys to remain unchanged on alloying, that is, the rigid-band model, in which the Fermi level is allowed to move in accordance with a change in electron concentration **e/a**.

Let us consider gamma-brasses existing over 57 to 70 at .%Zn in the Cu-Zn alloy system. One would wonder if the application of the rigid-band model over 57–70 at .%Zn concentration range can be justified, since areas under both Cu- and Zn-3d bands in the DOS definitely change in proportion to the solute concentration. However, as emphasized as the "golden rule" in Chapter 5, Section 5.6, a difference in the d-band profiles between the two competing phases at a given electron concentration would not affect the valence-band structure energy difference, once it is fully filled. Thus, we can safely proceed with our discussion by ignoring the composition dependence of the Zn- and Cu-3d band profiles and focusing only on a change in the DOS in the very vicinity of the Fermi level upon varying the solute concentration relative to the host Cu_5Zn_8. This allows us to plot the DOS as a function of electron concentration in place of energy and to assign $e/a = 21/13$ to the Fermi level for the host.

One more element of caution should be exercised in the discussion of phase stability over a solid solution range. The Fermi radius in the left-hand side of Equation 4.1 is obviously given by $k_F = [3\pi^2 (e/a)/V_a]^{1/3}$, where V_a is the volume per atom. By rewriting it as $k_F = [3\pi^2 (e/uc)/V_0]^{1/3}$, where V_0 is the volume per unit cell and e/uc is the number of electrons per unit cell, we can incorporate the effect of vacancies introduced into the unit cell through the electron concentration parameter e/uc. Betterton et al. [23] observed an increase in vacancy concentration with increasing Al and Ga concentrations in Cu-Al and Cu-Ga gamma-brasses. Following the pioneering work by Bradley and Taylor in 1937 for the B2-type Ni-Al alloys [24], Betterton et al. conjectured that vacancies are introduced so as to counterbalance an increase in e/a, thereby maintaining a constant e/uc and, in turn, the matching condition of Equation 4.1 over a whole solid solution range. This means that for the present purpose we should use e/uc as an electron concentration parameter rather than e/a.

As discussed in the preceding sections, the number of atoms in the unit cell and the e/a value for Cu_5Zn_8 can be fixed at 52 and 21/13, respectively. Hence, the Fermi level in the DOS for Cu_5Zn_8 is now replaced by e/uc equal to $52 \times (21/13) = 84$. An integration of the DOS in Figure 7.1a below and above the Fermi level can generate a conversion relation from the energy to e/uc. In this way, we can immediately construct the e/uc dependence of the DOS, as shown in Figure 7.11 for Cu_5Zn_8 [18]. It is seen that the pseudogap range is extended over $80 < e/uc < 88$ for the solid solution range of gamma-brass in the Cu-Zn alloy system. We consider the upper limit of e/uc equal to 88 to be reasonable in view of the fact that the Brillouin zone bounded by {330} and {411} zone planes of the gamma-brass structure is capable of accommodating totally 90 electrons per unit cell.* This is hereafter referred to as $(e/uc)_0 = 90$. If it exceeds 90, electrons have to overlap across the next Brillouin zone, needing extra electronic energies.

---

* The volume of the Brillouin zone bounded by {330} and {411} zone planes is given as $V_B = 45(2\pi/a)^3$ [25]. Since $2(L/2\pi)^3$ electrons can be accommodated in a unit volume of reciprocal space, we can fit $2(L/2\pi)^3 \times V_B = 90(L/a)^3$ electrons per volume $V = L^3$ or 90 electrons per unit cell in the zone.

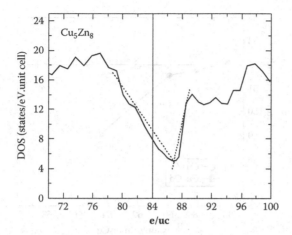

FIGURE 7.11   DOS in the vicinity of the Fermi level as a function of **e/uc** for Cu_5Zn_8 gamma-brass. A vertical line refers to the Fermi level. A dotted line approximates slopes of the pseudogap across the Fermi level. [From U. Mizutani, T. Noritake, T. Ohsuna, and T. Takeuchi, *Phil. Mag.* 90 (2010) 1985.]

Figure 7.12 shows the solute concentration dependence of the number of atoms per unit cell, N, for gamma-brasses in both Cu-Zn and Cu-Cd alloy systems, as determined from the lattice constant measured using x-ray powder diffractometer and the density by Archimedes' principle using a Si single crystal as a reference [18]. It clearly shows that N remains at the value of 52 up to about 61.5 at.%Zn but exhibits a gradual decrease with further increase in Zn concentration in the Cu-Zn alloy system. In contrast, the value of N in Cu-Cd gamma-brass system is almost 52 only at the lowest Cd concentration of 56 at.% and decreases fairly rapidly with increasing Cd concentration. It is surprising that almost one vacancy per unit cell is introduced at 61.5 at.%Cd, at which an ordered alloy Cu_5Cd_8 were to be formed.

We can now easily calculate the value of **e/uc** by taking a product of the number of atoms per unit cell, N, and an average **e/a** value by assuming valencies of Cu, Zn, and Cd to be one, two, and two, respectively. Indeed, the valency of Cu was confirmed from the Hume-Rothery plot discussed in Section 7.4 to be unity in Cu_5Zn_8. The results are plotted in Figure 7.13 as a function of the solute concentration for both alloy systems. Let us first discuss the data for Cu-Zn gamma-brasses. The value of **e/uc** is distributed over 82 to 86.3. A comparison between Figures 7.11 and 7.13 immediately tells us that the solid solution range of gamma-brass over 58 up to 70 at.% Zn is found inside the pseudogap. Therefore, we conclude that gamma-

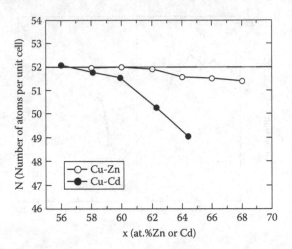

FIGURE 7.12 Solute concentration dependence of the number of atoms in the unit cell, N, for $Cu_{100-x}Zn_x$ and $Cu_{100-x}Cd_x$ gamma-brasses. [From U. Mizutani, T. Noritake, T. Ohsuna, and T. Takeuchi, *Phil. Mag.* 90 (2010) 1985.]

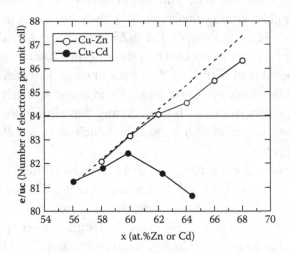

FIGURE 7.13 Solute concentration dependence of the number of itinerant electrons in the unit cell, **e/uc**, for $Cu_{100-x}Zn_x$ and $Cu_{100-x}Cd_x$ gamma-brasses [from U. Mizutani, T. Noritake, T. Ohsuna, and T. Takeuchi, *Phil. Mag.* 90 (2010) 1985]. A dashed straight line represents a change in **e/uc**, when N = 52 is assumed over the solid solution range.

brasses over the whole solute concentration range in the Cu-Zn alloy system are stabilized by the Hume-Rothery mechanism. However, the role of vacancies in this system is less clear, since a dashed line representing the e/uc calculated under the assumption of N = 52 throughout the solid solution range still remains below 88 at the termination of the solid solution range, that is, at 68 at .%Zn.

Another interesting phenomenon to be emphasized in Cu-Zn gamma-brasses is the persistence of N = 52 in the Cu-rich concentration range over 57 to 62 at .%Zn. Though the lowest value of e/uc is still within the pseudogap (see Figure 7.11), the Fermi surface may become a bit too small to satisfy the matching condition $(2k_F)^2 = |G|^2$. Morton observed the long-period super-lattice structure in the neighborhood of e/a = 1.58 in the Cu-Zn alloy system [26,27] and interpreted its formation in terms of the Sato-Toth theory [28,29], in which the split Brillouin zones constructed from split diffraction spots allow to maintain the matching condition to smaller e/a values with the generation of shorter reciprocal lattice vectors [30]. We will discuss more about this issue upon discussing Ni-Zn gamma-brasses in Chapter 8, Section 8.5.

The situation in Cu-Cd gamma-brass alloys is less clear, since defect-free gamma-brasses apparently do not exist [2–4,18]. Indeed, the value of e/uc remains well below 84 over the whole composition range, as shown in Figure 7.13. Such unique features are most likely caused by the possession of a large atomic size ratio between Cu and Cd, that is, $r_{Cd}/r_{Cu} = 1.23$ in comparison with $r_{Zn}/r_{Cu} = 1.09$ in Cu-Zn alloys. As a matter of fact, the α-phase primary solid solution range is extremely narrow in agreement with the Hume-Rothery 15% size rule. Moreover, neither bcc nor the CsCl-type B2 compound exists in the Cu-Cd alloy system. This is entirely different from the phase diagram in the Cu-Zn alloy system, in which successively appearing α-, β-, and γ-brasses are known to obey the Hume-Rothery electron concentration rule. Nevertheless, we have a fairly wide solid solution range for the gamma-brass. This suggests that the gamma-brass structure is so flexible in accommodating vacancies in its large unit cell that a larger amount of vacancies in the Cu-Cd system may be more easily introduced. We tend to believe that the lowering of the electronic energy by forming a pseudogap across the Fermi level may be large enough to overwhelm the size effect in stabilizing gamma-brass phase in the Cu-Cd alloy system in spite of a rather excessive accommodation of vacancies. Otherwise, gamma-brass would not appear as a stable phase due to such an unfavorable size ratio.

First-principles band calculations for the Cu-Cd gamma-brass are not feasible because there exists no stoichiometric compound in this system, free from chemical disorder and vacancies. For brevity, we may use the DOS for Cu_5Zn_8 gamma-brass, since chemical nature of constituent elements would be similar to each other. According to Figure 7.13, the value of the **e/uc** falls in the range of 80 to 82.4, which is well inside the pseudogap shown in Figure 7.11. As shown in Figure 7.9, a large negative magnetic susceptibility is observed after subtracting ionic contribution from the measured one. The results are found to fall on a rapidly declining universal curve with increasing the **e/a** value, along with the data for gamma-brasses in other alloy systems like Cu-Zn, Ag-Zn, and Ag-Cd [18,20]. This is interpreted as the possession of a large Landau diamagnetism [31] and can be taken as an indirect evidence for the existence of the Fermi level inside the pseudogap over the whole solid solution range of Cu-Cd gamma-brasses. In summary, we consider that all gamma-brasses over a whole solute concentration range in both Cu-Zn and Cu-Cd alloy systems are stabilized by a FsBz-induced pseudogap and, hence, obey the Hume-Rothery stabilization mechanism.

7.7 Cu-Al AND Cu-Ga GAMMA-BRASSES IN SOLID SOLUTION RANGES

In the same manner as in Figure 7.11, we can set the Fermi level at **e/uc** = 84 and rescale the DOS of Cu_9Al_4 in Figure 7.1b with respect to the parameter **e/uc** by integrating the DOS below and above the Fermi level. The results are shown in Figure 7.14. It can be seen that a pseudogap extends over the **e/uc** range from 81 to 88. This is quite reasonable since it is lower than $(e/uc)_0 = 90$. The solute concentration dependence of the number of atoms per unit cell, N, and that of the resulting **e/uc** are plotted in Figures 7.15 (a) and (b), respectively. The Cu-Al gamma-brasses are distributed over the range 84 < **e/uc** < 89. A comparison between Figures 7.14 and 15b indicates that its solid solution range is essentially inside the pseudogap. As indicated by a dashed line in Figure 7.15b, the value of **e/uc** would have gone well beyond 90 and the gamma-brass field would be much narrowed, if vacancies were not to be introduced with increasing Al concentration. Hence, we say that the Hume-Rothery stabilization mechanism works over the whole Al composition range in the Cu-Al gamma-brass system as a result of significant introduction of vacancies into the unit cell at higher Al concentration range.

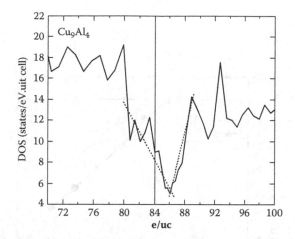

FIGURE 7.14 DOS in the vicinity of the Fermi level as a function of **e/uc** for Cu_9Al_4 gamma-brass. A vertical line refers to the Fermi level in Cu_9Al_4 gamma-brass. A dotted line approximates slopes of the pseudogap across the Fermi level. [From U. Mizutani, T. Noritake, T. Ohsuna, and T. Takeuchi, *Phil. Mag.* 90 (2010) 1985.]

Bradley et al. [32] revealed for the first time in 1938 that lowering the number of atoms in the unit cell well below N = 52 occurs and causes a cubic symmetry to break, when the Al concentration exceeds 36 at.%. Westman [33] claimed that the 38.9 at.%Al gamma-brass can be described in terms of the rhombohedral symmetry with space group $R3m$, a subgroup of $P\bar{4}3m$. More detailed structural studies were carried out by Kisi and Browne in 1991 [7]. They revealed that 31.3 to 34.0 at.%Al alloys were cubic with space group $P\bar{4}3m$, while 36.8 and 38.8 at .%Al alloys were rhombohedrally distorted with space group $R3m$. As shown in Figure 7.16, Mizutani et al. [18] observed the split of the {444} peak into {444} and {$\bar{4}44$} sub-peaks in the x-ray diffraction spectrum upon transformation into the $R3m$ structure for samples with x = 36, 37, and 39.9 at .%Al. This is consistent with the emergence of the rhombohedrally distorted phase in Cu-Al gamma-brasses with x ≥ 36.

One more unique feature is addressed in relation to the $R3m$ structure in Al-rich Cu-Al gamma-brasses. Kisi and Browne [7] pointed out that vacancy is absent in the cubic phase but is introduced only into sites IT in the cluster "b" upon the $R3m$ transformation (see Chapter 6, Figure 6.3 and also Appendix 2, Figure A2.10 and Section A2.2.4.7). They also observed that interatomic distance IT-IT in both clusters "a" and "b" increases with increasing Al concentration within the cubic phase.

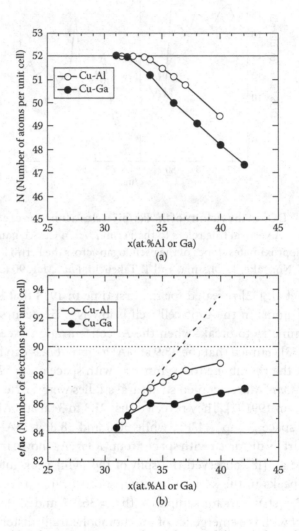

FIGURE 7.15 (a) Solute concentration dependence of the number of atoms in the unit cell, N, and (b) that of the number of itinerant electrons in the unit cell, **e/uc**, for $Cu_{100-x}Al_x$ and $Cu_{100-x}Ga_x$ gamma-brasses. A dashed straight line indicates a change in **e/uc**, when N = 52 is assumed over the solid solution range. A dotted horizontal line indicates $(e/uc)_0 = 90$. [From U. Mizutani, T. Noritake, T. Ohsuna, and T. Takeuchi, *Phil. Mag.* 90 (2010) 1985.]

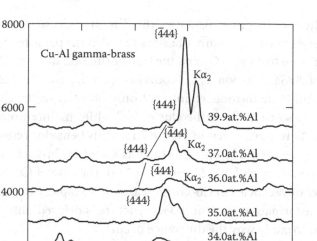

FIGURE 7.16 The (444) diffraction peak for $Cu_{100-x}Al_x$ gamma-brasses. The peak is split into $\{\overline{4}44\}$ and $\{444\}$ upon transformation into space group $R3m$. The former includes six-fold degenerate lines $(\overline{4}44, 4\overline{4}4, 44\overline{4}, \overline{4}4\overline{4}, \overline{4}\overline{4}4, \overline{4}\overline{4}\overline{4})$, while the latter two-fold $(444, \overline{4}\overline{4}\overline{4})$. A difference in multiplicity explains the observed difference in intensity. A subpeak on the right-hand side shoulder is due to Cu-$K\alpha_2$ radiation. [From U. Mizutani, T. Noritake, T. Ohsuna, and T. Takeuchi, *Phil. Mag.* 90 (2010) 1985.]

However, the interatomic distance IT-IT in the cluster "b" is shortened upon transformation as a result of the introduction of 1.2 vacancies into sites IT in "b." This is now compared with the situation in Cu-Ga gamma-brasses.

According to the structure analysis of Cu_9Ga_4 gamma-brass, it was identified to be isostructural to Cu_9Al_4 with space group $P\overline{4}3m$ [34]. However, the best refinements showed the existence of chemical disorder in sites CO of the two clusters "a" and "b" shown in Figure 6.3b. More detailed structure analysis was recently reported for five gamma-brasses over 32.0 to 40.0 at .%Ga concentration, using Spring-8 synchrotron radiation beam with the wavelength of 0.05 nm [18]. The Rietveld structure refinements for a series of Cu-Ga gamma-brasses revealed the persistence of the $P\bar{4}3m$ structure up to 38 at.%Ga and the transformation into the $R3m$ structure at 40 at.%Ga. As soon as Ga concentration departs from x = 31, vacancies begin to be introduced into not only sites CO in the clusters "a" and "b" but also sites IT in the cluster "b" while maintaining space group $P\bar{4}3m$. As is clearly seen in Figure 7.15a, this behavior reflects the downward departure of N from 52 as early as x = 32 in Cu-Ga in sharp contrast to the persistence of N = 52 up to x = 34 in the case of Cu-Al. Though the number of vacancies in Cu-Ga increases more rapidly than that in Cu-Al, the transformation into the $R3m$ structure is delayed until x = 40. It is of interest to study why the difference occurs.

Figure 7.17 shows the Ga concentration dependence of the interatomic distance IT-IT corresponding to Ga-Ga and Cu-Cu pairs in the two clusters "a" and "b," respectively. The number of vacancies over sites IT in the cluster "b" is also incorporated in Figure 7.17. It can be seen that it

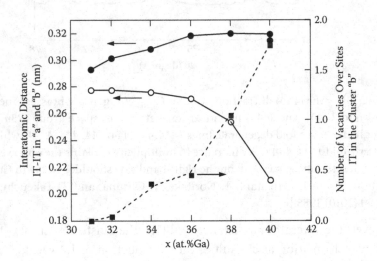

FIGURE 7.17 Ga concentration dependence of the interatomic distance IT-IT corresponding to Ga-Ga and Cu-Cu pairs on sites IT in the clusters "a" (●) and "b" (○), respectively, and the number of vacancies over sites IT in the cluster "b" (■). The Rietveld analysis was made with space group $P\bar{4}3m$ for x = 31 to 38 and with space group $R3m$ for x = 40 under the constraint of one vacancy at site IT1 and fractional occupancies of Cu atoms at three remaining sites IT2 in the cluster "b." [From U. Mizutani, T. Noritake, T. Ohsuna, and T. Takeuchi, *Phil. Mag.* 90 (2010) 1985.]

increases with an increase in the Ga content and reaches unity at x = 38. While the distance IT-IT in the cluster "a" gradually increases, the distance IT-IT in the cluster "b" sharply decreases above about 36 at .%Ga in harmony with a sharp increase in the number of vacancies in sites IT in the cluster "b." In contrast, a decrease in the interatomic distance CO-CO in the cluster "a" is found to be much more moderate, though the number of vacancies over sites CO in the cluster "a" increases as significantly as that over sites IT in the cluster "b" with increasing Ga concentration. Thus, it is clear that the transformation into the $R3m$ structure is essentially triggered by an increase in vacancies over sites IT in the cluster "b" up to 1.7 at 40 at .%Ga.

The way of distributing vacancies over the unit cell between Cu-Al and Cu-Ga gamma-brasses is quite different. The difference would originate from that in the atomic size ratio of constituent elements, though it is small: $r_{Al} / r_{Cu} = 1.12$ and $r_{Ga} / r_{Cu} = 1.10$. As mentioned in [7], vacancies in Cu-Al gamma-brasses are exclusively introduced into sites IT in the cluster "b" along the <111> direction and promote the transformation into the structure with space group $R3m$. However, a slightly smaller size ratio in Cu-Ga gamma-brass would be responsible for distributing vacancies not only over sites IT in the cluster "b," but also over CO in both clusters "a" and "b."

First-principles band calculations for Cu-Ga gamma-brasses are again formidable because of the absence of a stoichiometric compound free from chemical and geometrical disorder. Thus, we are obliged to rely on the DOS for Cu_9Al_4 because of the similarity in chemical nature of the constituent elements. As shown in Figure 7.15, values of **e/uc** in the Cu-Ga gamma-brasses are confined in a rather narrow range of 84 to 87 owing to the introduction of a large amount of vacancies. Judging from the **e/uc** dependence of the DOS for Cu_9Al_4 in Figure 7.14, we believe that the Fermi level would reside well inside the pseudogap throughout the whole Ga concentration range. A difference in the driving force into the $R3m$ transformation between Cu-Al and Cu-Ga alloy systems might be also related to the difference in the **e/uc** range. The value of **e/uc** exceeds 87 at 37 at .%Al, which is already near the upper edge of the pseudogap and, thereby, possibly rendering some modification in structure to stabilize the system. Instead, a slightly smaller size ratio in Cu-Ga alloy system allows to distribute vacancies more evenly over sites to effectively suppress the value of **e/uc** below 87 up to x = 40.

7.8 SUMMARY

We have repeatedly emphasized the need to discuss the Hume-Rothery electron concentration rule for CMAs by performing first-principles band calculations with a particular attention to two different electron concentration parameters e/a and **VEC**. We have introduced in this chapter techniques called the *FLAPW-Fourier method* and the *Hume-Rothery plot* to separate the parameter e/a from **VEC** and applied it to Cu_5Zn_8 and Cu_9Al_4 gamma-brasses. We could prove that a pseudogap at the Fermi level for both compounds is commonly caused by the FsBz interactions associated with the set of {330} and {411} lattice planes in agreement with the theory due to Mott and Jones in 1936 on the basis of the free electron model and could also explain why these two gamma-brasses are stabilized at e/a = 21/13.

The FLAPW-Fourier method could extract $|G|^2 = 18$ as a *critical* reciprocal lattice vector and the Hume-Rothery plot revealed $(2k_F)^2$ to be close to 18 in both Cu_5Zn_8 and Cu_9Al_4 gamma-brasses. This is taken as a demonstration for the validity of the matching condition given by Equation 4.1 by means of first-principles band calculations.

The solid solution range is discussed, using a criterion such that gamma-brass phase remains stable at finite temperatures if the Fermi level is inside a pseudogap, as revealed in ordered Cu_5Zn_8 and Cu_9Al_4 gamma-brasses. The parameter **e/uc** is employed as a practical electron concentration parameter for off-stoichiometric gamma-brasses. The parameter **e/uc** is expected to be constrained in the neighborhood of 84 given by the product of e/a = 21/13 and N = 52 over a whole solid solution range, since $|G|^2$ in equation (4.1) remains constant and equal to 18 regardless of solute concentrations for group (I) gamma-brasses.

Figure 7.18 is specifically prepared to show **e/uc** as a function of the number of valence electrons per unit cell **VE/uc** for group (I) gamma-brasses over a whole solid solution range. The data for Ni-Zn gamma-brasses discussed in Chapter 8, Section 8.5, are also included. It is clear that **e/uc** is distributed in the neighborhood of 84 and is always lower than $(e/uc)_0 = 90$ for all group (I) gamma-brasses. The value of **e/uc** is confined over the range from 80 to 90, reflecting the validity of the matching condition given by Equation 4.1. We will discuss this issue more universally in Chapter 10.

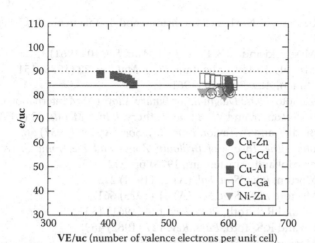

FIGURE 7.18 Number of itinerant electrons per unit cell, **e/uc**, as a function of the number of valence electrons per unit cell, **VE/uc**, for several group (I) gamma-brasses, including the data for Ni-Zn gamma-brasses. All the data are confined within the range 80 < **e/uc** < 90. The value of **VE/uc** is given by the product of **VEC** and N.

REFERENCES

1. P. Villars, *Pearson's Handbook, Crystallographic Data* (ASM, Materials Park, OH, 1997).
2. A.J. Bradley and C.H. Gregory, *Phil. Mag.* 12 (1931) 143.
3. O.V. Heidenstam, A. Johansson, and S. Westman, *Acta Chem. Scand.* 22 (1968) 653.
4. J.K. Brandon, R.Y. Brizard, P.C. Chieh, R.K. McMillan, and W.B. Pearson, *Acta Cryst.* B30 (1974) 1412.
5. A.J. Bradley and P. Jones, *J. Inst. Metals.* 51 (1933) 131.
6. L. Arnberg and S. Westman, *Acta Cryst.* A34 (1978) 399.
7. E.H. Kisi and J.D. Browne, *Acta Cryst.* (1991) B47 835.
8. R. Asahi, H. Sato, T. Takeuchi, and U. Mizutani, *Phys. Rev.* B 71 (2005) 165103.
9. E. Wimmer, H. Krakauer, M. Weinert, and A.J. Freeman, *Phys. Rev.* B 24 (1981) 864.
10. M. Weinert, E. Wimmer, and A.J. Freeman, *Phys. Rev.* B 26 (1982) 4571.
11. H.J.F. Jansen and A.J. Freeman, *Phys. Rev.* B 30 (1984) 561.
12. L. Hedin and B.I. Lundqvist, *J. Phys.* C 4 (1971) 2064.
13. H.J. Monkhorst and J.D. Pack, *Phys. Rev.* B 13 (1976) 5188.
14. C. Jepsen and O.K. Andersen, *Solid State Commun.* 9 (1971) 1763.
15. G. Lehmann and M. Taut, *Phys. Status Solid* 54 (1972) 469.
16. U. Mizutani, *Introduction to the Electron Theory of Metals* (Cambridge University Press, Cambridge, 2001).
17. R. Asahi, H. Sato, T. Takeuchi, and U. Mizutani, *Phys. Rev.* B 72 (2005) 125102.

18. U. Mizutani, T. Noritake, T. Ohsuna, and T. Takeuchi, *Phil. Mag.* 90 (2010) 1985.
19. T.B. Massalski and W.B. King, *Prog. Mat. Sci.* 10 (1961) 1.
20. T.B. Massalski and U. Mizutani, *Prog. Mat. Sci.* 22 (1978) 151.
21. J.M. Ziman, *Adv. Phys.* 10 (1961) 1.
22. H. Okamoto, *Phase Diagrams for Binary Alloys* (ASM International, OH, 2000).
23. J.O. Betterton, Jr. and W. Hume-Rothery, *J. Inst. Metals*, 80 (1951–52) 459.
24. A.J. Bradley and A. Taylor, *Proc. Roy. Soc.* [A] 159 (1937) 56.
25. H. Jones, *The Theory of Brillouin Zones and Electronic States in Crystals* (North-Holland, Amsterdam, 1975) pp. 212–215.
26. A.J. Morton, *Phys. Stat. Sol.* (A) 23 (1974) 275.
27. A.J. Morton, *Phys. Stat. Sol.* (A) 31 (1975) 661.
28. H. Sato and R.S. Toth, *Phys. Rev.* 124 (1961) 1833.
29. H. Sato and R.S. Toth, *Phys. Rev.* 127 (1962) 469.
30. A.J. Morton, *Phys. Stat. Sol.* (A) 44 (1977) 205.
31. H. Jones, *Proc. Roy. Soc. Lond.* A144 (1934) 225.
32. A.J. Bradley, H.J. Goldschmidt, and H. Lipson, *J. Inst. Metals*, 63 (1938) 149.
33. S. Westman, *Acta Chem. Scand.*, 19 (1965) 2369.
34. R. Sokhuyzen, J.K. Brandon, PC. Chieh, and W.B. Pearson, *Acta Cryst.* B30 (1974) 2910.

CHAPTER 8

Stabilization Mechanism of Gamma-Brasses Characterized by Increasing Orbital Hybridizations

Role of d-States-Mediated-FsBz Interactions

8.1 GAMMA-BRASSES IN GROUP (II)

According to Table 6.1, at least 11 gamma-brass alloys belong to group (II) with space group of either $I\bar{4}3m$ or $P\bar{4}3m$ in binary alloy systems. They are isostructural to those based on noble metals in group (I). We naturally wonder whether the stabilization mechanism remains unchanged between them. Indeed, many discussions have been repeatedly made with a belief that group (II) gamma-brasses also obey the Hume-Rothery electron concentration rule and are stabilized at $e/a = 21/13$ [1–5]. One of the difficulties in allowing such unsupported belief certainly originated from the fact that the e/a value of the transition metal (TM) element has been controversial and has remained unsolved. In Chapter 8, we try to shed more light on the stabilization mechanism of group (II) gamma-brasses by making full use of the FLAPW–Fourier method introduced in Chapter 7, Sections 7.3 and 7.4.

First, we have to choose group (II) gamma-brasses suitable for performing first-principles band calculations among those listed in Table 6.1. The best way is to find a stoichiometric ordered compound free from any chemical disorder and vacancies, as was encouraged by the existence of well-ordered Cu_5Zn_8 and Cu_9Al_4 in group (I) discussed in Chapter 7. According to the phase diagram [6], both Al_8V_5 and Mn_3In apparently exist only as a line compound. The composition ratio 8:5 in Al_8V_5 is consistent with a criterion for the formation of such an ordered gamma-brass (see Chapter 6, Section 6.2). However, Brandon et al. [7] reported the presence of chemical disorder in sites IT and OH in their structure analysis using a single-crystal (see Appendix 2, Section A2.2.1.12). The situation in Mn_3In is more serious. The composition ratio 3:1 is incompatible with structures forming an ordered gamma-brass. Indeed, Brandon et al. [8] revealed a substantial chemical disorder in Mn_3In (see Appendix 2, Section A2.2.2.10). To the best of our knowledge, an ordered gamma-brass has not been reported to exist in group (II). We are, thus, forced to construct a model structure to perform first-principles band calculations. The Al_8V_5 and TM_2Zn_{11} (TM = Ni, Co, Pd) gamma-brasses were selected, since the model structure could be constructed with a minimal sacrifice to eliminate chemical disorder from the observed structure.

8.2 TM-Zn (TM = Ni, Pd, Co, AND Fe) GAMMA-BRASSES

8.2.1 Construction of the Model Structure

Gamma-brass phase field is extended over 15–30 at.%TM in the family of the TM-Zn (TM = Mn, Fe, Co, Ni, Pd, Pt, Ir) alloy systems, all of which belong to group (II), as listed in Table 6.1. Among them, the Rietveld structure analysis is performed only at some particular compositions for those with TM = Fe, Co, Ni, Pd, and Ir [9–14]. For example, neutron diffraction studies are reported for the single phase $Ni_{17.7}Zn_{82.3}$ gamma-brass sample [10]. Its space group is identified as $I\bar{4}3m$. Among various models tested, the best fit is obtained, when Ni atoms are filled only into sites OT without any vacancies, resulting in Ni_2Zn_{11} or 15.38 at.%Ni alloy. The same conclusion is drawn for Ir_2Zn_{11}, where Ir atoms are exclusively filled into sites OT [12]. Though chemical disorder is claimed to be also present on sites OH in Pd-Zn [11] and on sites IT in Fe-Zn [13], both sets of data are consistent with a full occupancy of Pd or Fe atoms on sites OT. In the case of $Co_{20}Zn_{80}$ gamma-brass, the best refinement is achieved for the structure, in which Co atoms enter more preferentially into sites OT and Zn atoms into sites IT [14]. Its space group is deduced to be $I\bar{4}3m$.*

To construct an ordered gamma-brass suitable for first-principles band calculations, we consider it to be most appropriate to fill the TM element only into sites OT and Zn atoms into all remaining sites IT, OH, and CO, resulting in the chemical formula TM_2Zn_{11} (TM = Fe, Co, Ni, and Pd). The fractional coordinates of all atoms in the unit cell and the lattice constant are taken from those determined experimentally for alloys with the nearby compositions.† Fortunately, the composition of TM_2Zn_{11} (TM = Fe, Co, Ni, and Pd) gamma-brasses is almost marginal but is still within a solid solution range in the respective phase diagrams [6]. More details about the atomic structure of gamma-brasses are summarized in Appendix 2.

8.2.2 Electronic Structure Calculations and Stabilization Mechanism

8.2.2.1 Ni_2Zn_{11} and Pd_2Zn_{11} Gamma-Brasses

To begin with, we briefly discuss energy dispersion relations along the direction ΓN for the gamma-brass structure in the free electron model, which are shown in Figure 8.1. Let us direct our attention to electronic states at the symmetry points N, at which many parabolic bands cross with one another. They can be indexed in terms of the square of the reciprocal lattice vector $|G|^2$. As shown in Figure 3.2, the symmetry point N refers to the center of the {110} zone of the Brillouin zone of a bcc lattice. As noted in Chapter 7, Section 7.3, all zones, in which two integers in the set of Miller indices {hkl} are odd, pass through the symmetry points N upon reduction to the first zone. For example, degenerate electronic states at the lowest energy of −9.5 eV are easily assigned as $|G|^2 = 2$ associated with the set of {110} lattice planes. Further crossings of free-electron parabolic bands taking place one after another with increasing energy can be immediately assigned as an increasing order of $|G|^2 = 2, 6, 10, 14, 18, 22, 26, 30, 34, \ldots$ corresponding to the center of {110}, {211}, {310}, {321}, {330}, {411}, … zone planes of the Brillouin zone. The position of the Fermi level is determined

---

* The possession of space group $I\bar{4}3m$ is confirmed by the convergent beam electron diffraction studies for $Co_{20}Zn_{80}$ gamma-brass alloy [14]. This rules out the report on space group $P\bar{4}3m$ [6].

† The atomic coordinates of a crystal structure are usually expressed as fractional coordinates, that is, as fractions of the a, b, and c unit vectors. For example, an atom with fractional coordinates (0.5, 0.5, 0.5) would lie half way along each unit cell edge and is positioned at the center of the unit cell.

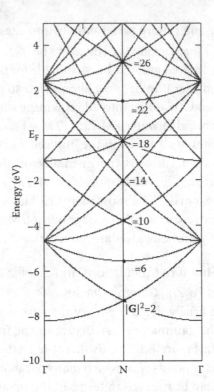

FIGURE 8.1 Energy dispersion relations along the direction ΓN for a canonical gamma-brass structure having $e/a = 21/13$ and $N = 52$ in the free electron model [from U. Mizutani et al., *Phys. Rev. B* 74 (2006) 235119]. The number along the energy axis at the symmetry point N refers to the value of $|\mathbf{G}|^2$. Note that the ratio of E_F over $E_{|G|^2=18}$ is 1.033, regardless of the choice of a lattice constant.

by filling electrons equal to $e/a = 21/13$ into the free-electron valence-band. We see that the electronic states near the Fermi level are dominated by those characterized by $|\mathbf{G}|^2 = 18$ in the free electron model. In Chapter 8, we deal with gamma-brasses, in which electronic states like $|\mathbf{G}|^2 = 18$ and its neighbors near the Fermi level are heavily perturbed by the TM-3d band involved.

The energy dispersion relations and DOS for Ni_2Zn_{11} and Pd_2Zn_{11} gamma-brasses are shown in Figures 8.2 and 8.3, respectively [14]. A large DOS in the energy range over -7 and -9 eV is due to the Zn-3d band in both cases. Since the amount of Ni or Pd element is only 15.38 at.%, their d bands are rather small. We find that both Ni-3d and Pd-4d bands are still well immersed below the Fermi level. We are ready to check if the FsBz interaction involving $|\mathbf{G}|^2 = 18$ characteristic of the gamma-brass

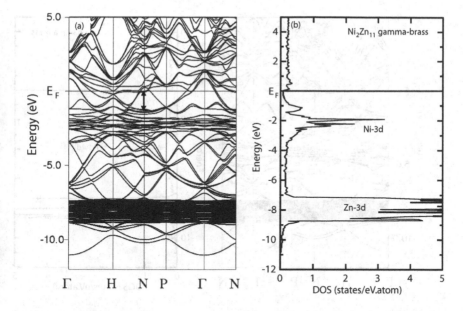

FIGURE 8.2   (a) Energy dispersion relations and (b) DOS calculated using the FLAPW method for Ni_2Zn_{11} gamma-brass. A double-headed arrow in (a) indicates a gap at the symmetry point N across the Fermi level. [From R. Asahi et al., *Phys. Rev.* B 72 (2005) 125102.]

structure is still effective enough to generate a pseudogap at the Fermi level, or is affected by the Ni-3d or Pd-4d band.

Figures 8.4a,b show energy spectra of the squared Fourier coefficients summed over equivalent zone planes,

$$\sum \left| C_{k+G}^i \right|^2$$

of the FLAPW wave function 7.1 outside the MT sphere at the symmetry points N at energy-eigen values ranging from −2 to +1 eV across the Fermi level. Here, $|G|^2$ is fixed at representative values of 14, 18, and 22. This technique was already introduced as the FLAPW-Fourier method in Chapter 7, Section 7.3. It is clear from Figures 8.4a,b that electronic states associated with $|G|^2$ = 18 predominantly appear at the bottom and the top of a pseudogap marked with an arrow. On the other hand, the $|G|^2$ = 14 states appear only below −1 eV, while the $|G|^2$ = 22 states only above +0.4 eV. This is in accordance with the behavior expected from the

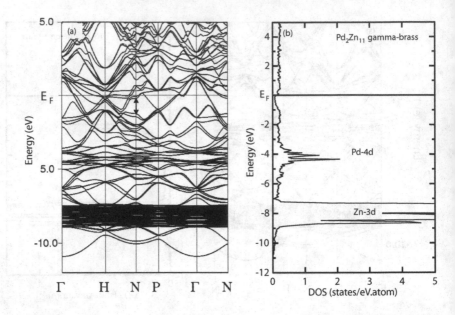

FIGURE 8.3   (a) Energy dispersion relations and (b) DOS calculated using the FLAPW method for Pd_2Zn_{11} gamma-brass. A double-headed arrow in (a) indicates a gap at the symmetry point N across the Fermi level. [From R. Asahi et al., *Phys. Rev.* B 72 (2005) 125102.]

free electron model in a sense that the higher the energy, the larger is the value of $|\mathbf{G}|^2$ involved. The results shown in Figure 8.4 are already decisive enough to conclude that a pseudogap in both Ni_2Zn_{11} and Pd_2Zn_{11} gamma-brasses is induced by the FsBz interactions involving the set of {330} and {411} lattice planes with $|\mathbf{G}|^2 = 18$.

In order to ascertain further our conclusion above, we show the FLAPW-Fourier spectra at the symmetry points N at two energies corresponding to lower and upper edges of the pseudogap as a function of the square of the Miller indices,

$$\sum h^2 \text{ or } |\mathbf{G}|^2$$

in Figures 8.5 and 8.6 on a semi-logarithmic scale for both Ni_2Zn_{11} and Pd_2Zn_{11}, respectively [14]. The Fourier coefficient in both cases is extremely large at $|\mathbf{G}|^2 = 18$, being well consistent with the data in Figures 8.4a,b. The ratio of the intensity at $|\mathbf{G}|^2 = 18$ over the next intense one is higher

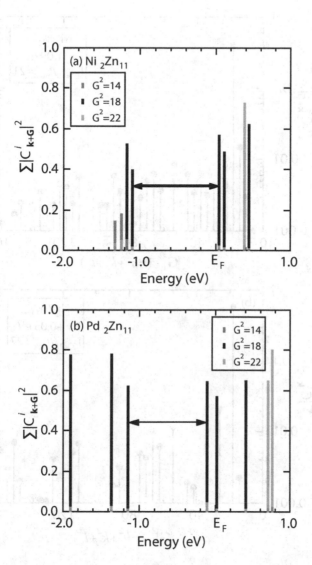

FIGURE 8.4 Energy dependence of the Fourier component $\Sigma \left| C_{\mathbf{k+G}}^{i} \right|^{2}$ of the FLAPW-wave function (7.1) outside the MT sphere at the symmetry points N for (a) Ni_2Zn_{11} and (b) Pd_2Zn_{11}. The value of $\left| \mathbf{G} \right|^{2}$ is fixed at three representative values 14, 18, and 22. The $\left| \mathbf{G} \right|^{2}$ = 26, 30, and 34 states are only weakly present below −1.0 eV and near the Fermi level.

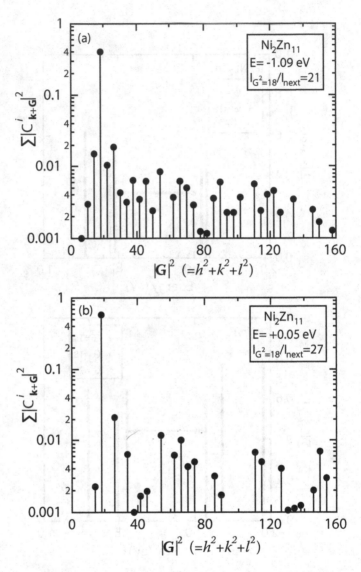

FIGURE 8.5 $|\mathbf{G}|^2 (= h^2 + k^2 + l^2)$ dependence of the Fourier component $\Sigma \left| C_{k+G}^i \right|^2$ of the FLAPW wave function (7.1) outside the MT sphere at energy eigen-values of (a) −1.09 and (b) +0.05 eV corresponding to the bottom and top of the pseudogap, respectively, at the symmetry points N for Ni_2Zn_{11} gamma-brass [from R. Asahi et al., *Phys. Rev.* B 72 (2005) 125102]. $I_{G^2=18} / I_{next}$: ratio of intensity at $|\mathbf{G}|^2 = 18$ over the next intense one. The ordinate is on a logarithmic scale.

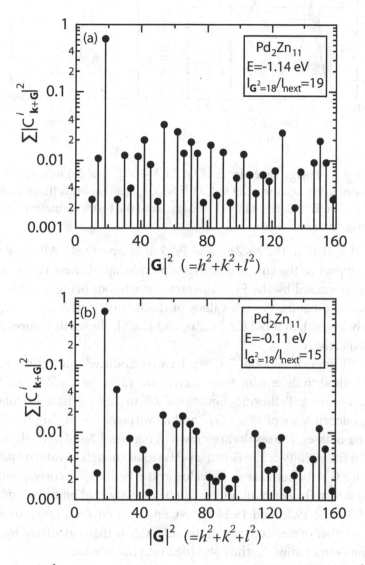

FIGURE 8.6 $|\mathbf{G}|^2 \ (= h^2 + k^2 + l^2)$ dependence of the Fourier component $\Sigma |C_{k+G}^i|^2$ of the FLAPW wave function (7.1) outside the MT sphere at energy eigen-values of (a) −1.14 and (b) −0.11 eV corresponding to the bottom and top of the pseudogap, respectively, at the symmetry points N for Pd$_2$Zn$_{11}$ gamma-brass [from R. Asahi et al., *Phys. Rev. B* 72 (2005) 125102]. $I_{G^2=18} / I_{next}$: ratio of intensity at $|\mathbf{G}|^2 = 18$ over the next intense one. The ordinate is on a logarithmic scale.

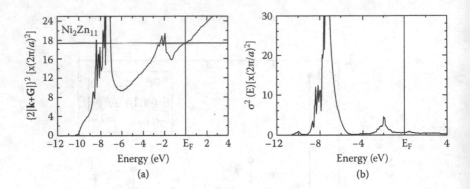

FIGURE 8.7 Energy dependence of (a) $\{2(|\mathbf{k}+\mathbf{G}|)\}^2$ and (b) its variance $\sigma^2(E)$ calculated using Equations 7.3 to 7.5 for Ni_2Zn_{11} gamma-brass [from R. Asahi et al., *Phys. Rev.* B 72 (2005) 125102]. The square of the Fermi diameter is read off at the Fermi level.

than 21 and 15 in the Ni_2Zn_{11} and Pd_2Zn_{11}, respectively. All these results lend support to the conclusion that a pseudogap in these two systems is definitely caused by the FsBz interactions without being affected by the respective d-bands, though values of the intensity ratio $I_{|G|^2=18} / I_{next}$ are slightly lower than those for Cu_5Zn_8 and Cu_9Al_4 shown in Figures 7.3 and 7.4, respectively.

In Chapter 7, Section 7.4, we have introduced the Hume-Rothery plot method to determine the effective **e/a** value for Cu_5Zn_8 and Cu_9Al_4 gamma-brasses. Following Equations 7.3 to 7.5, we now calculate the energy dependence of $\{2|\mathbf{k}+\mathbf{G}|\}^2$ and its variance for Ni_2Zn_{11} and Pd_2Zn_{11} gamma-brasses. The results are shown in Figures 8.7 and 8.8 [14]. The variance in the vicinity of the Fermi level is small enough to validate the value of $\{2|\mathbf{k}+\mathbf{G}|\}^2$ in both cases. Its value at the Fermi level corresponding to the square of the Fermi diameter can be directly read off from Figures 8.7 and 8.8 to be 19.36 and 19.27 for Ni_2Zn_{11} and Pd_2Zn_{11}, respectively. The total number of electrons per atom $(e/a)_{total}$ is then calculated by inserting the Fermi radius k_F thus obtained into the relation

$$\left(e/a\right)_{total} = \frac{8\pi k_F^3}{3N}$$

where the number of atoms per unit cell, N, is assumed to be equal to 52 and k_F is in units of

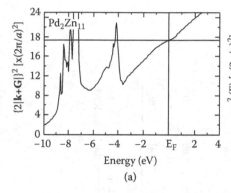
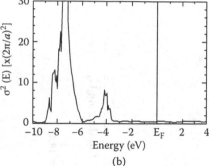

FIGURE 8.8 Energy dependence of (a) $\{2(|\mathbf{k}+\mathbf{G}|)\}^2$ and (b) its variance $\sigma^2(E)$ calculated using Equations 7.3 to 7.5 for Pd_2Zn_{11} gamma-brass [from R. Asahi et al., *Phys. Rev.* B 72 (2005) 125102]. The square of the Fermi diameter is read off at the Fermi level.

$$\frac{2\pi}{a}$$

The value of $(e/a)_{total}$ is found to be 1.72 and 1.70 for Ni_2Zn_{11} and Pd_2Zn_{11}, respectively. Since the pseudogap is confirmed to be induced by the FsBz interactions involving only $|\mathbf{G}|^2 = 18$, we conclude the value of $(e/a)_{total}$ = 1.70-1.72 to be still acceptable as gamma-brasses obeying the e/a = 21/13 rule. The valency of the TM element or $(e/a)_{TM}$ is easily obtained by taking the valency of the partner element Zn to be two. The $(e/a)_{TM}$ values for Ni and Pd turn out to be 0.15 and 0.07, respectively. All these results are summarized in Table 8.1.

If the arguments above are accepted, Ni_2Zn_{11} and Pd_2Zn_{11} gamma-brasses fit better into group (I) classification rather than into group (II).

8.2.2.2 Co_2Zn_{11} and Fe_2Zn_{11} Gamma-Brasses

The energy dispersion relations and DOS for Co_2Zn_{11} gamma-brass are shown in Figures 8.9 a,b, respectively [14]. Now the Fermi level is found in the Co-3d band, which is extended over energies from −2.5 to +0.5 eV across the Fermi level. Thus, it is of great importance to study how the d-band extending over the Fermi level affects the FsBz interactions associated with $|\mathbf{G}|^2 = 18$.

TABLE 8.1 $(2k_F)^2$, Critical $|G|^2$, $(e/a)_{total}$ and $(e/a)_{TM}$ for Group (II) Gamma-Brasses

	Al_8V_5	Fe_2Zn_{11}	Co_2Zn_{11}	Ni_2Zn_{11}	Pd_2Zn_{11}	Cu_5Zn_8	Cu_9Al_4	Ag_5Li_8
$(2k_F)^2\left(\times\left(\dfrac{2\pi}{a}\right)^2\right)$	21.0	20.0	19.5	19.36	19.27	18.47	18.45	13.4
$|G|^2\left(\times\left(\dfrac{2\pi}{a}\right)^2\right)$	14, 18, 22, 26, 30	18, 22	18, 22	18	18	18	18	—
$(e/a)_{total}$	1.94	1.80	1.73	1.72	17.0	1.60	1.60	1.0
$(e/a)_{TM}$	0.23	0.70	0.26	0.15	0.07	0.96	0.97	1.0
ref	[15]	[15]	[15]	[15]	[15]	[14]	[14]	[20]

Note: The data for group (I) Cu_5Zn_8 and Cu_9Al_4 gamma-brasses are also incorporated. Here, a *"critical"* $|G|^2$ means the square of the reciprocal lattice vector responsible for the formation of a pseudogap across the Fermi level. The most critical $|G|^2$s are underlined.

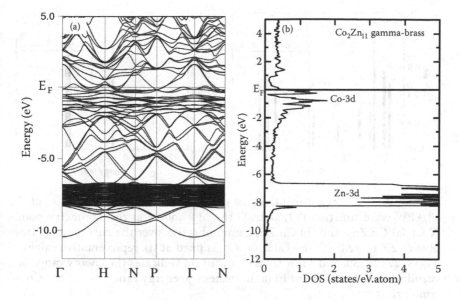

FIGURE 8.9 (a) Energy dispersion relations and (b) DOS calculated using the FLAPW method for Co_2Zn_{11} gamma-brass. [From R. Asahi et al., *Phys. Rev.* B 72 (2005) 125102.]

Figure 8.10a shows the energy spectrum of the FLAPW-Fourier component,

$$\sum \left| C_{k+G}^i \right|^2$$

for Co_2Zn_{11} gamma-brass over the energy range from -2.5 to $+2$ eV, which is wide enough to cover the Co-3d band. Here, $|G|^2$ is fixed at values from 14 to 34 at the symmetry points N of the bcc Brillouin zone. To emphasize its uniqueness, the energy spectrum for Cu_5Zn_8 gamma-brass in group (I) is also constructed over the same energy range and shown in Figure 8.10b. The energy range of the Co-3d band is marked with a dotted arrow in (a). Sizable Fourier components over the range of $|G|^2 = 14$ to 34 are widely distributed inside the Co-3d band in Co_2Zn_{11} gamma-brass. This obviously reflects the presence of flat energy dispersions inside the Co-3d band (Figure 8.9a). Especially, the Fourier components of $|G|^2 \geq 22$ normally appear above the Fermi level but remain significant even below about -1

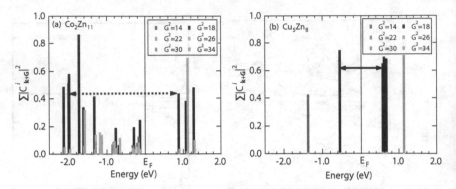

FIGURE 8.10 Energy dependence of the Fourier component $\Sigma |C_{k+G}^i|^2$ of the FLAPW-wave function (7.1) outside the MT sphere at the symmetry points N for (a) Co_2Zn_{11} and (b) Cu_5Zn_8 gamma-brasses over the energy range from about -2.5 to $+2.0$ eV. The value of $|G|^2$ is fixed at six representative values of 14, 18, 22, 26, 30, and 34. A solid arrow in (b) indicates the energy range of a pseudogap. A dotted arrow in (a) indicates the energy range, where the Co-3d band exists.

eV in Co_2Zn_{11}, indicating that the FLAPW wave function outside the MT sphere in this energy range is strongly perturbed by the Co-3d states and cannot be described simply in terms of the $|G|^2 = 14$ wave like in Cu_5Zn_8, as can be seen in Figure 8.10b.

What about the stabilization mechanism for Co_2Zn_{11} gamma-brass? First of all, we must note that the Co-3d band shown in Figure 8.9b is apparently separated by a deep pseudogap at about -0.41 eV into bonding and antibonding subbands due mainly to orbital hybridizations between Co-3d and Zn-4p states. Thus, one may argue that orbital hybridization effect would be responsible for its stabilization. But a large part of the antibonding subband is already below the Fermi level. This makes it difficult to explain its stabilization mechanism solely in terms of the orbital hybridization effect.

In the case of Cu_5Zn_8 discussed in Chapter 7, Section 7.3, the FsBz interaction associated with $|G|^2 = 18$ could account for the formation of a pseudogap across the Fermi level. This is clearly indicated in the energy spectrum shown in Figure 8.10b. In the case of Co_2Zn_{11}, we see from Figure 8.10a that the $|G|^2 = 18$ states (marked in black) are split into bonding states below about -1.7 eV and antibonding states above about $+0.9$ eV. Furthermore, the intensity of the bonding states due to the $|G|^2 = 18$ wave at about -1.7 to -2 eV is much stronger than that of the corresponding antibonding states at about +1.0 eV (see data marked in black in Figure 8.10a). Therefore, we take this as evidence that the FsBz interactions associated with $|G|^2 = 18$ are effective enough to stabilize Co_2Zn_{11} gamma-brass. Similarly, we find that the $|G|^2 = 22$ bonding states (marked in gray) also contribute to stabilizing Co_2Zn_{11}.

At this stage, we must note a striking difference in the behavior of splitting of the electronic states from that in Cu_5Zn_8 gamma-brass. A large part of bonding states in Co_2Zn_{11} gamma-brass is deeply shifted to energies over −1.7 to −2.1 eV near the bottom of the Co-3d band, while antibonding states to energies above about +0.9 eV near the top of the Co-3d band, resulting in a wider "pseudogap" in between them (see the dotted arrow in Figure 8.10a). The formation of a widely separated bonding and antibonding states may be understood in such a way that the FsBz interactions, involving the strongest $|G|^2 = 18$ and the next strongest $|G|^2 = 22$, occur as if to avoid the Co-3d band, which apparently pushes the resulting bonding and antibonding states to the respective bottom and top due to the orthogonality condition of the sp-like wave function with the d-like one. We call this phenomenon the *d-states-mediated-splitting* or *d-states-mediated-FsBz-interactions.**

An overall feature in the FLAPW–Fourier energy spectrum shown like in Figures 8.4 and 8.10 would be mostly reflected in the sp-partial DOS, since the energy spectrum is constructed from the wave function outside the MT sphere. Figure 8.11a shows the sp-partial DOS of Co_2Zn_{11} gamma-brass over the energy range from −4 to +3 eV in comparison with its total DOS in Figure 8.11b, which is duplicated from Figure 8.9b. The Co-3d bonding and antibonding subbands are marked in the total DOS in (b) with symbols (A), (B), and (C). Among them, we realized that the peaks (B) and (C) are greatly suppressed in the sp-partial DOS in (a). In other words, sp-electrons apparently form a pseudogap over −1 to +1 eV, though the peak "A" still remains near the Fermi level. We consider the formation of the pseudogap in the sp-partial DOS to be brought about by the d-states-mediated-splitting involving $|G|^2 = 18$ and 22 states and to be essential in stabilizing the Co-Zn gamma-brass. More details will be discussed upon dealing with the stability of Al_8V_5 gamma-brass.

---

* The d-states-mediated-splitting is unique to CMAs involving d-states in the valence band. It is specifically called d-states-mediated-FsBz-interactions, when splitting occurs across the Fermi level.

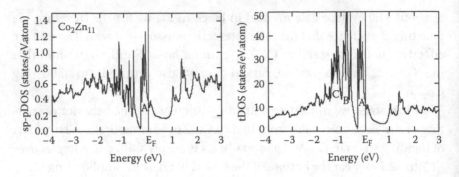

FIGURE 8.11 (a) sp-partial DOS for Co_2Zn_{11} gamma-brass and (b) its total DOS duplicated from Figure 8.9 (b). A: Co-3d antibonding subband, B: Co-3d non-bonding or bonding subband, C: Co-3d bonding subband.

The energy dispersion relations and DOS for Fe_2Zn_{11} gamma-brass are depicted in Figures 8.12a and b, respectively. The electronic structure is found to be quite similar to that of Co_2Zn_{11} discussed above. Hence, only a brief description will be needed. The energy spectrum of the FLAPW-Fourier components,

$$\sum \left| C_{k+G}^i \right|^2$$

for Fe_2Zn_{11} is shown in Figure 8.13a in comparison with that for Cu_9Al_4 in (b). Because of the presence of the dispersionless Fe-3d band over energies from -2 to $+1$ eV, many Fourier components over the range from $|G|^2 = 14$ to 34 remain finite below the Fermi level. More important is that splitting of the $|G|^2$-dependent electronic states occurs as if the Fe-3d band were to be avoided, forming bonding states near the bottom of the Fe-3d band and antibonding states near its top. This is the d-states-medi-ated-FsBz interactions mentioned above and must be responsible for the stabilization of the structure. In this way, we believe that the stabiliza-tion mechanism for TM_2Zn_{11} (TM = Fe and Co) results from the FsBz interactions involving, at least, two $|G|^2$s equal to 18 and 22 as a result of the mixing with the TM-3d states, that is, the d-states-mediated-FsBz-interactions. We are now ready to determine the valency of Fe and Co by performing the Hume-Rothery plot.

The Hume-Rothery plot is performed for Co_2Zn_{11} and Fe_2Zn_{11} gamma-brasses. The energy dependence of $\{2|k+G|\}^2$ and its variance are shown

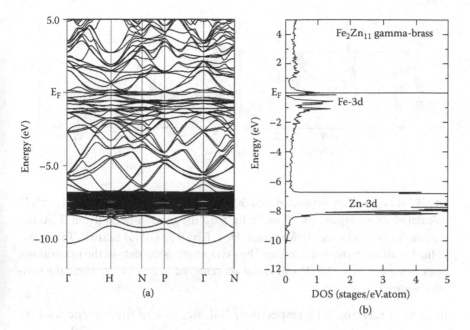

FIGURE 8.12 (a) Energy dispersion relations and (b) DOS calculated using the FLAPW method for Fe_2Zn_{11} gamma-brass. [From R. Asahi et al., *Phys. Rev.* B 72 (2005) 125102.]

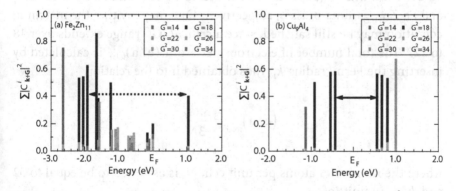

FIGURE 8.13 Energy spectrum of the Fourier component $\Sigma |C_{k+G}^i|^2$ of the FLAPW-wave function (7.1) outside the MT sphere at the symmetry points N for the (a) Fe_2Zn_{11} and (b) Cu_9Al_4 gamma-brasses. The value of $|G|^2$ is fixed at six representative values of 14, 18, 22, 26, 30, and 34. A solid arrow in (b) indicates the energy range of a pseudogap. A dotted arrow in (a) indicates the energy range, where the Fe-3d band exists.

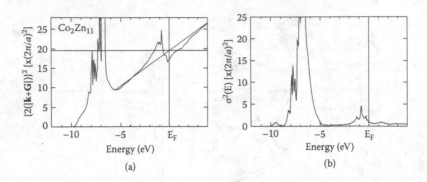

FIGURE 8.14 Energy dependence of (a) $\{2(|\mathbf{k}+\mathbf{G}|)\}^2$ and (b) its variance $\sigma^2(E)$ calculated using Equations 7.3 to 7.5 for Co_2Zn_{11} gamma-brass [from R. Asahi, H. Sato, T. Takeuchi, and U. Mizutani, *Phys. Rev.* B 72 (2005) 125102]. The square of the Fermi diameter is determined by extrapolating the data in the two regions: one centered at E = −3.5 eV and the other centered at E = +4 eV, where the variance is small.

in Figures 8.14 and 8.15, respectively [14]. Because of the presence of Co– and Fe-3d bands across the Fermi level, the variance is large in the energy range, over which they are spread. Hence, one cannot directly read off the value of $\{2|\mathbf{k}+\mathbf{G}|\}^2$ at the Fermi level. As shown in Figures 8.14a and 8.15a, a straight line is drawn through the data points, where the variance is low: one in the range from −3.5 to −4 eV and the other in the range from +4 to +5 eV. The value of $(2k_F)^2$ is determined to be 19.5 and 20.0 for Co_2Zn_{11} and Fe_2Zn_{11} gamma-brasses, respectively. One may say that the matching condition may be still satisfied, since a *critical* $|\mathbf{G}|^2$ range extends over 18 to 22.* The total number of electrons per atom $(e/a)_{total}$ is calculated by inserting the Fermi radius k_F thus obtained into the relation

$$\left(e/a\right)_{total} = \frac{8\pi k_F^3}{3N}$$

where the number of atoms per unit cell, N, is assumed to be equal to 52 and k_F is in units of

---

* The present analysis is limited only to the symmetry points N. If other symmetry points are included, the $|\mathbf{G}|^2$ = 20 corresponding to the set of {420} lattice-planes will be also counted as *critical*.

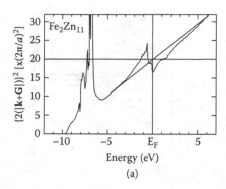
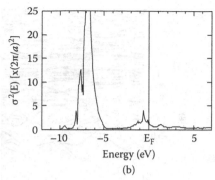

FIGURE 8.15 Energy dependence of (a) $\{2(|\mathbf{k}+\mathbf{G}|)\}^2$ and (b) its variance $\sigma^2(E)$ calculated using Equations 7.3 to 7.5 for Fe_2Zn_{11} gamma-brass [from R. Asahi, H. Sato, T. Takeuchi, and U. Mizutani, *Phys. Rev. B* 72 (2005) 125102]. The square of the Fermi diameter is determined by extrapolating the data in the two regions: one centered at $E = -3.5$ eV and the other centered at $E = +5$ eV.

$$\frac{2\pi}{a}$$

The value of $(e/a)_{total}$ is found to be close to 1.73 and 1.80 for Co_2Zn_{11} and Fe_2Zn_{11}, respectively. The valency of the TM element, $(e/a)_{TM}$, is easily obtained by assigning the valency of the partner element Zn as two. The $(e/a)_{TM}$ values for Co and Fe turn out to be 0.26 and 0.70, respectively. All relevant numerical data are summarized in Table 8.1.

8.3 Al_8V_5 GAMMA-BRASS

8.3.1 Construction of the Model Structure

As mentioned in Section 8.1, Brandon et al. [7] revealed that Al_8V_5 gamma-brass is isostructural to Cu_5Zn_8 gamma-brass with space group $I\bar{4}3m$ and contains 52 atoms in the unit cell with the lattice constant $a = 0.9234$ nm. It is found that four sites on IT are shared by two Al and two V atoms, four sites on OT by four V atoms, six sites on OH by four V atoms and two Al atoms and twelve sites on CO by twelve Al atoms. For first-principles band calculations, four Al and six V atoms are exclusively filled into sites IT and OH, respectively. This is made possible without changing the overall composition of Al_8V_5. The experimentally determined fractional coordinates of all 52 atoms in the unit cell and the lattice constant [7] are employed in both LMTO-ASA and FLAPW band calculations discussed below [15].

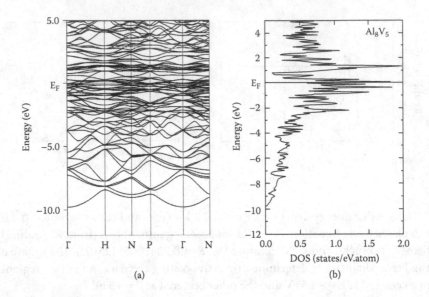

(a)   (b)

FIGURE 8.16 (a) Energy dispersion relations and (b) DOS calculated using the FLAPW method for Al_8V_5 gamma-brass. [From U. Mizutani et al., *Phys. Rev. B* 74 (2006) 235119.]

8.3.2 Electronic Structure Calculations and Stabilization Mechanism

The energy dispersion relations and DOS calculated using the FLAPW method for Al_8V_5 gamma-brass are shown in Figure 8.16 [15]. The dispersionless and bunched electronic states in the energy range from –3 to +2 eV across the Fermi level are certainly due to the V-3d band. It apparently consists of two V-3d subbands separated by a deep pseudogap at about +0.5 eV. Since the concentration of the transition metal element V is 38.5 at.% and is much higher than that in TM_2Zn_{11} (TM = Ni, Pd, Co, Fe) discussed in Section 8.2, the V-3d band forms a wider and higher DOS across the Fermi level.

As mentioned in Chapter 4, Section 4.10, the LMTO-ASA method is best suited to study the orbital hybridization effect between the V-3d and Al-3p states. The DOSs before and after zeroing the Al-3p/V-3d orbital hybridization terms in the LMTO-ASA wave function are depicted in Figures 8.17a,b, respectively [15]. Firstly, the DOS is well consistent with that derived from the FLAPW method shown in Figure 8.16b. More important is that a pseudogap at +0.5 eV disappears, when the V-3d/Al-3p orbital hybridization terms are intentionally deleted from the LMTO-ASA wave function. Therefore, we say that the V-3d states are mainly coupled with the Al-3p states to split the V-3d band into the bonding and antibonding sub-

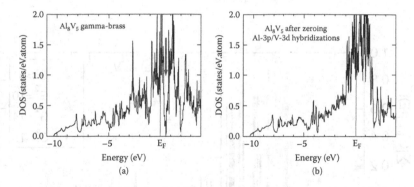

FIGURE 8.17 (a) Total DOS and (b) the DOS obtained after zeroing V-3d/Al-3p orbital hybridization terms in the LMTO-ASA wave function for Al_8V_5 gamma-brass. A deep pseudogap at about +0.5 eV above the Fermi level in (a) disappears in (b). [From U. Mizutani et al., *Phys. Rev.* B 74 (2006) 235119.]

bands, resulting in the pseudogap at about +0.5 eV. We consider the V-3d/Al-3p orbital hybridizations to be mainly responsible for stabilizing Al_8V_5 gamma-brass, since the Fermi level falls at such a position that the bonding subband is almost fully filled, while the antibonding subband is completely unoccupied. As a result, Al_8V_5 gamma-brass may well be regarded as being typical of an orbital hybridization-induced pseudogap system.

We are now interested in examining whether or not the stabilization mechanism specific to the gamma-brass structure, particularly, the FsBz interaction involving $|G|^2 = 18$, remains important in Al_8V_5 gamma-brass, in which the V-3d band widely spreads across the Fermi level. The interference effect of electron waves with different sets of lattice planes in the presence of the V-3d band is studied by means of the FLAPW-Fourier method. The energy spectrum of

$$\sum \left| C_{k+G}^i \right|^2$$

is constructed over $|G|^2$ values from 6 to 50 in the same manner as was done to construct Figures 8.4, 8.10, and 8.13 for group (II) gamma-brasses so far studied. The results for Al_8V_5 gamma-brass are shown in Figure 8.18 by dividing the range of $|G|^2$ into three regimes: (a) $6 \le |G|^2 \le 18$, (b) $22 \le |G|^2 \le 30$, and (c) $34 \le |G|^2 \le 50$.

Let us first direct our attention to the electronic states of $|G|^2 = 6$ corresponding to the symmetry points N on the {211} zone planes, which are

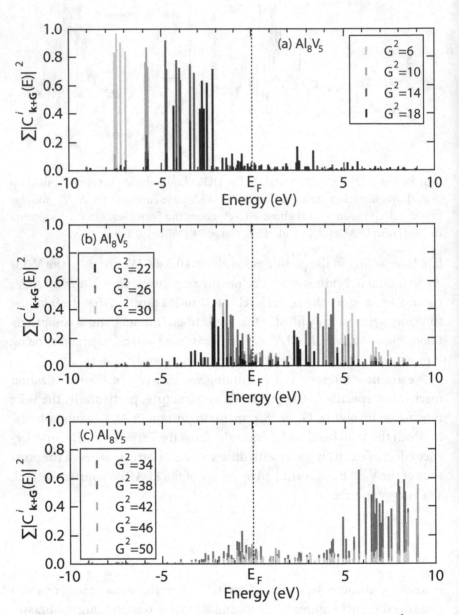

FIGURE 8.18 Energy dependence of the Fourier component $\Sigma \left| C_{k+G}^{i} \right|^{2}$ of the FLAPW-wave function outside the MT sphere at the symmetry points N for Al_8V_5 gamma-brass. The data are divided into three regimes, depending on the magnitude of $\left| G \right|^{2}$.

indicated by light gray bars in Figure 8.18a. The spectrum is sharply centered at around E = −7.4 eV. A comparison with the electronic states at the symmetry points N shown in both Figures 8.1 and 8.16a allows us to identify them as 6B and 6AB, which are weakly separated due to a small but finite form factor at $|G|^2 = 6$ without being affected by the V-3d states. However, the energy spectrum in Figures 8.18a–c begins to be widely spread, when $|G|^2$ is increased beyond 10. This is certainly due to the mixture of the V-3d states.

We find from Figure 8.18 that electronic states of $|G|^2 \geq 14$ are split into bonding states below about −2 eV and antibonding states above about +2 eV, giving rise to a pseudogap in between them. It is important to stress that the resulting bonding and antibonding states are densely distributed at the bottom and top of the V-3d band, respectively. This was already pointed out in connection with the stability of Co_2Zn_{11} and Fe_2Zn_{11} in Section 8.2.2.2 and named the d-states-mediated-splitting. In particular, the $|G|^2 = 18$ states marked with black in Figure 8.18a are split into intense bonding states below about −2 eV and weak antibonding states above about +3 eV at the bottom and top of the V-3d band, respectively.

In Figure 8.18b involving $22 \leq |G|^2 \leq 30$, the electronic states are almost equally split into bonding states below the Fermi level and the antibonding states above about +2 eV. The ratio of bonding states over antibonding states sharply decreases with increasing $|G|^2$. In Figure 8.18c involving $34 \leq |G|^2 \leq 50$, only a very small fraction of bonding states remains below the Fermi level while most antibonding states appear above about +5 eV. Undoubtedly, an energy gain from the regime (c) can be neglected. We can say that the interference of electrons extending outside the MT sphere with sets of lattice planes involving $14 \leq |G|^2 \leq 30$, in particular $|G|^2 = 18$ and 22, participates in the formation of bonding states near the bottom of the V-3d band. If all of the spectra from (a) to (c) are superimposed, one can realize that the V-3d-states-mediated-FsBz-interactions form a "pseudogap" in the energy range, where the V-3d band exists. However, a resulting pseudogap is masked in the total DOS in Figure 8.16b because of the presence of the V-3d states mainly residing inside the MT sphere.

The sp-partial DOS is calculated using the FLAPW method and plotted in Figure 8.19 [16]. As is expected, it clearly exhibits a pseudogap over the range from −2 to +2 eV in good agreement with the data in Figure 8.18. We consider that, among $|G|^2$'s ranging from 14 to 30, the $|G|^2 = 18$ wave characteristic of the gamma-brass structure plays the largest contribution to lowering the electronic energy by forming the bonding states below

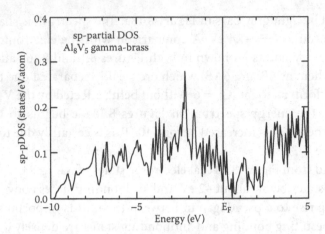

FIGURE 8.19 sp-electron partial DOS calculated using the FLAPW method for Al_8V_5 gamma-brass. [From U. Mizutani et al., Ab Initio test of the Hume-Rothery electron concentration rule for gamma-brasses, chapter 15 in *Diffuse Scattering in the 21st Century: Emerging Insights into Materials Structure and Behavior* (edited by R.I. Barabash, G.E. Ice, and P.E.A. Turchi, Momentum Press, New Jersey, 2009), pp. 283–301.]

–2 eV near the bottom of the V-3d band and that the $|G|^2 = 22$ wave is also important, since significant bonding states are formed below the Fermi level. Therefore, we conclude that Al_8V_5 gamma-brass is stabilized through the FsBz interactions mainly involving $|G|^2$s over 18 to 22 as a result of the d-states-mediated-splitting. Hence, its stabilization mechanism can be consistently described as the extension of that for Co_2Zn_{11} and Fe_2Zn_{11} discussed in Section 8.3.2.

We have so far limited our discussion to electronic states only at the symmetry points N. In order to prove the V-3d-states-mediated-FsBz interactions to occur effectively and equally at symmetry points other than the points N, we evaluated the $|G|^2$-dependence of two different energies of the system defined by the following two equations at the symmetry points N, Γ, and H (see Chapter 3, Figure 3.2b). The first one is given by

$$E_{av}\left(\left|\mathbf{G}\right|^2\right) = \frac{\displaystyle\sum_{E=E_0}^{\infty} \left|C_{\mathbf{k}+\mathbf{G}}(E)\right|^2 E}{\displaystyle\sum_{E=E_0}^{\infty} \left|C_{\mathbf{k}+\mathbf{G}}(E)\right|^2} \qquad (8.1)$$

where energy E weighted by state intensity $|C_{k+G}(E)|^2$ is summed from the bottom of the valence band, E_0, to an infinity, including both occupied and unoccupied states as a function of the square of the reciprocal lattice vector, $|G|^2$, at a given symmetry point. The $E_{av} - |G|^2$ relation thus obtained represents an energy dispersion of an electron obtained by averaging over both bonding and antibonding states formed below and above the Fermi level, respectively, at the symmetry points N, Γ, and H. The second one is given by

$$E_{oc}\left(|G|^2\right) = \frac{\sum\limits_{E=E_0}^{E_F} |C_{k+G}(E)|^2 E}{\sum\limits_{E=E_0}^{E_F} |C_{k+G}(E)|^2} \tag{8.2}$$

where the summation is limited up to the Fermi level so that the $E_{oc} - |G|^2$ relation represents the energy dispersion of an electron obtained by averaging only over the occupied states in the valence band at three different symmetry points. As shown in Figure 8.20, the $|G|^2$-dependence of both $E_{av}(|G|^2)$ and $E_{oc}(|G|^2)$ is quite universal, being independent of the choice of three different symmetry points, N, Γ, and H. Moreover, $E_{av}(|G|^2)$ is found to be close to the free electron behavior $E \propto |k+G|^2$, whereas $E_{oc}(|G|^2)$ is consistently lower than $E_{av}(|G|^2)$.[*] This can be taken as an additional proof that the V-3d-states-mediated-FsBz-interactions significantly contribute to lowering the electronic energy in Al_8V_5 gamma-brass.

The Hume–Rothery plot for Al_8V_5 gamma-brass is shown in Figure 8.21. The variance is large over the energy range −3 to +3 eV, since the V-3d band exists. Hence, the data below −3 eV and above +3 eV has to be extrapolated to the Fermi level. The square of the Fermi diameter $(2k_F)^2$ is determined to be 21.0 by taking the value at the position C obtained by averaging the two intersecting points A and B. Obviously, the accuracy in determining $(2k_F)^2$ for Al_8V_5 gamma-brass is the least reliable among those discussed earlier. The total number of electrons per

---

[*] Note that both $E_{av}(|G|^2)$ and $E_{oc}(|G|^2)$ represent an averaged energy per electron. The contributions from states of $|G|^2 \geq 24$ to the valence-band structure energy can be essentially neglected because of their low populations below the Fermi level.

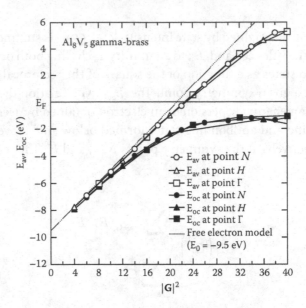

FIGURE 8.20 $|G|^2$-dependence of the energy $E_{av}(G)$ and $E_{oc}(G)$ per electron calculated using Equations 8.1 and 8.2 at the symmetry points N, Γ, and H for Al_8V_5 gamma-brass. A thin dotted line represents the free electron model. [From U. Mizutani et al., *Phys. Rev. B* 74 (2006) 235119.]

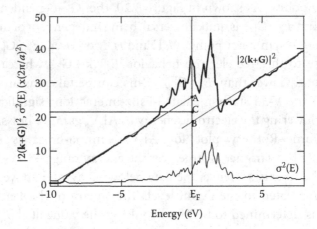

FIGURE 8.21 Energy dependence of (a) $\{2(|\mathbf{k}+\mathbf{G}|)\}^2$ and (b) its variance $\sigma^2(E)$ calculated using Equations 7.3 to 7.5 for Al_8V_5 gamma-brass. The square of the Fermi diameter at the point C is derived by taking an average of the intercepts at points A and B, which are obtained by extrapolating two straight lines to the Fermi level: one drawn through the data points below -3 eV and the other drawn through the data points above $+3$ eV. [From U. Mizutani et al., *Phys. Rev. B* 74 (2006) 235119.]

atom $(e/a)_{total}$ is calculated by inserting the Fermi radius k_F thus obtained into the relation

$$\left(e/a\right)_{total} = \frac{8\pi k_F^3}{3N}$$

where the number of atoms per unit cell, N, is equal to 52 and k_F is in units of

$$\frac{2\pi}{a}$$

The value of $(e/a)_{total}$ is no longer close to 21/13 (= 1.615) but is increased to 1.94. We see that the value tends to be progressively increased as the atomic number of the TM element involved becomes lower, i.e., Co_2Zn_{11}, Fe_2Zn_{11}, and Al_8V_5. The valency of the transition metal element, $(e/a)_{TM}$, is easily calculated by assigning that of the partner element Al as three. The value of $(e/a)_{TM}$ for the V atom turns out to be 0.23. A positive e/a value for the V atom means that it still serves as raising the charge density outside the MT sphere. All relevant data are included in Table 8.1.

8.4 Ag-Li GAMMA-BRASS

According to Table 6.1, there still exists another class of the gamma-brass, which is classified into group (III). The Ag-Li gamma-brass has been identified to be isostructural to the prototype Cu_5Zn_8 [17,18], though it consists of only monovalent elements Ag and Li without involving any TM elements. Hume-Rothery [19] mentioned in his book that "no combination of univalent elements can give the characteristic electron atom ratio of 21/13 although, if lithium were divalent, the above composition would be nearly that required for the 21/13 ratio." Hume-Rothery apparently tended to believe that the e/a = 21/13 rule would be universally applied to all gamma-brasses having the same complex structure. The study of the stabilization mechanism as well as the e/a determination for the Ag-Li gamma-brass is of great importance to gain a deeper insight into the Hume-Rothery electron concentration rule for CMAs. We include this unique gamma-brass in Chapter 8, since we will discuss its stabilization mechanism in terms of the Ag-4d-states-mediated-splitting.

8.4.1 Construction of the Model Structure

As mentioned in Chapter 6 and also Appendix 2, Section A2.2.1.13, chemical disorder slightly exists in the experimentally determined crystal structure of $Ag_{36}Li_{64}$ gamma-brass [18]. The model structure is constructed simply by ignoring small amount of Li atoms on sites OT and OH, where otherwise Ag atoms are located. This is made with a minimal sacrifice from the best-refined structure [18]. The ordered Ag_5Li_8 gamma-brass with the lattice constant of 0.99066 nm is employed for the FLAPW band calculations [20]. It may be noted that 61.5 at .%Li concentration in Ag_5Li_8 is slightly off from the minimum Li concentration of 63.5 at .%Li in the gamma-brass phase field in the equilibrium phase diagram [6].

8.4.2 Electronic Structure Calculations

Figures 8.22a,b show the FLAPW-derived energy dispersion relations and DOS for Ag_5Li_8 gamma-brass, respectively [20]. The bunched bands in the binding energies centered at −4.5 eV in (a) are obviously due to the Ag-4d band. The structure of dispersion relations in the energy region over +1 to +3 eV above the Fermi level is very similar to that found in the vicinity of the Fermi level for Cu_5Zn_8 gamma-brass shown in Figure 7.2a. The sparse dispersions obviously give rise to a pseudogap in the corresponding DOS, as is clearly seen in the insert to Figure 8.22b. It is worthwhile mentioning, at this stage, that an unusually sharp peak is observed in the DOS at −5.15 eV near the bottom of the Ag-4d band. We will come back to this phenomenon in Section 8.4.3, where its stabilization mechanism is discussed.

As discussed in Chapter 7, Sections 7.3 and 7.4, the FLAPW-Fourier analysis for Cu_5Zn_8 and Cu_9Al_4 gamma-brasses could identify a pseudogap at the Fermi level to originate from the FsBz interactions associated with the set of {330} and {411} lattice planes and deduce the effective **e/a** values to be essentially equal to 21/13 for both of them. A resemblance of the energy dispersion relations over the range +1 to +3 eV in Ag_5Li_8 gamma-brass with that across the Fermi level in Cu_5Zn_8 strongly suggests the FsBz interactions involving $|G|^2 = 18$ to be responsible for the formation of a pseudogap at 2 eV above the Fermi level. Needless to say, however, its presence above the Fermi level can have nothing to do with the stabilization of Ag_5Li_8 gamma-brass phase.

The Hume-Rothery plot was performed for Ag_5Li_8 gamma-brass [20]. The energy dependence of $|2(\mathbf{k}+\mathbf{G})|^2$ and its variance are shown in

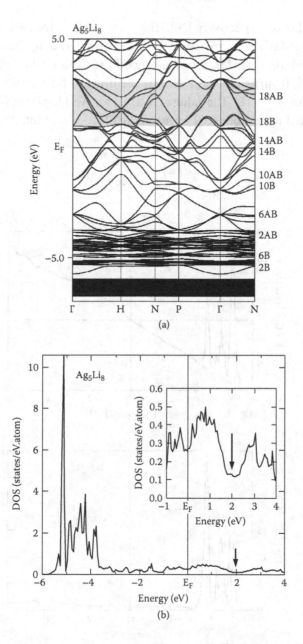

FIGURE 8.22 (a) Energy dispersion relations and (b) DOS calculated using the FLAPW method for Ag_5Li_8 gamma-brass [from U. Mizutani et al., *J. Phys.: Condens. Matter* 20 (2008) 275228]. The dispersion relations inside a rectangle highlighted by light gray resemble those in Cu_5Zn_8 gamma-brass across the Fermi level (see Chapter 7, Figure 7.2a). An insert to (b) represents the DOS in the vicinity of a pseudogap marked with an arrow.

Figures 8.23a,b, respectively [20]. The variance is extremely small in the neighborhood of the Fermi level. Thus, we can safely determine the square of the Fermi diameter $(2k_F)^2$ by reading off the value of the ordinate at the Fermi level. It turns out to be 13.4 in the units of $(2\pi/a)^2$, which is much smaller than $|\mathbf{G}|^2 = 18$. The value of $(\mathbf{e}/\mathbf{a})_{\text{total}}$ for Ag_5Li_8 gamma-brass is easily calculated to be 1.00 ± 0.02 by inserting $(2k_F)^2 = 13.4$ into the relation

$$\left(\mathbf{e}/\mathbf{a}\right)_{\text{total}} = \frac{8\pi k_F^3}{3N}$$

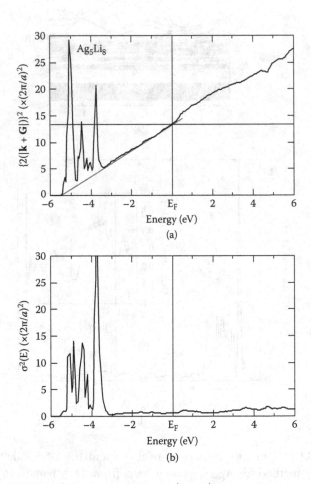

FIGURE 8.23 Energy dependence of (a) $\{2(|\mathbf{k}+\mathbf{G}|)\}^2$ and (b) its variance $\sigma^2(E)$ calculated using Equations 7.3 to 7.5 for Ag_5Li_8 gamma-brass. The square of the Fermi diameter is read off at the Fermi level. [From U. Mizutani et al., *J. Phys.: Condens. Matter* 20 (2008) 275228.]

where the number of atoms in the unit cell, N, is equal to 52 [18]. The valency of Li is confirmed to be unity, since that of Ag must be unity in the metallic state. We consider the present analysis to rule out clearly the Hume-Rothery postulate on divalency for Li [19].

8.4.3 Stabilization Mechanism

At first, we try to identify the role of the zone planes satisfying the matching condition given by Equation 4.1. As was described in the preceding section, the square of the Fermi diameter $(2k_F)^2$ in units of $(2\pi/a)^2$ is deduced to be 13.4 for Ag_5Li_8 from the Hume-Rothery plot. This immediately tells us that a set of {321} lattice planes with $|\mathbf{G}|^2 = 14$ must be a candidate interfering with electrons at the Fermi level. The NFE band calculations are performed to study its effect on the DOS. As shown in Figure 8.24, the form factor V_G is extremely large only at $|\mathbf{G}|^2 = 18$ corresponding to the {330} and {411} zone planes but is extremely small at $|\mathbf{G}|^2 = 14$ for Ag_5Li_8 gamma-brass [20].

The effect of eliminating form factors at $|\mathbf{G}|^2 = 18$ and 14 on the DOS is shown in Figures 8.26a,b, respectively, over the energy range −2 to +2 eV across the Fermi level [20]. A pseudogap marked by an arrow reproduces well that derived from the FLAPW method shown in Figure 8.22b. The pseudogap almost completely disappears when $V_{|\mathbf{G}|^2=18}$ is set to zero, confirming that it is definitely caused by the interaction involving the set of {330} and {411} lattice planes. Instead, the elimination of the form factor $V_{|\mathbf{G}|^2=14}$ hardly affects the DOS across the Fermi level. Thus, we conclude

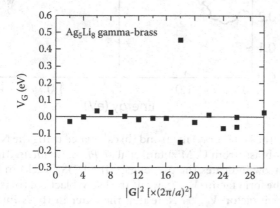

FIGURE 8.24 Form factor V_G as a function of the square of the allowed reciprocal lattice vector $|\mathbf{G}|^2$ for Ag_5Li_8 gamma-brass. [From U. Mizutani et al., *J. Phys.: Condens. Matter* 20 (2008) 275228.]

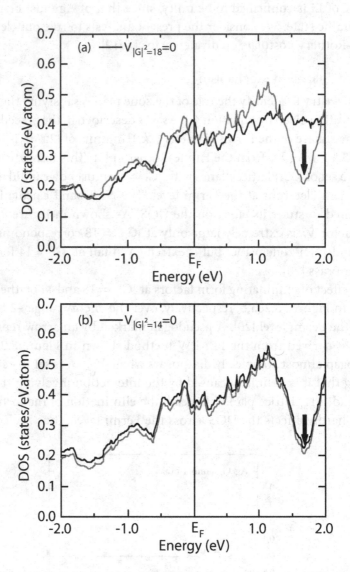

FIGURE 8.25 The DOS (gray) in (a) and (b) calculated using the NFE method for Ag_5Li_8 gamma-brass [from U. Mizutani et al., *J. Phys.: Condens. Matter* 20 (2008) 275228]. An arrow indicates the pseudogap. The DOS (black) in (a) is obtained after zeroing the form factor $V_{|G|^2=18}$, while the DOS (black) in (b) is obtained after zeroing the form factor $V_{|G|^2=14}$. Note that the latter in (b) is intentionally displaced upwards by 0.02 states/eV.atom to avoid an overlap with the DOS (gray).

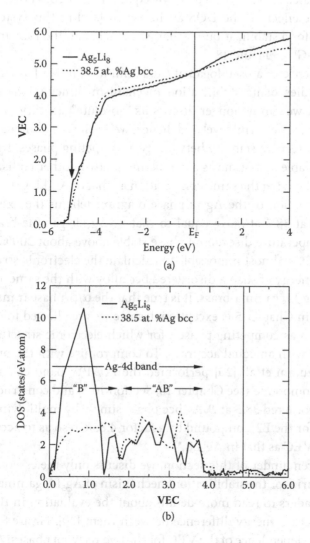

FIGURE 8.26 (a) **VEC** as a function of energy and (b) the DOS as a function of **VEC** for Ag$_5$Li$_8$ gamma-brass (solid line) and 38.5 at.%Ag bcc model structure (dotted line) obtained by renormalizing the DOS of AgLi B2-compound so as to accommodate the same number of **VEC** as that in Ag$_5$Li$_8$ [from U. Mizutani et al., *J. Phys.: Condens. Matter* 20 (2008) 275228]. An arrow in (a) indicates the presence of a flat-band. The vertical line represents the Fermi level, below which 4.85 electrons per atom are accommodated for both phases. Symbols "B" and "AB" in (b) represent the Ag-4d bonding and antibonding subbands, respectively.

that the matching condition given by equation (4.1) does not produce any measurable effect on the DOS at the Fermi level in this system and we must seek for a stabilization mechanism other than the FsBz interactions involving $|\mathbf{G}|^2$ = 14 and 18.

The absence of a pseudogap at the Fermi level in Ag_5Li_8 gamma-brass makes studies of its stabilization mechanism difficult. The reason for this is that we can no longer discuss its "absolute" stability in terms of a pseudogap at the Fermi level and, hence, we are forced to move to the discussion on relative stability between two competing phases. In principle, we are not able to assume as a competing phase another realistic ordered compound having the same composition as that of Ag_5Li_8 gamma-brass.

An inspection of the Ag-Li phase diagram tells us the existence of a bcc phase at 38.5 at .%Ag equal to that in Ag_5Li_8 gamma-brass. This is a high temperature disordered phase stable above about 200°C. Honestly speaking, it is almost impossible to calculate the electronic structure and the total-energy of such a disordered bcc alloy with the same accuracy as those of Ag_5Li_8 gamma-brass. It is true that the α/β-phase transformation discussed in Chapter 5 is exceptional, since we are allowed to choose fcc- and bcc-Cu as competing phases, for which electronic structures can be calculated with an equal accuracy. To compromise with the present situation, Mizutani et al. [20] performed the FLAPW band calculations for AgLi B2-compound (see Chapter 10, Section 10.3) and constructed a DOS for the disordered 38.5 at .%Ag bcc phase simply by multiplying the DOS obtained for the B2-compound by a factor 4.85/6.0 so as to accommodate the same **VEC** as that in Ag_5Li_8.*

In the remainder of this section, we discuss only the essence of a possible scenario for the stabilization mechanism of Ag_5Li_8 gamma-brass and ask the readers to read more details about the evaluation in the valence-band structure energy difference between them [20]. Figure 8.26a shows the energy dependence of the **VEC** for the two relevant phases. As a unique feature in Ag_5Li_8 gamma-brass, we can point out a sharp jump in **VEC** occurring at $E = -5.15$ eV, as marked by an arrow. A jump in **VEC** occurs without changing energy and its magnitude reaches as large as 1.25. This dramatically large jump in **VEC** can be safely attributed to the existence of

---

* Since Ag and Li atoms donate eleven and one electrons per atom to the valence band, the total numbers of electrons per atom, **VEC**, filled into the DOS for Ag_5Li_8 gamma-brass and AgLi B2-compound are quite different from each other, that is, $(11 \times 5 + 1 \times 8)/13 = 4.85$ and $(11 + 1)/2 = 6.0$, respectively.

almost flat energy dispersions at $E = -5.15$ eV in Figure 8.22a, and a delta-function-like peak in the DOS in Figure 8.22 (b).

The **VEC** dependence of the DOS is plotted in Figure 8.26b for the two phases. As is clear, Ag-4d bands for both phases can accommodate approximately four electrons per atom and are roughly divided into bonding "B" and anti-bonding "AB" states over the ranges $0 \leq \textbf{VEC} \lesssim 2.0$ and $2.0 \lesssim \textbf{VEC} \lesssim 4.0$, respectively. We can immediately find an unusually large growth of the Ag-4d "B" sub-band in Ag_5Li_8 gamma-brass. Such an abnormal growth of the "B" sub-band is absent in Figure 8.27, where the similar data for Cu_5Zn_8 gamma-brass are shown along with those for a disordered 38.5 at .%Cu-Zn bcc alloy, which are similarly constructed from the electronic structure of the CuZn B2-compound [21]. The "golden rule" discussed in relation to the phase competition between fcc- and bcc-Cu in Chapter 5, Section 5.6 holds true in the Cu-Zn system but obviously breaks down in the Ag-Li system [20]. This means that a sudden rise in the **VEC** slope marked by an arrow in Figure 8.26a for Ag_5Li_8 gamma-brass is truly unusual and is most likely responsible for the stabilization of Ag_5Li_8 gamma-brass.

Now we briefly discuss a possible origin for the formation of a flat-band near the bottom of the Ag-4d band in Ag_5Li_8 gamma-brass (neither in AgLi, CuZn B2-compounds nor in Cu_5Zn_8 gamma-brass). In this regard, we direct our attention to the evidence that the (211) x-ray diffraction peak is extremely strong and is comparable to the strongest peak (330) + (411) only in Ag_5Li_8 gamma-brass (see Appendix 2, Figure A2.8). This is in sharp contrast to that in other gamma-brasses like Cu_5Zn_8 or even Ag_5Zn_8 containing the same amount of Ag as Ag_5Li_8 (see its diffraction spectrum in Figure A2.1). Noritake et al. [18] attributed the occurrence of a huge (211) diffraction peak in Ag_5Li_8 to the predominant occupation of Ag atoms in the set of {211} lattice planes.

Figure 8.28 shows the FLAPW–Fourier energy spectra associated with the $|\textbf{G}|^2 = 6, 10, 14,$ and 18 components of the FLAPW wave function outside the MT sphere at the symmetry points N for Ag_5Li_8 gamma-brass. As is clear from the argument in Section 8.4.2, the $|\textbf{G}|^2 = 18$ electronic states are split into bonding and antibonding states due to the interference with the set of {330} and {411} lattice planes and are responsible for the formation of a pseudogap at about +2 eV above the Fermi level (see Figure 8.22). On the other hand, the NFE-like behavior without involving any splitting is confirmed for $|\textbf{G}|^2 = 10$ and 14. Indeed, they are well away from Ag-4d band. More interesting to be noted is the occurrence of the Ag-4d-states-

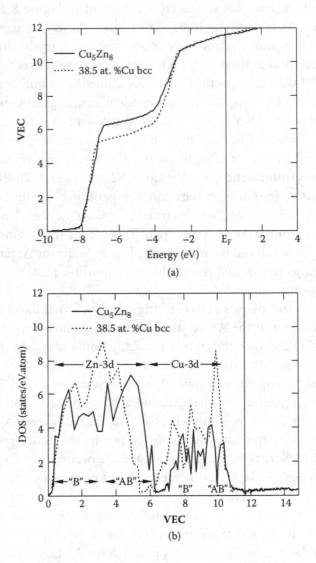

FIGURE 8.27 (a) **VEC** as a function of energy and (b) the DOS as a function of **VEC** for Cu_5Zn_8 gamma-brass (solid line) and 38.5 at.%Cu bcc model structure (dotted line) obtained by renormalizing the DOS of the CuZn B2-compound so as to accommodate the same number of **VEC** as that in Cu_5Zn_8.

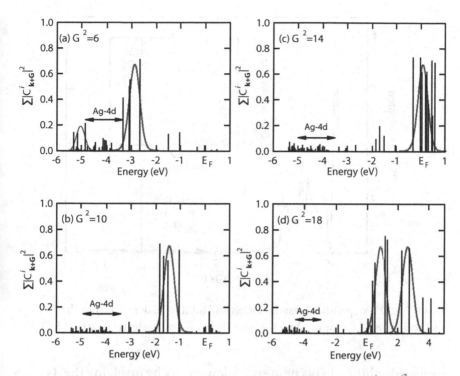

FIGURE 8.28 Energy dependence of the Fourier component $\Sigma \left| C^i_{k+G} \right|^2$ of the FLAPW-wave function (7.1) outside the MT sphere at the symmetry points N for Ag_5Li_8 gamma-brass. The value of $\left| G \right|^2$ is fixed at four representative values of 6, 10, 14, and 18. An arrow indicates the energy range of the Ag-4d band. A Gaussian-like peak is drawn as a guide to eyes to approximate the distribution of the relevant $\left| G \right|^2$-dependent electronic states.

mediated-splitting associated with the $\left| G \right|^2 = 6$ states, which are widely split into bonding and antibonding states as if to avoid the Ag-4d band.

We show in Figure 8.29 the s-, p-, and d-states partial DOSs derived from the FLAPW method for Ag_5Li_8 gamma-brass. Obviously, the flat-band at $E = -5.15$ eV discussed above is mainly composed of Ag-4d states. However, as shown in the insert to Figure 8.29, both s- and p-states also participate in forming the flat-band. Measuring from the bottom of the valence band, we can roughly estimate the kinetic energy of itinerant sp-electrons sustaining the flat-band to be 2.5 eV. Its insertion into the free electron equation $\lambda [nm] = 1.226 / \sqrt{E[eV]}$ yields the wavelength $\lambda = 0.78$ nm. This is in good agreement with the lattice spacing $2d = 0.78$ nm in the set of {211} lattice planes. In this way, the Bragg condition is satisfied and the stationary waves can be formed. We consider the Ag-4d-states-

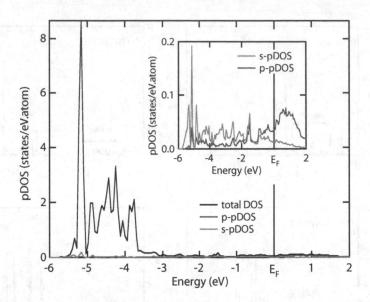

FIGURE 8.29 s-, p-, and d-partial DOSs and total DOS derived from the FLAPW method for Ag_5Li_8 gamma-brass. Insert shows s- and p-electron partial DOSs.

mediated-splitting to occur at such a low energy by involving the $|\mathbf{G}|^2 = 6$ states and to result in a profound condensation of a large number of Ag-4d electrons at −5.15 eV in promoting the stabilization of this compound. We can definitely say that the stabilization of Ag_5Li_8 gamma-brass has little to do with the FsBz interaction involving the set of {330} and {411} lattice planes. The discussion on phase stability without involving a pseudogap at the Fermi level is quite difficult. But it may be possible that the stability of this unique gamma-brass is essentially brought about by the set of Ag-rich {211} lattice planes, which is also characteristic of the gamma-brass structure.

8.5 Ni-Zn AND Co-Zn GAMMA-BRASSES IN SOLID SOLUTION RANGES

As listed in Table 6.1, a rather wide solid solution range exists in the gamma-brass phase field in both Ni-Zn and Co-Zn alloy systems. In Chapter 7, Sections 7.6 and 7.7, we studied the solute concentration dependence of the number of vacancies introduced into the unit cell and discussed why Cu-Zn, Cu-Cd, Cu-Al, and Cu-Ga gamma-brasses have a rather wide solid solution range in terms of the FsBz interactions. Figures 8.30a,b show the

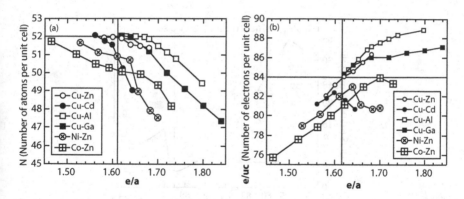

FIGURE 8.30   e/a dependence of (a) the number of atoms in the unit cell, N, and (b) the number of itinerant electrons in the unit cell, e/uc, for Cu-Zn, Cu-Cd, Cu-Al, Cu-Ga, Ni-Zn, and Co-Zn gamma-brasses [from U. Mizutani et al., *Phil. Mag.* 90 (2010) 1985]. The e/a values of Ni and Co are taken to be +0.15 and +0.26, respectively. [From R. Asahi, H. Sato, T. Takeuchi, and U. Mizutani, *Phys. Rev.* B 72 (2005) 125102.]

e/a dependence of the number of atoms per unit cell, N, and that of the number of electrons per unit cell, e/uc, for Ni-Zn and Co-Zn gamma-brasses over a whole composition range, along with the data for group (I) gamma-brasses already shown in Figures 7.12, 7.13, and 7.15 [22]. Here the e/a values for Ni and Co are taken to be +0.15 and +0.26, respectively, as discussed in Section 8.2.2. It can be seen that the number of atoms in the unit cell decreases, while the value of e/uc systematically increases with increasing e/a, regardless of the alloy systems chosen. This indicates the soundness of the $(e/a)_{TM}$ values determined from the Hume-Rothery plot for Ni and Co.

We mentioned in Section 8.2.2.1 that Ni-Zn gamma-brasses exhibit a clear pseudogap immediately below the Fermi level and its origin is successfully interpreted in terms of the FsBz interaction associated with $|G|^2$ = 18. Encouraged by this finding, we replot the DOS as a function of e/uc in Figure 8.31 by locating the Fermi level at e/uc = 0.15 * 8 + 2 * 44 = 89.2 and integrating the DOS below and above the Fermi level. The experimentally derived e/uc values shown in Figure 8.30b are distributed over 78 to 83 across its whole solid solution range, as marked with an arrow in Figure 8.31. The Fermi level is situated inside the pseudogap over the whole concentration range. Thus, we conclude that Ni-Zn gamma-brasses over the solid solution range are also subjected to the Hume-Rothery

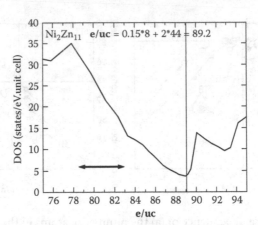

FIGURE 8.31 **e/uc** dependence of the DOS for the Ni_2Zn_{11} model structure [from U. Mizutani et al., *Phil. Mag.* 90 (2010) 1985]. A vertical line refers to its Fermi level. An arrow indicates the solid solution range of the Ni-Zn gamma-brass.

stabilization mechanism, though the present rigid-band approach relies on the DOS obtained from the idealized model structure Ni_2Zn_{11}.*

Another interesting feature in the Ni-Zn gamma-brasses is the occurrence of a long-period super-lattice structure in the Ni-rich region containing more than 21at .%Ni. We discussed in Chapter 7, Section 7.6, the super-lattice formation in Cu-rich Cu-Zn gamma-brasses. A similar phenomenon has been observed in Ni-rich Ni-Zn gamma-brasses [22,23]. The formation of the long-period super-lattice structure is believed to be electronic in origin [24,25]. Figure 8.32a shows the Zn concentration dependence of the mean period of the super-lattice structure, which is reproduced from the data by Morton [23]. The two sets of data for the Cu-Zn and Ni-Zn alloy systems show no correlations, when plotted as a function of the Zn concentration. However, the data almost fall on a universal curve, when plotted against **e/uc**, as shown in Figure 8.32b. This may be taken as another evidence for the validity for an appropriate assignment of the **e/a** value for Ni.

The situation in the Co-Zn gamma-brasses is not straightforward. The DOS calculated for the model structure Co_2Zn_{11} was already shown in Figure 8.9b. The Fermi level is located near the edge of the Co-3d antibonding subband in Co_2Zn_{11}. It is important to realize that the DOS-**e/uc**

---

* As is clear from Figure 8.30a, gamma-brasses in the Zn-rich concentration range contain a large number of vacancies. This is the reason why the solid solution range marked with an arrow is located far below **e/uc** = 89.2 in Figure 8.31.

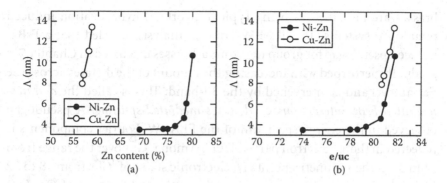

FIGURE 8.32 (a) Zn concentration and (b) **e/uc** dependences of the periodicity Λ of the long-period super-lattice structure observed in Cu-rich Cu-Zn gamma-brasses (O) and Ni-rich Ni-Zn gamma-brasses (●) [from U. Mizutani et al., *Phil. Mag.* 90 (2010) 1985; A.J. Morton, *Phys. Stat. Sol.* A 44 (1977) 205.22, 23]. The **e/ uc** values are re-evaluated by using the **e/a** = 0.15 for Ni [from U. Mizutani, T. Noritake, T. Ohsuna and T. Takeuchi, *Phil. Mag.* 90 (2010) 1985].

plot is meaningful, only when the energy region of our interest is away from the d-band like in Cu_5Zn_8, Cu_9Al_4, and Ni_2Zn_{11} gamma-brasses discussed earlier. In other words, the construction of the DOS-**e/uc** plot for the Co-Zn gamma-brass, where the Fermi level falls inside the d-band, is obviously unsuccessful [22].

The FLAPW band calculations have been so far limited only to the model structure Co_2Zn_{11}, that is, 15.3 at .%Co. Information about the electronic structure of more Co-rich gamma-brasses is needed, since its solid solution range extends over 15 to 29 at .%Co [22]. We consider the Rietveld structure analysis for the Co-rich gamma-brass to be of importance in order to construct a model structure at the composition as close to 30.8 at .%Co corresponding to Co_4Zn_9 as possible. Its FLAPW–Fourier analysis will provide more crucial information about the stabilization mechanism of group (II) Co-Zn gamma-brasses.

8.6 SUMMARY

In Chapter 8, we have studied whether the Hume-Rothery electron concentration rule continues to hold or must be modified or even breaks down in gamma-brasses, where one of the constituent elements is chosen from the 3d-transition metal (TM) elements, and where both constituent elements are monovalent. We chose TM_2Zn_{11} (TM = Ni, Pd, Co, and Fe) and Al_8V_5 gamma-brasses from group (II) and Ag_5Li_8 gamma-brass from group (III), as listed in Table 6.1. Among them, group (II) gamma-brasses are identified as being typical of orbital hybridization-induced pseudogap systems. The FLAPW-Fourier analysis revealed that a FsBz-induced pseudogap for group (I) gamma-brasses discussed in Chapter 7 is gradually perturbed with increasing the amount of the d-states across the Fermi level and is intervened by the 3d-band. This is called the *d-states-mediated-FsBz interactions* or *d-states-mediated-splitting*. A pseudogap observed in the energy spectrum of the FLAPW–Fourier components is reflected in the sp-electron partial DOS. Among $|\mathbf{G}|^2$s over the range from 14 to 30 at the symmetry points N, electronic states of $|\mathbf{G}|^2$s from 18 to 22 make the most significant contribution to the stabilization of Co_2Zn_{11}, Fe_2Zn_{11}, and Al_8V_5 gamma-brasses in group (II). It is emphasized that the range of *critical* $|\mathbf{G}|^2$s participating in forming a pseudogap across the Fermi level is widened due to the d-states-mediated-FsBz interactions. We are, therefore, led to conclude that the Hume-Rothery stabilization mechanism remains effective in group (II) gamma-brasses. More details about this issue will be discussed in Chapter 10.

The Hume-Rothery plot revealed that the **e/a** value gradually deviates from **e/a** = 21/13 in group (II), as the atomic number of the TM element decreases. We claimed that the **e/a** = 21/13 rule essentially holds for Ni_2Zn_{11} and Pd_2Zn_{11} gamma-brasses but the departure from the **e/a** = 21/13 rule is gradually increased with decreasing the atomic number of the TM element involved. The **e/a** value for Ag_5Li_8 gamma-brass is deduced to be unity within the accuracy of 1.00 ± 0.02. The interference phenomenon to cause a deep pseudogap occurs at about 2 eV above the Fermi level and, hence, the matching condition definitely breaks down in Ag_5Li_8 gamma-brass. Its stabilization mechanism is discussed in terms of the formation of an extremely flat band at the bottom of the Ag-4d band. The formation of stationary waves due to the Ag-4d-states-mediated-splitting associated with $|\mathbf{G}|^2 = 6$ is suggested as its possible origin.

REFERENCES

1. W. Ekman, *Z. Phys. Chem. Abt.* B 12 (1931) 57.
2. W.L. Bragg, *Nature*, 131 No. 3317 (1933) 749.
3. H. Witte, *Metallwirtschaft* 16 (1937) 237.
4. G.V. Raynor, *Prog. Met. Phys.* 1 (1940) 1.
5. W.B. Pearson, *The Crystal Chemistry and Physics of Metals and Alloys* (Wiley-Interscience, New York, 1972), p. 109.
6. H. Okamoto, *Phase Diagrams for Binary Alloys* (ASM International, OH, 2000).
7. J.K. Brandon, W.B. Pearson, P.W. Riley, C. Chieh, and R. Stokhuyzen, *Acta Cryst.* B33 (1977) 1088.

8. J.K. Brandon, H.S. Kim, and W.B. Pearson, *Acta Cryst.* B35 (1979) 1937.
9. P. Villar (Ed.), *Pearson's Handbook*, Vol. 2 (ASM International, OH, 1997).
10. A. Johansson, H. Ljung, and S. Westman, *Acta Chem. Scand.* 22 (1968) 2743.
11. V.-A. Edström and S. Westman, *Acta Chem. Scand.* 23 (1969) 279.
12. L. Arnberg and S. Westman, *Acta Chem. Scand.* 26 (1972) 513.
13. J.K. Brandon, R.Y. Brizard, P.C. Chieh, R.K. McMillan, and W.B. Pearson, *Acta Cryst.* B30 (1974) 1412.
14. R. Asahi, H. Sato, T. Takeuchi, and U. Mizutani, *Phys. Rev.* B 72 (2005) 125102.
15. U. Mizutani, R. Asahi, H. Sato, and T. Takeuchi, *Phys. Rev.* B 74 (2006) 235119.
16. U. Mizutani, R. Asahi, H. Sato, and T. Takeuchi, Ab Initio test of the Hume-Rothery electron concentration rule for gamma-brasses, chapter 15 in *Diffuse Scattering in the 21st Century: Emerging Insights into Materials Structure and Behavior* (edited by R.I. Barabash, G.E. Ice, and P.E.A. Turchi, Momentum Press, New Jersey, 2009), pp. 283–301.
17. L. Arnberg and S. Westman, *Acta Chem. Scand.* 26 (1972) 1748.
18. T. Noritake, M. Aoki, S. Towata, T. Takeuchi, and U. Mizutani, *Acta Cryst.* B 63 (2007) 726.
19. W. Hume-Rothery, *Atomic Theory for Students of Metallurgy* (Institute of Metals, Monograph and Report Series No.3, The Institute of Metals, London, 1962), p. 306.
20. U. Mizutani, R. Asahi, H. Sato, T. Noritake, and T. Takeuchi, *J. Phys.: Condens. Matter* 20 (2008) 275228.
21. U. Mizutani, R. Asahi, T. Takeuchi, H. Sato, O.Y. Kontsevoi, and A.J. Freeman, *Z. Kristallogr.* 224 (2009) 17.
22. U. Mizutani, T. Noritake, T Ohsuna, and T. Takeuchi, *Phil. Mag.* 90 (2010) 1985.
23. A.J. Morton, *Phys. Stat. Sol.* A 44 (1977) 205.
24. H. Sato and R.S. Toth, *Phys. Rev.* 124 (1961) 1833.
25. H. Sato and R.S. Toth, *Phys. Rev.* 127 (1962) 469.

CHAPTER 9

Stabilization Mechanism of 1/1-1/1-1/1 Approximants

9.1 ELECTRONIC STRUCTURE OF 1/1-1/1-1/1 APPROXIMANTS

Since the discovery of the Al-Mn quasicrystal by Shechtman et al. in 1984 [1], quasicrystals have been established as a new family of compounds characterized by the lack of translational symmetry but having rotational symmetries forbidden in normal crystals. Their stabilization mechanisms have become one of the most exciting recent topics. However, first-principles band calculations based on the Bloch theorem in the reciprocal space are unfortunately not applicable in this case because of the infinitely large unit cell. Instead, one can perform first-principles band calculations for the family of *approximants*, since the lattice periodicity is assured, no matter how large is the unit cell. In particular, the electronic structure of the lowest-order 1/1-1/1-1/1 approximant, containing some 130 to 170 atoms in the unit cell, has been extensively studied in the past. This is indeed a promising approach to deepen our understanding of the electronic structure of a quasicrystal, since we know from the cut-and-projection method (see Chapter 6, Section 6.1) that the local atomic structure between a quasicrystal and its approximant is essentially the same. In this chapter, we exclusively focus on the exploration of the stabilization mechanism of a 1/1-1/1-1/1 approximant itself as another class of CMAs in relation to the Hume-Rothery electron concentration rule.

As noted in Chapter 1, Section 1.2, Traverse et al. [2] in 1988 revealed from the soft x-ray emission spectra a large depression in the DOS at the Fermi level in the Al-Mn quasicrystal relative to its amorphous and crystalline counterparts. Their pioneering work already hinted at the presence of a pseudogap in quasicrystals. At that time, however, people had intuitively expected that first-principles band calculations were beyond a practical level of computations, even for 1/1-1/1-1/1 approximants. A breakthrough was brought about by Fujiwara in 1989 [3], who performed LMTO-ASA band calculations for the Al-Mn approximant, containing some 138 atoms in the unit cell. This work revealed a deep pseudogap at the Fermi level and suggested it to be most likely responsible for the stability of such CMAs containing icosahedral clusters similar to those in quasicrystals. Since then, first-principles LMTO-ASA electronic structure calculations have been extensively carried out to affirm whether a pseudogap is a universal feature of all approximants and whether it becomes more pronounced, as the order of the approximants is increased toward that of a quasicrystal. Readers may consult the recent developments on this topic in books and in review articles [4–6].

As discussed in Chapter 4, Section 4.10, the LMTO-ASA method has been recognized as a fast but efficient scheme for first-principles electronic structure calculations. It owes its speed to the relatively small basis set, which consists essentially of atomic orbitals of constituent elements. This is the reason why the LMTO-ASA band calculation method is capable of determining the electronic structure even for CMAs with a giant unit cell, as in the case of 1/1-1/1-1/1 approximants. Moreover, it allows to extract the effect of orbital hybridizations among neighboring atoms and their involvement in the formation of a pseudogap. On the other hand, the FLAPW method is more suited when the contribution arising from a specific FsBz interaction needs to be examined in order to understand more clearly the physics behind the Hume-Rothery electron concentration rule (see Chapter 4, Section 4.12). Since the FLAPW wave function outside the MT sphere in a crystal is given as a sum of plane waves over more than two thousand reciprocal lattice vectors, solving a secular equation becomes a hard task for CMAs containing more than one hundred atoms in the unit cell. This is probably the reason why, in the past, theoretical studies with respect to the Hume-Rothery electron concentration rule for CMAs have not made much progress beyond the free electron model.

A large number of electronic structure calculations have already been reported on 1/1-1/1-1/1 approximants. Representative works reported on families of RT- and MI-type 1/1-1/1-1/1 approximants are summarized in Table 9.1 [3,7–17]. As emphasized in Chapter 6, chemical disorder and defects were revealed at various sites in many of them. They must be removed to perform first-principles band calculations, and, hence, the construction of a "model structure" becomes inevitable. We consider it important to criticize its appropriateness in comparison with a structure obtained by refining the measured structural data. However, there was a tendency for theoreticians working on band calculations to describe only briefly the model structure they constructed.

Efforts to probe more specifically the origins of the Hume-Rothery electron concentration rule were indeed fairly limited in the past. Among them, Trambly de Laissardière et al. [18] discussed the Hume-Rothery electron concentration rule by constructing the Anderson Hamiltonian, which was composed of two terms describing the motion of nearly free sp-electrons on one hand and d-impurities on the other hand to treat electrons in Al-TM based alloys. Obviously, their approach is based on the model Hamiltonian instead of first-principles band calculations. In this chapter, we aim at elucidating the stabilization mechanism of only a few 1/1-1/1-1/1 approximants on the basis of first-principles LMTO-ASA and FLAPW band calculations with an emphasis on the role of the FsBz interactions in the formation of a pseudogap. They include Al-Mg-Zn and Al-Li-Cu approximants from the RT-type family and Al-Cu-TM-Si (TM = Fe and Ru) and Al-Re-Si 1/1-1/1-1/1 approximants from the MI-type family. Nearly Free Electron (NFE) band calculations turned out to be very valuable to extract the FsBz interactions for systems like 1/1-1/1-1/1 approximants, where the application of the FLAPW method is rather limited.

9.2 Al-Mg-Zn 1/1-1/1-1/1 APPROXIMANT

9.2.1 Construction of the Model Structure

The Al-Mg-Zn 1/1-1/1-1/1 approximant is, we consider, the best suited to explore the FsBz interactions and to examine whether a pseudogap is indeed FsBz-induced, since its valence band can be well described in the NFE model except for the Zn-3d band near its energy bottom. All the structure analyses so far reported claim a large amount of chemical disorder between Al and Zn atoms [19–21]. The fractional occupancy at the center of the cluster (sites A: 2a) is small and can be essentially regarded as being vacant, regardless of the Al concentration [20,21]. Both Al and Zn atoms are randomly filled into 12 sites (B: 24g) on the first icosahedron

TABLE 9.1 First-Principles Band Calculations for 1/1-1/1-1/1 Approximants

Cluster Type	System (Number of Atoms in the Unit Cell)	Method	Reference	Cluster Type	System (Number of Atoms in the Unit Cell)	Method	Ref.
	$Al_{16}Mg_{39.5}Zn_{44.5}$ (162)	LMTO-ASA	[7] J. Hafner and M. Krajčí (1993)		$Al_{82.6}Mn_{17.4}$ (138)	LMTO-ASA	[3] T. Fujiwara (1989)
	$Al_{45}Mg_{40}Zn_{15}$ (160)	LMTO	[8] S. Roche and T. Fujiwara (1998)		Al-Cu-Fe-Si (139)	LMTO-ASA	[8] S.Roche and T. Fujiwara (1998)
	$Al_{30}Mg_{40}Zn_{30}$ (160)	LMTO-ASA	[9] H. Sato et al. (2001)		$Al_{68.8}Pd_{15.6}Mn_{15.6}$ (128)	LMTO-ASA	[13] M. Krajčí et al. (1995)
	$Al_{60}Li_{32.5}Cu_{7.5}$ (160)	LMTO-ASA	[10] T. Fujiwara and T. Yokokawa (1992)	MI	$Al_{68.8}Pd_{15.6}Re_{15.6}$ (128)	LMTO-ASA	[14] M. Krajčí and Hafner (1999)
RT	Al-Li-Cu	LMTO-ASA	[11] M. Windisch et al. (1994)		$Al_{68}Cu_7TM_{17}Si_8$ (TM = Fe and Ru) (144)	LMTO-ASA	[15] U. Mizutani et al. (2004)
	$Al_{52.5}Li_{32.5}Cu_{15}$ (160)	LMTO-ASA	[12] H. Sato et al. (2004)		$Al_{68}Cu_7TM_{17}Si_8$ (TM = Fe and Ru) (144, 138)	FLAPW	[16] U. Mizutani et al. (2009)
					$Al_{73.6}Re_{17.4}Si_9$ (138)	LMTO-ASA	[17] T. Takeuchi et al., (2003)

Note: Chemical formula is given in %.

with an approximate ratio of 1:4, 12 sites (C: 24g) on the second icosa-hedron with 3:2, and 24 sites (F: 48h) on the truncated icosahedron with 3:2, whereas no chemical disorder exists in the totally thirty-two sites D, E, G, and H for Mg atoms, resulting in 80-atom cluster (see Chapter 6, Figure 6.4 and Section 6.3).

Hafner and Krajčí [7] constructed a model structure of Al-Mg-Zn 1/1, 2/1, 3/2, and 5/3 cubic approximants by applying the cut-and-projection method for a periodic Penrose lattice, as described in Chapter 6, Section 6.1, and decorating it as proposed by Henley and Elser [22]. Briefly, Al atoms are placed on all vertices and Zn atoms on the mid-edge positions in all structural units, consisting of the two kinds of rhombohedra: prolate (PR) and oblate (OR) ones. Two Mg atoms are placed along the trigonal axis in each PR. A composition of the structure thus obtained is deduced to be $Al_{26}Mg_{64}Zn_{72}$, or $Al_{16}Mg_{39.5}Zn_{44.5}$ in %, containing 162 atoms per unit cell in contrast to 160 atoms per unit cell experimentally observed. This implies that the center of the first icosahedron (sites A: 2a) is filled with Al in their model. According to their composition, 12 Al atoms are filled into either sites B on the first icosahedron or sites C on the second icosahedron. The diffraction pattern calculated from their model struc-ture apparently agrees well with the measured one. Their composition is located at the Al-poor side of the solid solution range [20].

Following the Rietveld structure analysis for a series of $Al_xMg_{39.5}Zn_{60.5-x}$ (20.5≤x≤50.5) 1/1-1/1-1/1 approximants [20], Roche and Fujiwara [8] put Zn atoms only into sites B, and Al atoms into sites C and F, while Mg atoms into sites D, E, G, and H, resulting in $Al_{72}Mg_{64}Zn_{24}$ per unit cell or $Al_{45}Mg_{40}Zn_{15}$ in %. Their model structure is located at the Al-rich side of its solid solution range. Sato et al. [9] constructed a model structure with the composition $Al_{48}Mg_{64}Zn_{48}$ or $Al_{30}Mg_{40}Zn_{30}$ in % by filling Zn atoms into both sites B and C and Al atoms into sites F, while Mg atoms into sites D, E, G, and H. The experimentally determined lattice constant of 1.4355 nm for the $Al_{30}Mg_{40}Zn_{30}$ approximant was employed. Since the center of the cluster (sites A) is set to zero in both [8] and [9] in accordance with the experiment, the total number of atoms in the unit cell is 160. The Al con-centration in [9] is located in the middle of the solid solution range.

9.2.2 Electronic Structure Calculations

Prior to the discussions on the electronic structure calculations, we briefly note the experimental studies concerning the Hume-Rothery sta-bilization mechanism in Al-Mg-Zn 1/1-1/1-1/1 approximants. In 1995,

Takeuchi and Mizutani [23] attributed a reduction in the intensity of the observed x-ray photoemission valence band spectra near the Fermi level, relative to that of pure Al, and a clear reduction in the observed electronic specific heat coefficient relative to the free electron value to the existence of a pseudogap. The Fermi diameter was estimated by assuming valencies of Mg, Zn, and Al to be two, two, and three, respectively. Judging from observed x-ray diffraction peaks, they assumed reciprocal lattice vectors associated with the set of {543}, {710}, and {550} lattice planes to be the best candidate to satisfy the matching condition in Equation 4.1 and to be responsible for the formation of a pseudogap. At that time, the set of {543}, {710}, and {550} lattice planes with $|\mathbf{G}|^2 = 50$ was simply assumed to be *critical*.

Hafner and Krajčí [7] in 1993 calculated the electronic structure for their model structure $Al_{16}Mg_{39.5}Zn_{44.5}$ (in %) containing totally 162 atoms in the unit cell, using the LMTO-ASA method. The energy dispersion relations and DOS are reproduced in Figures 9.1a,b. A pseudogap is clearly seen across the Fermi level in the DOS. They briefly noted that the pseudogap was induced by closely spaced {631}, {710}+{550}, and {640} reciprocal lattice vectors without any detailed analysis. (Note that {543} is missing in [7].) Their statement above was probably made simply on the basis of the free electron model. Unfortunately, their dispersion relations along the direction ΓX or <200> in Figure 9.1a are too dense to extract any meaningful information about the pseudogap.

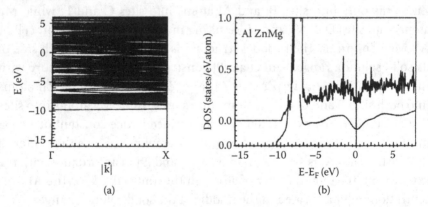

FIGURE 9.1   (a) Energy dispersion relations calculated for the model structure $Al_{26}Mg_{64}Zn_{72}$ using the LMTO-ASA method along the ΓX direction. (b) The corresponding DOS. A smooth DOS in (b) was obtained by recursion methods. [From J. Hafner and M. Krajčí, *Phys. Rev. B* 47 (1993) 11795.]

Roche and Fujiwara [8] in 1998 calculated the electronic structure of their model structure $Al_{72}Mg_{64}Zn_{24}$ by using the LMTO method. Their results on the energy dispersion relations in the vicinity of the Fermi level and the total DOS are shown in Figures 9.2a,b, respectively. A double minima structure is found inside a pseudogap formed immediately below the Fermi level. A width of the DOS pseudogap below the Fermi level is about 1 eV (see Chapter 2, Figure 2.7 in Section 2.3.). However, neighboring energy eigen-states at the symmetry points N across the Fermi level are separated from one another only about the order of 0.1 eV, as can be seen from Figure 9.2a. Thus, the DOS pseudogap of about 1 eV in width cannot be explained simply by studying energy dispersions at the symmetry points N across the Fermi level. Indeed, the situation is entirely different from that in Cu_5Zn_8 and Cu_9Al_4 gamma-brasses shown in Figure 7.2, where a gap of the order of 1 eV (i.e., 10 times larger) is opened across the Fermi level at the symmetry points N.

The LMTO-ASA band calculations were also performed for the $Al_{48}Mg_{64}Zn_{48}$ model structure in 2001 [9]. The energy dispersion relations and DOS are shown in Figures 9.3a,b, respectively. It is clear that a pseudogap is present at the Fermi level in the DOS and is split into two minima A and B separated by a small peak at the Fermi level. An energy separation at the symmetry points N across the Fermi level in Figure 9.3a is again found to be only of the order of 0.1 eV. These energy separations are too small to account for the width of the pseudogap in the DOS, which is again about 1 eV below the Fermi level, as can be seen from Figure 9.3b. This means that, in sharp contrast to the situation in Cu_5Zn_8 and Cu_9Al_4 gamma-brasses (see Chapter 7, Figure 7.2), the electronic structure analysis solely at the symmetry points N across the Fermi level can hardly explain the origin of a DOS pseudogap in the approximant. We consider an overall deficiency in electron populations along almost all directions to be likely responsible for its formation in the Al-Mg-Zn approximant, as can be seen in sparse dispersion relations across the Fermi level inside a rectangle highlighted by white coloring in Figure 9.3a.

9.2.3 Stabilization Mechanism

Figure 9.4a shows the LMTO-ASA derived energy dispersion relations along the <710> direction for the $Al_{48}Mg_{64}Zn_{48}$ model structure in comparison with those derived from the free electron model in Figure 9.4b [9]. A vertical line at

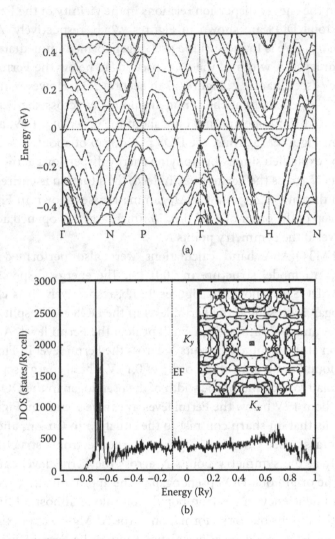

FIGURE 9.2   (a) Energy dispersion relations and (b) DOS calculated using the LMTO method for the model structure $Al_{72}Mg_{64}Zn_{24}$. A pseudogap immediately below the Fermi level is apparently split into double minima. An insert to Figure 9.2b shows its Fermi surface contour in the $k_x - k_y$ plane. [From S. Roche and T. Fujiwara, *Phys. Rev. B* 58 (1998) 11338.]

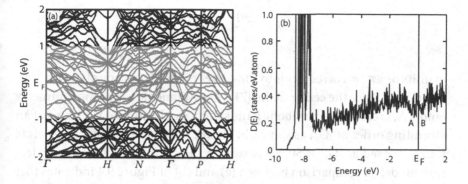

FIGURE 9.3 (a) Energy dispersion relations near the Fermi level and (b) DOS calculated using the LMTO-ASA method for the model structure $Al_{48}Mg_{64}Zn_{48}$ [from H. Sato, T. Takeuchi, and U. Mizutani, *Phys. Rev.* B 64 (2001) 094207]. A pseudogap in (b) is split into double minima (A) and (B) separated by a small peak.

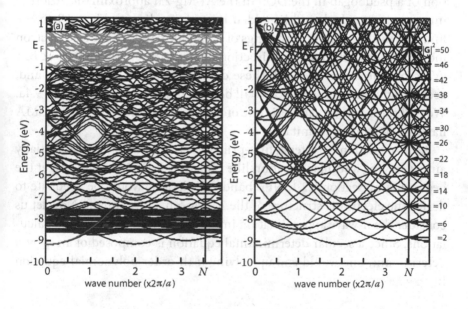

FIGURE 9.4 Energy dispersion relations along the direction <710> derived from (a) LMTO-ASA method and (b) the free electron model for the model structure $Al_{48}Mg_{64}Zn_{48}$ [from H. Sato, T. Takeuchi, and U. Mizutani, *Phys. Rev.* B 64 (2001) 094207]. The crossing of parabolic bands along the energy axis at the center N of the {710} zone planes in (b) can be identified with respect to the square of the relevant reciprocal lattice vector $|G|^2$. The parabolic bands passing the Fermi level correspond to $|G|^2 = 50$ in (b). A shaded rectangle in (a) indicates the region, where a pseudogap is formed in the DOS, as shown in Figure 9.3 (b).

$$k = \frac{\sqrt{50}}{2}(=3.535)$$

in units of $2\pi/a$ corresponds to the center of the $\{710\}$ zone planes in both (a) and (b). Since the center of the $\{710\}$ zone planes refers to the symmetry points N, crossings of parabolic bands there can be indexed in terms of an ascending order of $|\mathbf{G}|^2$, as shown in Figure 9.4b. It is seen that degenerate states indexed as $|\mathbf{G}|^2 = 50$ fall closest to the Fermi level in the free electron model. A comparison between (a) and (b) of Figure 9.4 indicates that degenerate states at the center of $\{710\}+\{543\}+\{550\}$ zones with $|\mathbf{G}|^2 = 50$ are only weakly lifted as a result of the interaction with the set of these lattice planes and that resulting electronic states remain in the vicinity of the Fermi level. This means that the two-wave approximation involving zones only with $|\mathbf{G}|^2 = 50$ is definitely too crude to explain the formation of a pseudogap in the DOS in the Al-Mg-Zn approximant. Another important remark should be added here. According to Figure 9.4b, we find that the frequency of crossings of parabolic bands in the free electron model exceeds 200 along the direction ΓN over the energy range $-1 \leq E \leq 1$ eV across the Fermi level. All these degenerate states will be lifted and, accordingly, energy dispersions will be flattened, as shown in Figure 9.4a. We will show below that this is one of the prominent features of the CMA having 160 atoms per unit cell.

As discussed in Chapter 4, first-principles band calculations essentially aim at solving a secular equation like Equation 4.62. As mentioned at the beginning of Section 9.2.1, NFE band calculations would be adequate to derive the electronic structure of the Al-Mg-Zn approximant. Now let us consider the n-wave approximation in the framework of NFE band calculations. Since a secular determinantal equation is composed of an n-by-n determinant, one would end up solving n-th order polynomial equation in an energy E:

$$a_n E^n + a_{n-1} E^{n-1} + \cdots + a_0 = 0 \qquad (9.1)$$

where the coefficient a_n is composed of unperturbed free electron energies and non-zero form factors.* In the two-wave approximation, Equation 9.1

---

* See Footnote on pg. 84, in Chapter 4, for the definition of the form factor.

is reduced to a quadratic equation and results in the well-known energy dispersion relation [24]:

$$E(\mathbf{k}) = \left(\frac{1}{2}\right)\left[-\left(\frac{a_1}{a_2}\right) \pm \sqrt{\left(\frac{a_1}{a_2}\right)^2 - 4\left(\frac{a_0}{a_2}\right)}\right]$$

$$= \left(\frac{1}{2}\right)\left[(E_0 + E_n) \pm \sqrt{(E_0 - E_n)^2 + 4V_n^* V_n}\right]$$

(9.2)

where V_n is the form factor or the Fourier component of the ionic potential associated with the reciprocal lattice vector $\mathbf{G} = (2\pi/a)\mathbf{n}$ and both $E_0 = \mathbf{k}^2$ and $E_n = (\mathbf{k} - \mathbf{G}_n)^2$ are unperturbed free electron energies centered at $\mathbf{G} = 0$ and $\mathbf{G} = \mathbf{G}_n$.

Suppose that NFE band calculations are carried out with more than 1000-wave approximation. Degenerate free electron parabolic bands indexed with $|\mathbf{G}|^2 = 50$ at the symmetry points N near the Fermi level in Figure 9.4b, for example, will be lifted in a complicated manner as a result of perturbations due to many nonzero form factors involved in Equation 9.1. However, as long as the form factors are small enough to validate the NFE approximation, the effect of each form factor on the dispersion relations is limited only in the vicinity of the crossing region of more than two parabolic bands. In the case of the two-wave approximation given by Equation 9.2, we know that the perturbation due to the form factor V_n develops only in the vicinity of $E_0 = E_n$ and otherwise the free electron parabolic band is preserved (see Figure 4.3(a)).

Figures 9.5a,b compare the wave number dependences of the form factor between Cu_5Zn_8 gamma-brass and $Al_{48}Mg_{64}Zn_{48}$ 1/1-1/1-1/1 approximant. It is clear that only the $|\mathbf{G}|^2 = 18$ form factor is extremely large and located very near the Fermi level in Cu_5Zn_8. Its magnitude reaches 0.8 eV while the rest is less than 0.1 eV in the range $0.9 \leq |\mathbf{G}|/2k_F \leq 1.0$. This is apparently responsible for the opening of a large gap reaching about 1 eV across the Fermi level at the symmetry points N in its dispersion relations (see Chapter 7, Figure 7.2a) and allowed us to interpret the DOS pseudogap formed across the Fermi level solely in terms of the gap opening at the symmetry points N. Instead, the form factor closest to the Fermi level in $Al_{48}Mg_{64}Zn_{48}$ 1/1-1/1-1/1 approximant is that at $|\mathbf{G}|^2 = 50$ associated with {543}, {710}, and {550} zone planes but its magnitude is only 0.2 eV. In

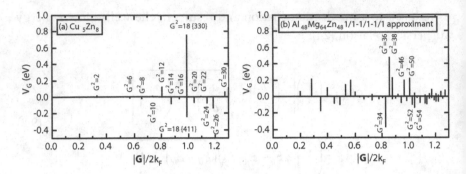

FIGURE 9.5 Form factors V_G in the NFE model as a function of the wave number normalized with respect to the Fermi diameter $2k_F$ for (a) Cu_5Zn_8 gamma-brass and (b) $Al_{48}Mg_{64}Zn_{48}$ 1/1-1/1-1/1 approximant. Note that form factors take finite values only at wave numbers corresponding to the reciprocal vector $|G|$ allowed for each system. Sizable form factors appearing in the vicinity of the Fermi level are indexed in terms of $|G|^2$. The values of $(2k_F)^2$ for Cu_5Zn_8 gamma-brass and $Al_{48}Mg_{64}Zn_{48}$ 1/1-1/1-1/1 approximant are taken to be 18.47 and 49.52, respectively, in units of $(2\pi/a)^2$, where a is the lattice constant.

addition, two form factors at $|G|^2$ = 46 and 50, being comparable in magnitude, exist over the narrow range $0.9 \le |G|/2k_F \le 1.0$.

To extract the role of form factors at $|G|^2$ = 50 and 46, which may be hereafter referred to as $V(50)$ and $V(46)$, in the formation of a pseudogap, we performed 1505-wave NFE electronic structure calculations for the $Al_{48}Mg_{64}Zn_{48}$ 1/1-1/1-1/1 approximant model structure (see [9]) in such a way that form factors other than critical ones are set to zero. Figures 9.6a,b show energy dispersion relations derived when only either $V(46)$ or $V(50)$ is nonzero, respectively. It is clear that each form factor lifts degenerate states only in the energy region above the respective $|G|/2k_F$ values shown in Figure 9.5b but preserves free electron bands below them. As can be seen from dispersion relations inside shaded rectangles in Figures 9.6a,b, the form factor $V(50)$ causes electronic states to be less populated over energies immediately above the Fermi level, whereas $V(46)$ over those centered at −0.5 eV below the Fermi level. Sparse and flat dispersion relations become more evident in Figure 9.6c, where both form factors $V(50)$ and $V(46)$ are simultaneously activated, along with low energy-lying form factors $V(2)$, $V(4)$, and $V(6)$.

Figures 9.7a–e show the effect of different combinations of form factors on the DOS near the Fermi level in 1505-wave NFE band calculations.

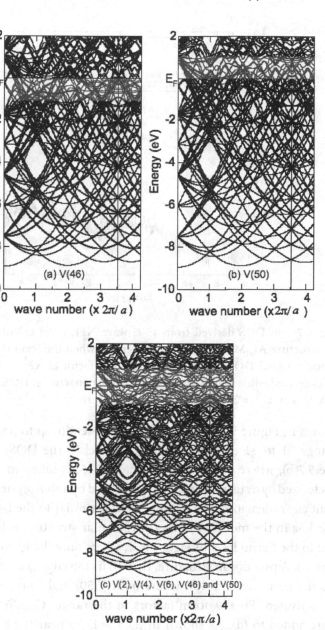

FIGURE 9.6 Energy dispersion relations along the direction <711> for the model structure $Al_{48}Mg_{64}Zn_{48}$ obtained when (a) only the form factor at $|\mathbf{G}|^2 = 46$ is nonzero and (b) only the form factor at $|\mathbf{G}|^2 = 50$ is nonzero in 1505-wave NFE band calculations. In (c), form factors at $|\mathbf{G}|^2 = 2, 4, 6, 46,$ and 50 are non-zero. An energy range, where sparse electronic states are produced by the respective form factors, is highlighted by a shaded rectangle.

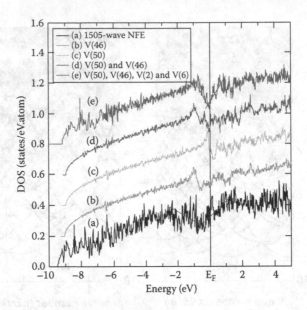

FIGURE 9.7 (a) DOS derived from 1505-wave NFE band calculations for the model structure $Al_{48}Mg_{64}Zn_{48}$. (b) DOS derived when the form factor at $|G|^2 = 46$ is non-zero. (c) DOS derived when the form factor at $|G|^2 = 50$ is nonzero. (d) DOS derived when both $|G|^2 = 46$ and 50 are nonzero. (e) DOS derived when form factors at $|G|^2 = 2, 4, 6, 46,$ and 50 are nonzero.

As shown in Figure 9.7a, we can confirm a pseudogap to be formed over the range −1 to +1 eV across the Fermi level in the DOS. The DOS in Figure 9.7(b), where only the form factor $V(46)$ is taken into account, is characterized by a cusp at about −1 eV, followed by a sharply declining slope without clear formation of a pseudogap. It is similar to the DOS shown in Figure 3.3a in the model of Jones (I). A similar structure is formed but is shifted to the Fermi level in Figure 9.7c, where only the form factor $V(50)$ is nonzero. A pseudogap structure, though its top edge is blurred, emerges across the Fermi level in (d), where both $V(50)$ and $V(46)$ are concomitantly activated. Finally, form factors in the range $|G|^2 \leq 6$ or $|G|/2k_F < 0.35$ are added to (d). As shown in (e), the DOS near the bottom of the valence band is heavily perturbed. More important is that the structure of the pseudogap becomes sharp and clear across the Fermi level.

From the analysis above, we can say that both form factors $V(50)$ and $V(46)$, or the sets of the Brillouin zone planes associated with $|G|^2 = 50$ and 46, are essential in the formation of a DOS pseudogap across the Fermi level in the Al-Mg-Zn 1/1-1/1-1/1 approximant, and its structure is matured through assistance of low energy-lying form factors in the range $|\mathbf{G}|^2 \leq 6$. This can be viewed as the formation of a more or less spherical Brillouin zone net in the reciprocal space, thereby causing the Bragg reflections to occur in almost all directions. In other words, the FsBz interactions involving such multi-zones must be responsible for the formation of a pseudogap in the Al-Mg-Zn 1/1-1/1-1/1 approximant containing as many as 160 atoms per unit cell.

9.3 Al-Li-Cu 1/1-1/1-1/1 APPROXIMANT

9.3.1 Construction of the Model Structure

The atomic structure of the Al-Li-Cu 1/1-1/1-1/1 approximant was independently studied by the two groups. Both neutron and x-ray diffraction measurements were carried out on powder sample with composition $Al_{94.6}Cu_{17}Li_{48.4}$ by Guryan et al. [25] and on a single-crystal with composition $Al_{88.62}Cu_{19.377}Li_{50.335}$ by Audier et al. [26]. They could successfully refine the structure with space group $Im\bar{3}$ with the absence of atoms at the center of the cluster (sites A: 2a). It contains totally 160 atoms in the unit cell with the lattice constant of 1.3891 nm [25] or 1.39056 nm [26]. Li atoms are filled into sites (D: 16f), (E: 24g), and (H: 12e) without measurable chemical disorder. Chemical disorder is the most substantial in sites (C: 24g), where Al and Cu atoms are almost evenly distributed. Sites B and F are shared by Al and Cu atoms with a proportion of 89:11 in its RT-type cluster (see Chapter 6, Figure 6.4). Such chemical disorder must be eliminated in band calculations.

In LMTO-ASA band calculations for the Al-Li-Cu 1/1-1/1-1/1 approximant by Fujiwara and Yokokawa [10], the model structure was constructed with space group $Pm3$ by filling Al and Cu atoms into sites C in the cluster "a" and "b" (see Chapter 6, Figure 6.4), respectively. They simply noted that Al and Li atoms are filled into remaining sites without further detailed description. Their resulting model structure of $Al_{96}Li_{52}Cu_{12}$ per unit cell or $Al_{60}Li_{32.5}Cu_{7.5}$ in % contains 160 atoms in the unit cell. Unfortunately, no reference to the atomic structure, from which they deduced their model structure, was provided in [10]. Windisch et al. [11] also constructed the model structure for the Al-Li-Cu 1/1-1/1-1/1 approximant by using a slightly modified Henley–Elser model [22] discussed in Section 9.2.1. It contains 160 atoms in the unit cell with space group $Im\bar{3}$. However, its chemical formula was not explicitly mentioned in [11]. Sato et al. [12] modified the atomic structure refined by Guryan et al. [25] to eliminate chemical disorder from the unit cell: Al atoms are exclusively filled into sites (B: 24g) on the first ico-sahedron, Li atoms into sites (D: 16f) and (E: 24g) on the dodecahedron, Cu atoms into sites (C: 24g) on outer icosahedron and Al atoms into remaining outer shells. A fractional occupancy at the center of the RT-cluster (sites A: 2a) is set to zero in agreement with the refined structure [25,26]. This leads to the model structure $Al_{84}Li_{52}Cu_{24}$ per unit cell or $Al_{52.5}Li_{32.5}Cu_{15}$ in % containing 160 atoms in the unit cell with space group $Im\overline{3}$.

9.3.2 Electronic Structure Calculations

Fujiwara and Yokokawa [10] revealed a deep pseudogap across the Fermi level for their model structure $Al_{96}Li_{52}Cu_{12}$ and discussed the mechanism leading to its formation in terms of the matching condition (4.1) by esti-mating the Fermi diameter simply by inserting the LMTO-ASA derived valence band width, i.e., E_F into the free electron relation $E_F = k_F^2$ in atomic units. A more important message from their work is that they judged from a narrow width of the Cu-3d peak only a small mixing between the Cu-3d and other orbitals. To confirm this, they calculated the DOS for two other model structures: one $Al_{84}Li_{52}Cu_{24}$ obtained by filling only Cu atoms into sites C in both clusters "a" and "b" and the other $Al_{108}Li_{52}$ by filling only Al atoms into them. They are led to conclude that Cu atoms play no role in the formation of the pseudogap, since the pseudogap structure remains unchanged among their three model struc-tures, including $Al_{108}Li_{52}$, where Cu is absent. As will be discussed later, their finding plays an important role to elucidate the mechanism for the formation of a pseudogap in this system.

Windisch et al. [11] reported the electronic structure of not only 1/1-1/1-1/1 but also higher-order Al-Li-Cu approximants up to 8/5-8/5-8/5. For the 1/1-1/1-1/1 approximant with a 160-atom cubic cell, the electronic structure was calculated self-consistently, using the standard LMTO technique in the atomic-sphere approximation (ASA). Their results are essentially consistent with that reported by Fujiwara and Yokokawa [10].

More detailed studies concerning the electronic structure and stabili-zation mechanism were reported by Sato et al., using the model structure $Al_{84}Li_{52}Cu_{24}$ discussed above [12]. The LMTO-ASA method was employed to study the orbital hybridization effect on the formation of a pseudogap, whereas NFE band calculations to study its origin from the viewpoint of the FsBz interactions. The total DOS derived from the LMTO-ASA is shown in Figure 9.8 and is compared with those obtained after intentionally zeroing (a), the Cu-3d/spd hybridization terms and (b) both Cu-3d/spd and Al-3p/

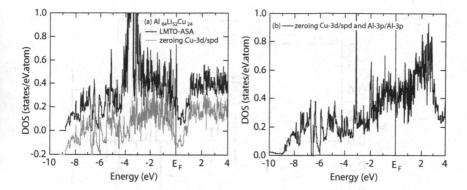

FIGURE 9.8 (a) Total DOS calculated using LMTO-ASA band calculations for the model structure $Al_{84}Li_{52}Cu_{24}$ (black line) and that obtained after deleting the Cu-3d/spd orbital hybridization terms (gray line). The latter is intentionally displaced downwards by 0.2 states/eV.atom to separate two DOSs from one another. (b) Total DOS obtained after deleting both the Cu-3d/all-spd and Al-3p/Al-3p orbital hybridization terms. [From H. Sato, T. Takeuchi, and U. Mizutani, *Phys. Rev. B* 70 (2004) 024210.]

Al-3p orbital hybridization terms. It is clear that the pseudogap remains unchanged in (a), but disappears when Al-3p/Al-3p terms are deleted in (b). This implies that the Cu-3d band has little to do with a pseudogap formation, in good agreement with Fujiwara and Yokokawa [10]. Instead, the role of Al-3p/Al-3p orbital hybridizations has been emphasized.

9.3.3 Stabilization Mechanism

The stabilization mechanism developed in Reference 12 will be described below from the point of view of both the FsBz interactions and orbital hybridizations. At first, the e/a value of their model structure $Al_{84}Li_{52}Cu_{24}$ is easily calculated to be 2.05, since valencies of Al, Li, and Cu are three, unity, and unity, respectively. An insertion of e/a = 2.05 and N = 160 into the matching Equation 4.1 immediately indicates that the zone {631} is the best candidate for the FsBz interaction responsible for the formation of a pseudogap in $Al_{84}Li_{52}Cu_{24}$. The form factor for the $Al_{84}Li_{52}Cu_{24}$ model structure is shown in Figure 9.9 as a function of the wave number normalized with respect to its Fermi diameter. It is clear that many form factors comparable in magnitude are distributed across the Fermi level, indicating the presence of the multi-zone effects most likely involving form factors at both $|G|^2$ = 46 and 50 in a similar manner to the situation in the Al-Mg-Zn 1/1-1/1-1/1 approximant discussed in Section 9.2.3.

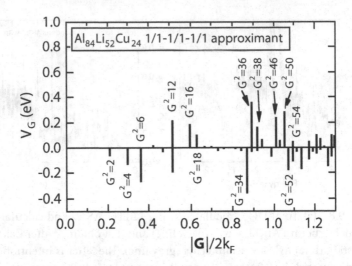

FIGURE 9.9 Form factor V_G in NFE band calculations as a function of the wave number normalized with respect to the Fermi diameter $2k_F$ for the model structure $Al_{84}Li_{52}Cu_{24}$ [from H. Sato, T. Takeuchi, and U. Mizutani, *Phys. Rev.* B 70 (2004) 024210]. Sizable form factors appearing in the vicinity of the Fermi level are indexed in terms of $|G|^2$. The value of $(2k_F)^2$ for $Al_{84}Li_{52}Cu_{24}$ is taken to be 45.20 in units of $(2\pi/a)^2$, where a is the lattice constant.

Sato et al. [12] carried out NFE band calculations for their model structure $Al_{84}Li_{52}Cu_{24}$. As emphasized in Section 9.3.2, the NFE model would be reasonable, since the Cu-3d band plays no essential role on the formation of a pseudogap and form factors are fairly small. After checking that the 1061-wave approximation is accurate enough to reproduce main features derived from LMTO-ASA band calculations, they tried to extract zone planes responsible for the formation of a pseudogap at the Fermi level. The DOS thus obtained is plotted in Figure 9.10a. A pseudogap is found at the Fermi level. The DOS derived after zeroing several important form factors is depicted in Figures 9.10b–d. An elimination of the form factor $V_{\{631\}}$ with $|G|^2 = 46$ significantly reduces the depth of the pseudogap but still leaves it to be noticeable (Figure 9.10b). A similar reduction in the depth of the pseudogap is also observed, when form factors $V_{\{543\}}$, $V_{\{710\}}$, and $V_{\{550\}}$, all of which are associated with $|G|^2 = 50$, are eliminated (Figure 9.10c). However, when both sets of form factors with $|G|^2 = 46$ and 50 are simultaneously deleted, we can almost perfectly erase the pseudogap, as shown in Figure 9.10d. This means that the Brillouin zones associated with both $|G|^2 = 46$ and 50 equally contribute to the formation of a pseudogap.

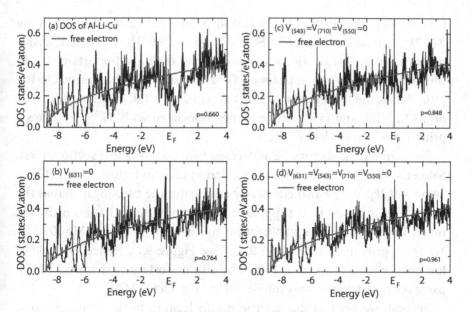

FIGURE 9.10 Effect of the deletion of form factors on the DOS calculated using 1061-wave NFE band calculations for the model structure $Al_{84}Li_{52}Cu_{24}$ [from H. Sato, T. Takeuchi, and U. Mizutani, *Phys. Rev. B* 70 (2004) 024210]. (a) A pseudogap is significant with $p = 0.660$, (b) the pseudogap remains sizable with $p = 0.764$ after zeroing the form factor V_{631}, (c) the pseudogap remains with $p = 0.848$ after zeroing form factors V_{543}, V_{710}, and V_{550}, and (d) the pseudogap essentially disappears with $p = 0.961$ after zeroing form factors V_{631}, V_{543}, V_{710}, and V_{550}. See the text for the definition of the parameter p.

To evaluate a diminishing pseudogap by the elimination of particular form factors in a more quantitative way, they introduced a parameter p, which is defined as a ratio of the number of states involved in the energy range $E_F < E < +1$ eV over that in the free electron states. As incorporated into Figure 9.10, the parameter p becomes close to unity only when two form factors at $|G|^2 = 46$ and 50 are simultaneously removed. In addition to zones passing through the symmetry points N studied above, they also studied the effect of deletion of zones, which pass the symmetry points Γ and H upon reduction to the first zone. Their contribution to the formation of the pseudogap is found to be of minor importance.

As is clear from the arguments above, the mechanism to form a pseudogap in the Al-Li-Cu 1/1-1/1-1/1 approximant is essentially the same as that in the Al-Mg-Zn 1/1-1/1-1/1 approximant discussed in Section 9.2.3. But the electron concentration e/a = 2.05 in the $Al_{84}Li_{52}Cu_{24}$ approximant is lower than e/a = 2.3 in the $Al_{48}Mg_{64}Zn_{48}$ approximant. This difference is reflected in the position of the two form factors $V(46)$ and $V(50)$ on the $|\mathbf{G}|/2k_F$ axis, as can be seen in Figures 9.5b and 9.9. The set of {631} lattice planes associated with $|\mathbf{G}|^2 = 46$ would serve the most important role in the former, since $V(46)$ appears at $|\mathbf{G}|/2k_F$ nearly equal to unity. On the other hand, the form factor $V(50)$ is at $|\mathbf{G}|/2k_F = 1.0$, and, hence, the set of {543}, {710}, and {550} lattice planes associated with $|\mathbf{G}|^2 = 50$ plays a principal role in the latter.

Two more remarks may be addressed before ending this section. First, Sato et al. [12] pointed out that the deletion of form factors having $|\mathbf{G}|^2 \leq 16$ or $|\mathbf{G}|/2k_F < 0.59$ also effectively eliminates the pseudogap across the Fermi level in the Al-Li-Cu approximant. Its importance has been already noted in discussing the stabilization mechanism of the Al-Mg-Zn 1/1-1/1-1/1 approximant in Section 9.2.3 (see Figure 9.7). Hence, the role of low energy-lying form factors must be also important to shape up a DOS pseudogap structure across the Fermi level in such CMAs.

Second, we need to discuss the origin leading to the formation of a pseudogap in the Al-Li-Cu approximant. We have claimed in Section 9.3.2 that the pseudogap is formed through the Al-3p/Al-3p orbital hybridization, while in Section 9.3.3 it is formed through the FsBz interactions involving $|\mathbf{G}|^2 = 46$ and 50. Is this self-consistent with each other? Sato et al. [12] calculated a charge distribution of electrons within the range of thermal energies (≈ 30 meV) below the Fermi level on the (200) plane and could reveal the highest charge density extending along the line connecting neighboring Al atoms. This is obviously attributed to the existence of the Al-3p/Al-3p bonding states. They further proved that this unique charge distribution is destroyed, when form factors at $|\mathbf{G}|^2 = 46$ and 50 are deleted in NFE band calculations. From this they concluded that these two mechanisms are not independent of one another but essentially describe the same phenomenon from the two different points of view: the Al-3p/Al-3p orbital hybridizations in the real space and the FsBz interactions in the reciprocal space.

9.4 Al-Cu-TM-Si (TM = Fe OR Ru) 1/1-1/1-1/1 APPROXIMANTS

9.4.1 Construction of the Model Structure

The atomic structure of $Al_{68}Cu_7(Fe_{1-x}Ru_x)_{17}Si_8$ (x = 0, 0.5 and 1) 1/1-1/1-1/1 approximants was experimentally determined by analyzing the powder diffraction spectra taken with the beam line of the synchrotron radiation, SPring-8, Japan, by means of the Rietveld method [27,28]. Space group is reduced to $Im\bar{3}$ as a result of random distributions of different chemical species in most of equivalent sites in the unit cell. The exception is that the transition metal element Fe or Ru is exclusively filled into 12 sites (TM: 24g) on the larger icosahedron without any chemical disorder for both $Al_{68}Cu_7Fe_{17}Si_8$ and $Al_{68}Cu_7Ru_{17}Si_8$ corresponding to x = 0 and 1, respectively, in the chemical formula above. As listed in Table 9.2, chemical disorder always exists in the distribution of Al and Cu atoms on the inner icosahedron (sites II: 24g) and also on the icosidodecahedron (sites MI1: 12d and sites MI2: 48h). The first shell is formed by 12 atoms on the inner icosahedron while the second shell by 30 atoms on the icosidodecahedron and 12 atoms on the larger icosahedron TM (see Chapter 6, Figure 6.5). In glue sites G1, G2, and G3 connecting neighboring MI-clusters, chemical disorder is the most significant. In particular, glue sites G1 are heavily disordered by a mixture of Cu, Si, and vacancies, while glue sites G2 and G3 are occupied by Si and Al with slightly different coordinates, thereby being often referred to as *split sites*. The lattice constant of $Al_{68}Cu_7Fe_{17}Si_8$ and $Al_{68}Cu_7Ru_{17}Si_8$ approximants turned out to be 1.248 and 1.2496 nm, respectively. The experimentally derived total number of atoms per unit cell is deduced to be 139, as listed in Table 9.2.

TABLE 9.2 Refined Atomic Structure for $Al_{68}Cu_7Fe_{17}Si_8$ 1/1-1/1-1/1 Approximant and Two Model Structures

Refined Structure with Space Group $Im\bar{3}$			Model Structures with Space Group $Pm\bar{3}$	
Site	Atoms	Occ.	Model 1	Model 2
II/24g	Al/Cu	0.887/0.113	24Al	24Al
MI1/12d	Al/Cu	0.908/0.092	12Al	12Al
MI2/48h	Al/Cu	0.964/0.036	48Al	48Al
TM/24g	Fe	1.0	24Fe	24Fe
G1/12e	Cu/Si/ vacancy	0.04/0.6/0.36	G1 in "a" 6 Cu	6Cu
			G1 in "b" 6 Si	Vacant
G2/24g	Si	0.378	12Al	12Si
G3/24g	Al	0.622	12Al	12Al
Chemical formula		$Al_{68}Cu_7Fe_{17}Si_8$(%)	$Al_{108}Cu_6Fe_{24}Si_6$ ($Al_{75}Cu_{4.2}Fe_{16.6}Si_{4.2}$ in %)	$Al_{96}Cu_6Fe_{24}Si_{12}$ ($Al_{69.6}Cu_{4.3}Fe_{16.6}Si_{8.7}$ in %)
Atoms/ unit cell		139	144	138
e/a		2.566	2.591	2.617

The model structure listed as "model 1" in Table 9.2 is constructed after a slight modification of the experimentally determined atomic structure discussed above [27]. The presence of Cu atoms with occupancies less than about 10 % on sites II, MI1, and MI2 is fully ignored. These sites together with glue sites G3 are filled only with Al atoms. To compensate for the deficient Cu, the model 1 assumes Cu atoms to fill into glue sites G1. Space group $Pm3$ is intentionally employed to put six Cu atoms on glue sites G1 in the MI-type cluster "a" and six Si atoms on G1 in the other cluster "b" (see Chapter 6, Figure 6.7). We obtain the chemical formula $Al_{108}Cu_6TM_{24}Si_6$ (TM = Fe and Ru) per unit cell or $Al_{75}Cu_{4.2}TM_{16.6}Si_{4.2}$ in %. The total number of atoms per unit cell is increased from the measured value of 139 to 144, since vacancies in the glue sites G1 are fully filled with atoms.

The structure "model 2" in Table 9.2 is also constructed to check how sensitively the electronic structure depends on the choice of the model. In the model 2, atom distribution on glue sites is fixed as follows: six sites G1 in the cluster "a" are filled with Cu atoms and those in the cluster "b" are left vacant, while 12 Si atoms are placed into sites G2. This alteration results in the chemical formula $Al_{96}Cu_6TM_{24}Si_{12}$ (TM = Fe and Ru) per unit cell or $Al_{69.6}Cu_{4.3}TM_{16.6}Si_{8.7}$ in %. Note that the number of atoms in the unit cell is 138, one atom lower than that in the refined structure.

9.4.2 Electronic Structure Calculations and Stabilization Mechanism
9.4.2.1 LMTO-ASA Band Calculations
The electronic structure is calculated for the $Al_{108}Cu_6Ru_{24}Si_6$ approximant (model 1) by means of the LMTO-ASA method [15]. The resulting DOS is shown in Figure 9.11 (a). It is characterized by a large d-band originating from Cu-3d and Ru-4d states over energies centered at about −4 eV and a deep pseudogap across the Fermi level. The sp-d hybridization terms, representing a mixture of Cu-3d and Ru-4d states with all other states, are deleted from LMTO-ASA wave function to study the effect of the sp-d orbital hybridizations on DOS and dispersion relations. The DOS thus obtained is incorporated into Figure 9.11b using the same energy scale as in (a), to allow a direct comparison with the original DOS. Obviously, the Cu-3d states centered at about −3.5 eV and the Ru-4d states centered at −1 eV are now well separated. Its sp-partial DOS is then calculated and shown in Figure 9.11c, where a free electron parabolic DOS is drawn as a guide. It is clear that a pseudogap remains finite at the Fermi level in the sp-d hybridization-free sp-partial DOS. This must be attributed to the FsBz interactions [15].

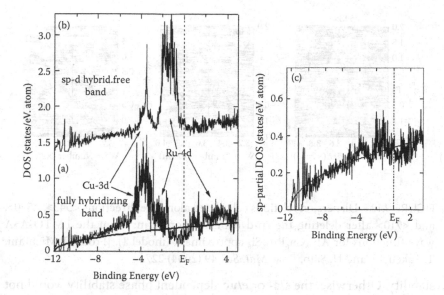

FIGURE 9.11 (a) Total DOS and (b) that obtained after deleting the sp-d hybridization terms in the LMTO-ASA wavefunction and (c) sp-d hybridization-free sp-partial DOS for $Al_{108}Cu_6Ru_{24}Si_6$ approximant (model 1). [From U. Mizutani, T. Takeuchi, and H. Sato, *Prog. Mat. Sci.* 49 (2004) 227.]

The energy dispersion relations are calculated along directions <543>, <550> and <710> after deleting all sp-d hybridization terms (Figure 9.11 (b)). The results are shown in Figure 9.12 along with free electron parabolic bands [15]. It can be seen that many free electron parabolic bands merge at

$$k = \frac{\sqrt{50}}{2} = 3.535$$

in units of $2\pi/a$ immediately below the Fermi level along the three directions and that electronic states involved are immediately identified as $|G|^2$ = 50 corresponding to the center of {543}, {550}, and {710} zone planes. These degenerate states are lifted in sp-d hybridization-free bands. It is claimed from the analysis above that the set of {543}, {550}, and {710} lattice planes must be responsible for the formation of a FsBz-induced pseudogap at the Fermi level in the sp-d hybridization-free sp-partial DOS shown in Figure 9.11c. We consider the analysis above important, since *a FsBz-induced pseudogap is present at the Fermi level but is simply hidden behind a much larger orbital hybridization-induced pseudogap.* A FsBz-induced pseudogap, though it is small in size, must play a key role in phase

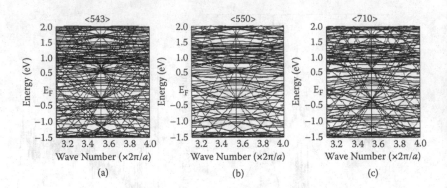

FIGURE 9.12 Dispersion relations calculated along the directions <543>, <550>, and <710> after deleting the sp-d hybridization terms from the LMTO-ASA wave functions for $Al_{108}Cu_6Ru_{24}Si_6$ approximant (model 1). [From U. Mizutani, T. Takeuchi, and H. Sato, *Prog. Mat. Sci.* 49 (2004) 227.]

stability. Otherwise, the **e/a**- or **e/uc**-dependent phase stability would not be observed (see Chapter 10, Section 10.7).

A pseudogap at the Fermi level in the strongly hybridizing band shown in Figure 9.11a is much deeper than the FsBz-induced one shown in Figure 9.11c. The latter is certainly amplified as a result of the sp-d hybridization effect, which contributes to splitting of the Ru-4d states centered at about −1 eV (Figure 9.11b) into bonding states over −2 to −4 eV and antibonding states above the Fermi level (Figure 9.11a). The reason why MI-type quasicrystals and their approximants in the Al-Cu-TM alloy system need Fe, Ru, or Os as the TM element from the same column in the periodic table, is clear. The unhybridized TM-d states need to be present at about −1 eV, as is the case for the Ru-4d states shown in Figure 9.11b. Its location at this energy is important because the sp-d hybridization creates bonding states at energies lower than −2 eV, while antibonding states above the Fermi level. Such electronic configurations can effectively contribute to further lowering an electronic energy of the system by deepening a FsBz-induced pseudogap at the Fermi level through orbital hybridizations. Instead, the role of Cu is less straightforward but certainly serves as adjusting the electron concentration so as to bring the Fermi level near the minimum in the pseudogap.*

---

* It seems to be more difficult to explain the role of Si. However, we are aware that the approximant can be stabilized by adding Si to the Al-Cu-TM (TM = Fe, Ru, and Os) quasicrystalline phase. It is interesting to note that Si is exclusively filled into the outermost glue sites, which would likely play a key role in restoring the lattice periodicity instead of continuing the icosahedral symmetry (see Table 9.2).

As is clear from the argument above, we could confirm the validity of the matching condition given by Equation 4.1 even for the Al-Cu-Ru-Si approximant characterized by an orbital hybridization-induced pseudogap. Since the set of {543}, {710}, and {550} lattice planes is deduced to be responsible for the formation of a FsBz-induced pseudogap, the value of $(e/a)_{total}$ can be easily estimated to be 2.57 from Equation 4.1 [15]. The valency of Ru is then calculated to be 0.76, provided that valencies of Cu, Al, and Si are unity, three, and four, respectively. As will be discussed in Section 9.4.2.2, these values well coincide with those deduced from a theoretically more rigorous Hume-Rothery plot based on the FLAPW-Fourier method, lending support to the above analysis based on LMTO-ASA band calculations.

9.4.2.2 FLAPW Band Calculations

The LMTO-ASA band calculations discussed in the preceding section are performed only for the model structure 1 with TM = Ru. In the present section, the FLAPW band calculations are employed to further elucidate the origin of the formation of a pseudogap in Al-Cu-TM-Si (TM = Fe and Ru) 1/1-1/1-1/1 approximants. First, we study if the choice of either Fe or Ru as the TM element or that of either the model 1 or 2 would cause any significant difference in the electronic structure.

Figure 9.13 shows energy dispersion relations for the model structure 2 with (a) TM = Fe and (b) TM = Ru. Both sets of data are quite similar to each other: there exist extremely bunched dispersion-less electronic states below about −1 eV and above about +1 eV, leaving sparsely populated electronic states across the Fermi level. This certainly gives rise to a pseudogap in their respective DOSs. Similar results are obtained for the model structure 1 with TM = Fe and Ru. From this, we can judge a difference in the electronic structure caused by the choice of the model structure 1 or 2 to be of minor importance in the rest of discussions.

The total DOS and s-, p-, and d-partial DOSs are calculated, using the FLAPW method for both $Al_{108}Cu_6Fe_{24}Si_6$ and $Al_{108}Cu_6Ru_{24}Si_6$ 1/1-1/1-1/1 approximants (model 1) [16]. As shown in Figure 9.14, the total DOS is quite consistent with LMTO-ASA derived DOS shown in Figure 9.10a. We can clearly see that both the Fe-3d and Ru-4d bands are split into bonding and antibonding subbands, resulting in a deep pseudogap across the Fermi level. As discussed in Section 9.4.2.1, this is largely caused by the Al-3p/TM-d orbital hybridizations and certainly contributes to the stabilization of these approximants.

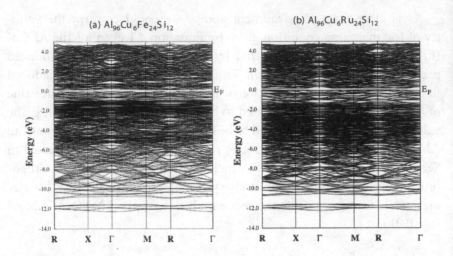

FIGURE 9.13 Energy dispersion relations calculated using the FLAPW method for (a) $Al_{96}Cu_6Fe_{24}Si_{12}$ and (b) $Al_{96}Cu_6Ru_{24}Si_{12}$ 1/1-1/1-1/1 approximants (model 2).

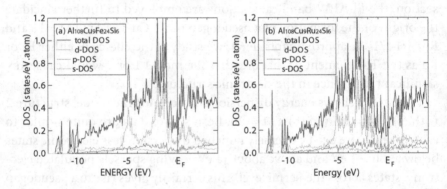

FIGURE 9.14 Total, s-, p-, and d-partial DOSs calculated using the FLAPW method for (a) $Al_{108}Cu_6Fe_{24}Si_6$ and (b) $Al_{108}Cu_6Ru_{24}Si_6$ 1/1-1/1-1/1 approximants (144 atoms per unit cell, model 1). [From U. Mizutani, R. Asahi, T. Takeuchi, H. Sato, O.Y. Kontsevoi, and A.J. Freeman, *Z. Kristallogr.* 224 (2009) 17.]

The *Hume–Rothery plot* is performed for these two approximants, using the model structure 1. Figures 9.15a,b show the energy dependence of $\left|2(\mathbf{k}+\mathbf{G})\right|^2$ and its variance for $Al_{108}Cu_6Fe_{24}Si_6$ and $Al_{108}Cu_6Ru_{24}Si_6$ approximants, respectively [16]. One can draw a straight line passing the origin as well as regions, where the variance is small, and deduce the intercept with the Fermi level to be equal to 50 for both cases. Since this represents the square of the Fermi diameter, the value of $(e/a)_{total}$ is immediately calculated to be 2.59 and 2.56 for TM = Fe and Ru, respectively. The values

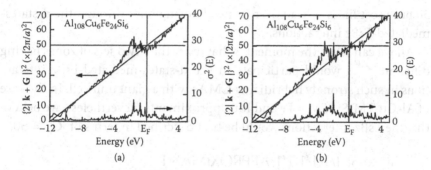

FIGURE 9.15 Energy dependence of $\left|2(\mathbf{k}+\mathbf{G})\right|^2$ and its variance calculated from the Hume-Rothery plot for (a) $Al_{108}Cu_6Fe_{24}Si_6$ and (b) $Al_{108}Cu_6Ru_{24}Si_6$ 1/1-1/1-1/1 approximants (model 1). [From U. Mizutani, R. Asahi, T. Takeuchi, H. Sato, O.Y. Kontsevoi, and A.J. Freeman, *Z. Kristallogr.* 224 (2009) 17.]

of e/a for Fe and Ru turn out to be equally 0.8, provided that valencies of Cu, Al, and Si are assigned to unity, three, and four, respectively. This is quite consistent with the results derived from LMTO-ASA band calculations discussed in Section 9.4.2.1.

The LMTO-ASA analysis in the preceding section claimed that the interaction of electrons with the set of {543}, {710}, and {550} lattice planes with $\left|\mathbf{G}\right|^2$ = 50 gives rise to a FsBz-induced pseudogap across the Fermi level in $Al_{68}Cu_7Ru_{17}Si_8$ approximant with space group $Im\bar{3}$. The value of $(2k_F)^2 = 50$ derived from the Hume-Rothery plot leads to the fulfillment of the matching condition given by Equation 4.1 and ascertains the existence of the Hume-Rothery stabilization mechanism even for the orbital hybridization-induced pseudogap system.

Energy dispersion relations shown in Figure 9.13 for both $Al_{96}Cu_6Fe_{24}Si_{12}$ and $Al_{96}Cu_6Ru_{24}Si_{12}$ model structures are almost dispersion-less and highly bunched below about –1 eV, where the bottom of the pseudogap is located. The extraction of the FsBz interactions in strongly hybridizing Al-Cu-TM-Si (TM = Fe and Ru) 1/1-1/1-1/1 approximants would be tough, since both multi-zone effect discussed in the RT-type approximants in Sections 9.2 and 9.3 and the d-states-mediated-splitting discussed in Chapter 8 are expected to occur simultaneously in a complicated manner. Because of this complexity in the electronic structure, the FLAPW-Fourier analysis has not yet been attempted, though we consider it to be of urgent necessity to extract specific FsBz interactions. Sizable Fourier components of the FLAPW wave function outside the MT sphere at the symmetry points M near the bottom of the pseudogap would not be localized only at $\left|\mathbf{G}\right|^2 =$

50 but most likely spread over its neighboring ones through the d-states-mediated-FsBz interactions.

All we can say at the moment is that more than two sets of zones having different $|\mathbf{G}|^2$ s would participate in the d-states-mediated-FsBz interactions in such strongly hybridizing CMAs with a giant unit cell. In the case of Al-Cu-TM-Si (TM = Fe and Ru) approximants, nevertheless, we believe that the FsBz interactions would be best described in terms of $|\mathbf{G}|^2 = 50$.

9.5 Al-Re-Si 1/1-1/1-1/1 APPROXIMANT

9.5.1 Atomic Structure Free from Chemical Disorder

Takeuchi et al. [17] determined the atomic structure of three single-phase $Al_{82.6-x}Re_{17.4}Si_x$ (x = 7, 9 and 12) 1/1-1/1-1/1 approximants by analyzing the powder diffraction spectra taken with the wavelength of 0.07 nm at the beam line BL02B2, synchrotron radiation facility, SPring-8, Japan. They could refine the data by assuming space group $Pm\bar{3}$. In particular, the data for x = 9, i.e., $Al_{73.6}Re_{17.4}Si_9$ can be best refined without assuming any chemical disorder on all sites in the MI-cluster and glue sites as well. Al atoms exclusively enter into totally 24 sites (IIa: 12j) and (IIb: 12k) on inner icosahedra in two MI-type clusters "a" and "b" at the center and corner of the unit cell, respectively (see Chapter 6, Figure 6.5). Similarly, 24 Re atoms are filled into sites (TMa: 12j) and (TMb: 12k). Further, 60 Al atoms are filled into sites (MI1a: 6e) and (MI1b: 6h), (MI2a: 24l) and (MI2b: 24l), while six Al and twelve Si atoms enter into glue sites (G2a: 6f) and (G2b: 12j), respectively. Finally, further twelve Al atoms are filled into common glue sites (G1: 12k). The ordered structure above leads to the chemical formula $Al_{102}Re_{24}Si_{12}$ containing totally 138 atoms in the unit cell with the lattice constant $a = 1.28603$ nm.

9.5.2 Electronic Structure Calculations

The $Al_{102}Re_{24}Si_{12}$ per unit cell or $Al_{73.6}Re_{17.4}Si_9$ in % is an exceptionally unique 1/1-1/1-1/1 approximant free from chemical disorder, at least, within the accuracy of their Rietveld refinement [17]. This provides us with a unique opportunity to allow first-principles band calculations for the 1/1-1/1-1/1 approximant by directly employing the experimentally refined atomic structure. The DOS was calculated using the LMTO-ASA method for $Al_{102}Re_{24}Si_{12}$ to analyze the Si concentration dependence of various physical properties such as the electronic specific heat coefficient, the Pauli-paramagnetic susceptibility and the temperature dependence of

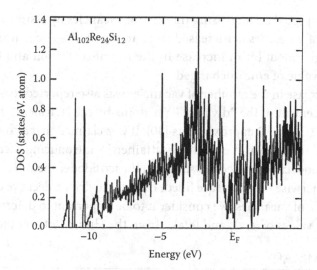

FIGURE 9.16   Total DOS calculated using the LMTO-ASA method for $Al_{102}Re_{24}Si_{12}$ 1/1-1/1-1/1 approximant. [From T. Takeuchi, T. Onogi, U. Mizutani, H. Sato, K. Kato, and T. Kamiyama, *Phys. Rev. B* 68 (2003) 184203.]

resistivity [17]. As shown in Figure 9.16, its total DOS is characterized by a deep pseudogap at the Fermi level. The origin of the pseudogap is most likely caused by Re-5d/Al-3p orbital hybridizations. Unfortunately, however, no analysis from the viewpoint of the FsBz interactions has been so far attempted for this compound.

As will be discussed in Chapter 10, Section 10.3, the effective **e/a** value of Mn is deduced to be 0.46. Assuming the valency of Re to be equal to that of Mn, we can easily calculate the value of $(e/a)_{total}$ to be 2.64 close to 2.58 for the Al-Cu-Ru-Si approximant as deduced from the Hume-Rothery plot in Section 9.4.2.2. This suggests that the set of {543}, {710}, and {550} lattice planes associated with $|G|^2 = 50$ would be most likely responsible for the formation of a FsBz-induced pseudogap, which is apparently hidden behind a huge orbital hybridization-induced one, as shown in Figure 9.16 (see Chapter 10, Section 10.7 and Figure 10.17.).

9.6 ROLE OF VACANCIES TO MAINTAIN A CONSTANT e/uc IN THE 1/1-1/1-1/1 APPROXIMANTS

Takeuchi et al. [29] evaluated the Fe concentration dependence of vacancies introduced into glue sites upon substitution of Fe for Mn atoms in a series of $Al_{73}Mn_{18-x}Fe_xSi_9$ ($0 \le x \le 12.5$) 1/1-1/1-1/1 approximants by performing the Rietveld structure analysis for the x-ray powder diffraction data.

They interpreted an increasing number of vacancies by assuming that the number of vacancies is increased so as to counterbalance an increase in e/a brought about by an increase in the Fe concentration and thereby to keep the value of e/uc unchanged.

An increase in the number of vacancies was also reported to occur upon the replacement of the "divalent" Yb atoms by the "trivalent" Y atoms in Cd-Yb-Y 1/1-1/1-1/1 approximants [30]. It was claimed that a constant e/uc in spite of an increase in e/a is likely attained by introducing vacancies into both the dodecahedral second shell and the icosidodecahedral fourth shell, where otherwise Cd atoms are filled. In order to draw a decisive conclusion on the role of vacancies, we consider it to be important to determine accurately the values of $(e/a)_Y$ and $(e/a)_{Yb}$ from the Hume-Rothery plot.

9.7 SUMMARY

Extensive studies have been undertaken of the stabilization mechanisms involved in two RT-type Al-Mg-Zn and Al-Li-Cu approximants, and two MI-type Al-Cu-TM-Si (TM = Fe and Ru) approximants, by utilizing both LMTO-ASA and FLAPW first-principles band calculations. In the case of RT-type Al-Mg-Zn and Al-Li-Cu approximants, NFE band calculations are quite effective in identifying the origin of a DOS pseudogap. An introduction of a single form factor at the Fermi level like that at $|G|^2 = 50$ in the Al-Mg-Zn approximant is found to perturb the bands only in the vicinity of the Fermi level (see Figure 9.6b). But this is apparently not enough to cause a pseudogap in the DOS. It is the simultaneous introduction of the two form factors at $|G|^2 = 50$ and 46 that is effective enough to produce a sizable DOS pseudogap across the Fermi level. This may be called a multi-zone effect. It is also of interest to note that the structure of a pseudogap becomes clearer, when electronic states perturbed by $V(50)$ and $V(46)$ are further modulated by long-wavelength excitations in the range $|G|^2 \leq 6$. A more quantitative analysis was made for the Al-Li-Cu approximant, where the situation is essentially the same as that in the Al-Mg-Zn approximant.

First-principles LMTO-ASA band calculations for the strongly hybridized Al-Cu-Ru-Si approximant indicate that a FsBz-induced pseudogap is formed as a result of interference of electrons with the set of {543}, {550}, and {710} lattice planes, but this is simply hidden behind a much larger orbital hybridization-induced pseudogap. A combination of the LMTO-ASA analysis and the *Hume-Rothery plot* based on FLAPW band calculations confirmed that the matching condition still holds and, hence, the

Hume-Rothery stabilization mechanism works even for an orbital hybridization-induced pseudogap system. A similar situation is expected to hold true in the MI-type Al-Re-Si approximant.

REFERENCES

1. D. Shechtman, I. Blech, D. Gratias, and J.W. Cahn, *Phys. Rev. Letters* 53 (1984) 1951.
2. A. Traverse, L. Dumoulin, E. Belin, and C. Sénémaud, in *Quasicrystalline Materials*, eds. Ch. Janot and J.M. Dubois (World Scientific, Singapore, 1988), pp. 399–408.
3. T. Fujiwara, *Phys. Rev.* B 40 (1989) 942.
4. Z.M. Stadnik, *Physical Properties of Quasicrystals* (Springer Verlag, Berlin, 1999).
5. U. Mizutani, T. Takeuchi, and H. Sato, *J. Phys.: Condens. Matter* 14 (2002) R767.
6. G.T. Laissardière, D. Nguyen-Manh, and D. Mayou, *Prog. Mat. Sci.* 50 (2005) 679.
7. J. Hafner and M. Krajčí, *Phys. Rev.* B 47 (1993) 11795.
8. S. Roche and T. Fujiwara, *Phys. Rev.* B 58 (1998) 11338.
9. H. Sato, T. Takeuchi, and U. Mizutani, *Phys. Rev.* B 64 (2001) 094207.
10. T. Fujiwara and T. Yokokawa, *Phys. Rev. Letters* 66 (1992) 333.
11. M. Windisch, M. Krajčí, and J. Hafner, *J. Phys.: Condens. Matter* 6 (1994) 6977.
12. H. Sato, T. Takeuchi, and U. Mizutani, *Phys. Rev.* B 70 (2004) 024210.
13. M. Krajčí, M. Windisch, J. Hafner, G. Kresse, and M. Mihalkovič, *Phys. Rev.* B 51 (1995) 17355.
14. M. Krajčí, and J. Hafner, *Phys. Rev.* B 59 (1999) 8347.
15. U. Mizutani, T. Takeuchi, and H. Sato, *Prog. Mat. Sci.* 49 (2004) 227.
16. U. Mizutani, R. Asahi, T. Takeuchi, H. Sato, O.Y. Kontsevoi, and A.J. Freeman, *Z. Kristallogr.* 224 (2009) 17.
17. T. Takeuchi, T. Onogi, T. Otagiri, U. Mizutani, H. Sato, K. Kato, and T. Kamiyama, *Phys. Rev.* B 68 (2003) 184203.
18. G. Trambly de Laissardière, D. Nguyen Manh, L. Magaud, J.P. Julien, F. Cyrot-Lackmann, and D. Mayou, *Phys. Rev.* B 52 (1995) 7920.
19. G. Bergman, J.L.T. Waugh, and L. Pauling, *Acta Crystallogr.* 10 (1957) 254.
20. U. Mizutani, W. Iwakami, T. Takeuchi, M. Sakata, and M. Takata, *Phil. Mag. Letters* 76 (1997) 349.
21. W. Sun, F.J. Lincoln, K. Sugiyama, and K. Hiraga, *Mat. Sci. Eng.* 294–296 (2000) 327.
22. C.L. Henley and V. Elser, *Phil. Mag. Letters* 53 (1986) 115.
23. T. Takeuchi and U. Mizutani, *Phys. Rev.* B 52 (1995) 9300.
24. U. Mizutani, *Introduction to the Electron Theory of Metals* (Cambridge University Press, Cambridge, 2001).
25. C.A. Guryan, P.W. Stephens, A.I. Goldman, and F.W. Gayle, *Phys. Rev.* B 37 (1988) 8495.

26. M. Audier, J. Pannetier, M. Leblanc, C. Janot, J-M. Lang, and B. Dubost, *Physica B* 153 (1988) 136.

27. U. Mizutani, T. Takeuchi, E. Banno, V. Fournee, M. Takata, and H. Sato, *Mater. Res. Soc. Symp. Proc.* (2001) 643 [Materials Research Society, K.13.1.1.].

28. T. Takeuchi and U. Mizutani, *J. Alloys Compounds*, 342 (2002) 416.

29. T. Takeuchi, T. Onogi, E. Banno, and U. Mizutani, *Mater. Trans.* 42 (2001) 933.

30. H. Takahashi, T. Takeuchi, U. Mizutani, J.Q. Guo, and A.-P. Tsai, *J. Non-Cryst. Solids* 334,335 (2004) 228.

The Interplay and Contrasts Involved in the Chemistry, Physics, and Crystal Structures of Alloys and Compounds

10.1 e/a OR VEC AS AN ELECTRON CONCENTRATION PARAMETER

As has been emphasized frequently in this monograph, the electron concentration plays a crucial role in control of phase stability as well as numerous physical properties of alloys. In Chapter 1, we have introduced two different notions of electron concentration: one is the e/a, appearing in connection with the Hume-Rothery electron concentration rule, and the other is the VEC, i.e., the number of electrons per atom, including the d-electrons being involved in the valence band. The VEC is obviously a key parameter in determining the Fermi level when first-principles band calculations are carried out to study the band structure. One must cautiously select either the e/a or the VEC as an electron concentration parameter, depending on the situation involved.

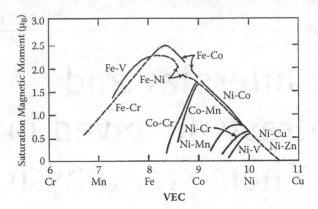

FIGURE 10.1 Slater–Pauling curve representing the **VEC** dependence of saturation magnetization in 3d-transition metal alloys. [From R.M. Bozorth, *Ferromagnetism,* (D van Nostrand, New York, 1951); U. Mizutani, *Introduction to the Electron Theory of Metals* (Cambridge University Press, Cambridge, 2001).]

Let us first consider cases, in which a universal behavior is evident when plotted against the **VEC**. Included in Figures 10.1 to 10.4 are the saturation magnetization known as the Slater–Pauling curve [1], the electronic specific heat coefficient for bcc alloys of 3d transition metals (TMs) [2], the superconducting transition temperature of TM alloys [3] and the themoelectric power in the Heusler (L2$_1$)-type Fe$_2$VAl, where Fe, V, and Al can be partially replaced by late TM elements like Co and Pt, early TM elements like Ti, Zr, and Mo, and polyvalent elements like Si and Ge, respectively, without changing the structure [4]. The reason why all these physical properties exhibit universal behaviors with respect to the **VEC** stems from the fact that they are quantities clearly related to the total DOS at the Fermi level. In Section 10.8, we shall discuss in more detail why a master curve is evident for the **VEC** dependence of the Seebeck coefficient in the Fe$_2$VAl system.

What about the **e/a** as an electron concentration parameter? Here, again, we have many examples in which a universal alloying behavior emerges when plotted against **e/a**. The axial ratio *c/a*, the magnetic susceptibility corrected for ionic contributions, and the electronic specific heat coefficient in noble metal alloys are typical examples, as already shown in Chapter 7, Figures 7.8 to 7.10 in Section 7.5. Recall that such a universal behavior collapses, if the **VEC** is employed in place of **e/a** in these cases. This can be easily understood, if the discussion in Section 7.2 is recalled: both Cu$_5$Zn$_8$ and Cu$_9$Al$_4$ gamma-brasses correspond to a common **e/a** value of 21/13, but yield different values of **VEC** of 11.615 and 8.538, respectively (see Section

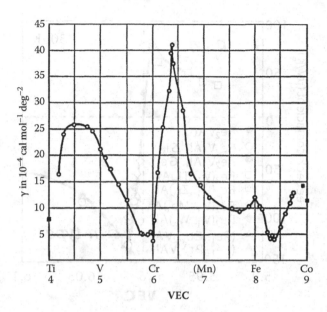

FIGURE 10.2 **VEC** dependence of the electronic specific heat coefficient in 3d-transition metal alloys. [From C.H. Cheng, K. Gupta, E.C. van Reuth, and P.A. Beck, *Phys. Rev.* 126 (1962) 2030.]

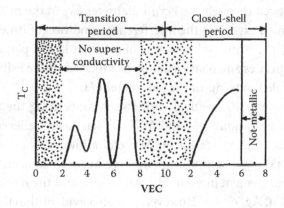

FIGURE 10.3 **VEC** dependence of superconducting transition temperatures throughout the periodic table. [From B.T. Matthias, *Phys. Rev.* 97 (1955) 74.]

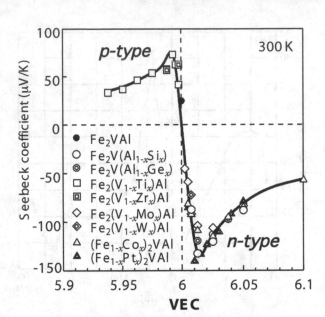

FIGURE 10.4 **VEC** dependence of the Seebeck coefficient in the Heusler (L2₁)-type Fe_2VAl alloys. [From Y. Nishino, *The Science of Complex Alloy Phases*, edited by T.B. Massalski and P.E.A. Turchi, TMS (The Minerals, Metals & Materials Society, 2005), pp. 325–344.]

7.2). The difference obviously arises from the fact that the partner element Zn possesses ten 3d-electrons in the valence band but Al does not. The reason why a critical electron concentration parameter must be **e/a** in dealing with the Hume-Rothery stabilization mechanism was already explained in detail in Chapter 4, Section 4.2 and Chapter 7, Sections 7.3, and 7.4: the **e/a** is introduced through the Fermi diameter $2k_F$ in the matching condition (4.1), which expresses the interference condition of itinerant electron waves near the Fermi level with sets of relevant lattice planes. Therefore, physical properties and phase stability dominated by the FsBz interactions should be scaled in terms of the parameter **e/a**.

In this Chapter 10, we deal with issues concerning the two different electron concentrations, **e/a** and **VEC**. It has been believed over many years that the value of **e/a** for noble metal alloys can be reasonably well assigned by taking a composition-average between nominal valencies of the polyvalent element like Zn, Mg, Al, Si, etc., and the monovalent noble metals like Cu, Ag, or Au. However, it is also evident that the determination of **e/a** becomes less clear for the TM element located to the left of the noble metals, since the Fermi level often falls then in the energy range where the d-band is only partially filled. A large number of arguments have been made in attempts to ascertain the e/a value for TM elements, but no clear-cut definition has yet emerged. Indeed, this has posed a great difficulty in interpreting the Hume-Rothery electron concentration rule for alloys containing TM elements. To begin with, we will review the existing proposals on e/a for TM elements.

10.2 EXISTING PROPOSALS ON e/a FOR THE TRANSITION METAL ELEMENT

The most frequently cited e/a values for TM elements are probably those proposed by Raynor in 1949 [5]. He directed his attention to $CrAl_7$, $MnAl_6$, $FeAl_3$, Co_2Al_9, and $NiAl_3$, all of which come into equilibrium with the Al-primary solid solution with the 3d-TM element, and pointed out that, except for $FeAl_3$, the proportion of Al decreases as the number of holes per atom in 3d-atomic orbitals of the TM atom decreases in the Pauling model. From this, he was intuitively guided to conclude that electrons donated by Al fill the holes of the 3d-orbitals and the extent to which electrons are absorbed by the TM atom determines the stability of these Al-TM compounds. If the e/a ratio for these compounds is calculated under the assumption that each Al atom donates three electrons per atom and that each TM atom absorbs them to fill the holes in the 3d-orbitals, an approximately constant e/a ratio is maintained, except in the case of $FeAl_3$: $CrAl_7 \rightarrow 2.05$, $MnAl_6 \rightarrow 2.05$, $FeAl_3 \rightarrow 1.58$, $Co_2Al_9 \rightarrow 2.12$, and $NiAl_3 \rightarrow 2.09$. This led Raynor [5] to postulate that they are indeed typical of electron compounds and that the effective e/a values for Cr, Mn, Fe, Co, and Ni in the Al-TM alloys can be taken to have negative values −4.66, −3.66, −2.66, −1.71 and −0.61, respectively.*

In 1988, Tsai et al. [6] discovered thermally stable quasicrystals in Al-Cu-TM (TM = Fe, Ru, and Os) alloy systems. They noticed that all these Al-based quasicrystals seem to correspond to the value of e/a close to 1.75, provided that negative valencies proposed by Raynor are assigned to TM elements involved. Using Raynor's negative e/a scheme for TM elements, they further discovered new quasicrystals in Al-Pd-TM (TM = Mn and Re) alloy systems [7]. From their experiments, they were convinced to conclude that all Al-based quasicrystals obey the Hume-Rothery electron concentration rule with the e/a values centered at about 1.8. Their findings

---

* For example, the effective e/a value for $MnAl_6$ is calculated to be $(e/a)_{total} = [(-3.66) + 6 \times 3]/7$ = 2.05. Unfortunately, a physical basis for the assignment of $(e/a)_{Fe} = -2.66$ is quite obscure. Its value becomes −0.6, if Fe_3Al is also assumed to be stabilized at $(e/a)_{total} = 2.1$. Raynor made no comments on why he assumed $(e/a)_{total} = 1.58$ instead of 2.1 for Fe_3Al [5].

were so dramatic that many researchers have been encouraged to employ the Hume-Rothery electron concentration rule coupled with Raynor's negative valency scheme for TM elements as a powerful guide to search for new quasicrystals [8–10]. Even a theoretical reasoning for the Raynor's postulate has been attempted [11,12].

There have also been proposals for assigning a positive e/a value for the TM element. For example, Haworth and Hume-Rothery [13] studied the maximum solubility limit of the α-phase in Cu-TM-Zn and Cu-TM-Al alloy systems and assigned the e/a value of 1.8, 1.0, 0.8, and 0.6 to Mn, Fe, Co, and Ni, respectively, in order for the solubility limit to meet a common e/a value of 1.4. Mizutani et al. [14] performed first-principles LMTO-ASA band calculations for Al-Cu-Ru-Si 1/1-1/1-1/1 approximant and revealed a deep pseudogap across the Fermi level. They showed that the matching of the Fermi surface with the set of {543}, {710}, and {550} lattice planes gives rise to a positive e/a value of 0.76 for Ru (see Chapter 9, Section 9.4.2.1). More recently, Ishimasa et al. [10,15] noted that stable quasicrystals in Sc-TM alloy systems they discovered always correspond to the e/a value in the neighborhood of 2.1, provided that positive e/a values proposed by Haworth and Hume-Rothery are employed for the TM element involved. It appears that Raynor's negative valency scheme has been more frequently applied to Al-TM alloys, while the Haworth and Hume-Rothery positive valency scheme to Cu-TM alloys.

Throughout the present volume, we have stressed the need for performing first-principles band calculations to lay foundations for the Hume-Rothery electron concentration rule for CMAs. The FLAPW-Fourier method, together with the Hume-Rothery plot, is devised to extract the *critical* reciprocal lattice vector responsible for the formation of a pseudogap via the FsBz interactions, and the resulting electron concentration parameter e/a. A tabulation of the effective e/a values for TM elements in the periodic table is certainly of great interest and could be helpful to future researchers. Fortunately, the Hume-Rothery plot can be performed for any compounds, regardless of the size of the unit cell. As mentioned in Chapter 1, Section 1.2, the e/a rule originally proposed by Hume-Rothery in 1926 pointed to a regularity that, in spite of differences in atomic composition, is clearly responsible for the fact that all three compounds Cu_5Sn, Cu_3Al, and CuZn crystallize into a common structure of the bcc phase having $e/a = 3/2$ [16]. Many researchers, including Hume-Rothery and Raynor, tended to believe that all B2-compounds, including those containing TM elements as a partner element, are also likely stabilized at e/a

= 3/2. Our objective in the next section is to examine if the **e/a** value for B2-compounds is really 3/2, irrespective of the presence or absence of the TM element.

10.3 HUME-ROTHERY PLOT FOR THE B2-COMPOUNDS

Let us first consider whether the Hume-Rothery plot provides the **e/a** value consistent with an averaged valency of constituent elements for noble metal alloys. Energy dispersion relations and the Hume-Rothery plot for the equiatomic AgLi B2-compound are shown in Figures 10.5a,b, respectively [17]. In spite of the location of the Ag-4d band far below the Fermi level, the variance $\sigma^2(E)$ in AgLi B2-compound remains large and a deviation of data points from a straight line in $\{2|\mathbf{k}+\mathbf{G}|\}^2$ is significant up to about +5 eV. This is in sharp contrast to that in Ag_5Li_8 gamma-brass shown in Figure 8.23, where the variance becomes low, as soon as the Ag-4d band terminates at about −3.0 eV. Thus, a square of the Fermi diameter in the B2-compound has to be determined by drawing a straight line connecting data points above about +10 eV and near the bottom of the valence band, as shown in Figure 10.5b. It is deduced to be 1.5 ± 0.1 in units of $(2\pi/a)^2$. The value of $(e/a)_{total}$ is then calculated to be 0.9 ± 0.1 from the relation

$$\left(e/a \right)_{total} = \frac{8\pi k_F^3}{3N}$$

where the number of atoms per unit cell, N, is equal to 2 and k_F is in units of $(2\pi/a)^2$. This agrees well with its average valency of monovalent Ag and Li within the accuracy of the analysis.

As is clear from the above argument, AgLi B2-compound must be excluded from the family of the **e/a** = 3/2 compounds. It is of interest, at this stage, to consider why the variance of AgLi B2-compound remains significant up to about +5 eV, whereas that of Ag_5Li_8 gamma-brass becomes well suppressed immediately after the Ag-4d band terminates at about −3.0 eV. A comparison of energy dispersion relations between these two compounds shown in Figures 10.5a and 8.22a immediately tells us that there is no essential difference in the position of the Ag-4d band: both are located over the energy range from −5.5 to −3.5 eV. However, we notice that dispersion relations in the B2-compound are much simpler and more scarce than those in the gamma-brass. The reason the latter is more crowded and less dispersive is because zone foldings more frequently occur when the unit cell becomes larger and larger. We consider this to be responsible

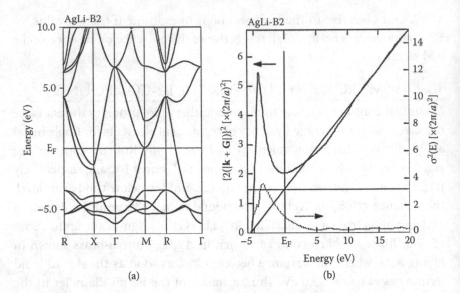

FIGURE 10.5 (a) Energy dispersion relations and (b) the Hume-Rothery plot for AgLi B2-compound. [From U. Mizutani, R. Asahi, H. Sato, T. Noritake, and T. Takeuchi, *J. Phys.: Condens. Matter* 20 (2008) 275228.]

for a quick restoration of the free electron behavior in gamma-brass, as soon as the Ag-4d band is over, but for a persistent departure from this, up to +5eV, in the B2-compound. Therefore, we can say that the higher the number of atoms in the unit cell, the more accurately the Fermi diameter can be determined from the Hume-Rothery plot. The determination of the Fermi diameter for structurally simple compounds like B2-compounds is generally less accurate than that in CMAs.

The DOS and the Hume-Rothery plot for AgMg B2-compound are depicted in Figures 10.6a,b, respectively [18]. There is no pseudogap at the Fermi level. The value of **VEC** at the Fermi level is obviously equal to (11 + 2)/2 = 6.5. From the Hume-Rothery plot, the square of the Fermi diameter is deduced to be 2.0 ± 0.1 in units of $(2\pi/a)^2$. The value of $(e/a)_{total}$ is accordingly calculated to be 1.48 ± 0.07, being consistent with the **e/a** = 3/2 rule. Judging from its high melting point of 820°C and a wide solubility range, Hume-Rothery [19] conjectured an electrochemical effect in AgMg to be particularly significant among B2-compounds. He even speculated a possible presence of AgMg molecules in liquid phase. Indeed, a difference in the electronegativities, $\Delta\chi$, between Ag and Mg metals is 0.7 and is much larger than 0.3 in the Cu-Zn alloy system, according to the Pauling electronegativity table (χ_{Mg} = 1.9, χ_{Ag} = 1.2, χ_{Zn} = 1.6, and

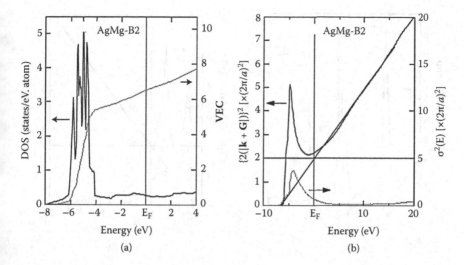

FIGURE 10.6   (a) DOS and (b) the Hume-Rothery plot for AgMg B2-compound. [From U. Mizutani, R. Asahi, T. Takeuchi, H. Sato, O.Y. Kontsevoi, and A.J. Freeman, *Z. Kristallogr.* 224 (2009) 17.]

χ_{Cu} = 1.9). Nevertheless, we revealed that the **e/a** value deduced from the Hume-Rothery plot is close to 3/2 and is apparently unaffected by the electrochemical effect.

Figures 10.7a,b show the DOS and the Hume-Rothery plot for CuZn B2-compound, respectively [18]. There is no pseudogap and **VEC** = (11 + 12)/2 = 11.5 at the Fermi level. Though the Cu-3d band terminates at about −2.0 eV, the variance continues to remain fairly large up to +5 eV above the Fermi level. A straight line is drawn by connecting data points above about +5 eV and those near the bottom of the valence band, where the variance is small. From the intercept at the Fermi level, the square of the Fermi diameter for CuZn is determined to be 2.0 ± 0.1 in units of $(2\pi/a)^2$. The **e/a** value is calculated to be 1.48 ± 0.07, in good agreement with an average nominal valency of 3/2. Thus, CuZn B2-compound can be correctly regarded as a 3/2 compound. The effective **e/a** value of Cu turned out to be +0.96, in good agreement with its nominal valency of unity, if the valency of two is assigned to Zn.

We have so far discussed B2-compounds involving the noble metals Cu or Ag and normal metals like Li, Mg, and Zn. The **e/a** value deduced from the Hume-Rothery plot is found to agree well with a composition-weighted average of valencies of constituent elements. What happens if the TM element is contained as one of constituent elements in B2-compound?

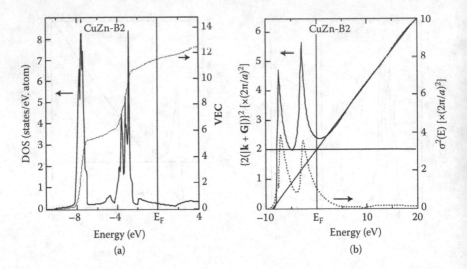

FIGURE 10.7 (a) DOS and (b) the Hume-Rothery plot for CuZn B2-compound. [From U. Mizutani, R. Asahi, T. Takeuchi, H. Sato, O.Y. Kontsevoi, and A.J. Freeman, *Z. Kristallogr.* 224 (2009) 17.]

Hume-Rothery [20] and Raynor [21] wondered whether the valency of Ni should be unity in NiZn, but zero in NiAl B2-compounds, provided that each of them is indeed an $e/a = 3/2$ electron compound, and valencies of Zn and Al are two and three, respectively.*

Figures 10.8a,b show the DOS and the Hume-Rothery plot for NiZn B2-compound [18]. The **VEC** reaches $(10 + 12)/2 = 11$ at the Fermi level. The Fermi level sits immediately following the peak in the DOS. The square of the Fermi diameter is determined to be 1.8 ± 0.1 in units of $(2\pi/a)^2$ from the intercept of a straight line fitted to the data, where the variance is small. The value of $(e/a)_{total}$ is accordingly determined to be 1.26 ± 0.07. It is clear that NiZn B2-compound can no longer be regarded as an $e/a = 3/2$ compound. The valency of Ni in NiZn is deduced to be 0.54, provided that the valency of Zn is two. What about NiAl B2-compound? Its DOS and the Hume-Rothery plot are shown in Figures 10.9a,b, respectively [18]. The DOS has no pseudogap at the Fermi level. The **VEC** reaches $(10 + 3)/2 = 6.5$ at the Fermi level. The square of the Fermi diameter is found to be 2.25 ± 0.05 in units of $(2\pi/a)^2$ and the value of $(e/a)_{total}$ to be

---

* Raynor expressed a dilemma in the determination of valency of Ni in NiZn and NiAl B2-compounds at the 1952 Abingdon conference, where both Mott and Jones participated [21]. No positive suggestions were apparently made by them. The problem has remained unsolved till now.

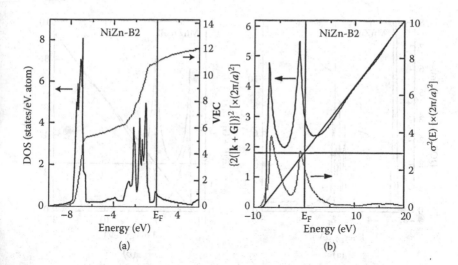

FIGURE 10.8 (a) DOS and (b) the Hume-Rothery plot for NiZn B2-compound. [From U. Mizutani, R. Asahi, T. Takeuchi, H. Sato, O.Y. Kontsevoi, and A.J. Freeman, *Z. Kristallogr.* 224 (2009) 17.]

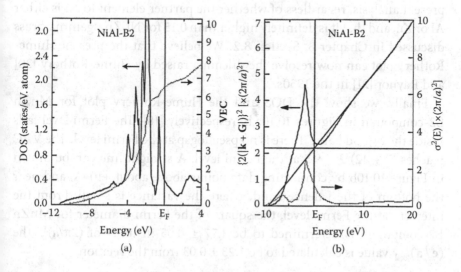

FIGURE 10.9 (a) DOS and (b) the Hume-Rothery plot for NiAl B2-compound. [From U. Mizutani, R. Asahi, T. Takeuchi, H. Sato, O.Y. Kontsevoi, and A.J. Freeman, *Z. Kristallogr.* 224 (2009) 17.]

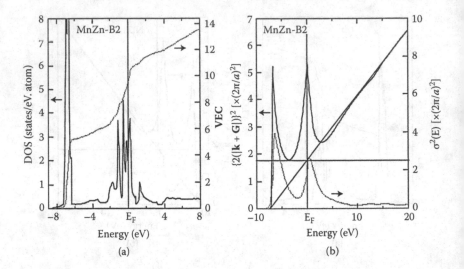

FIGURE 10.10 (a) DOS and (b) the Hume-Rothery plot for MnZn B2-compound. [From U. Mizutani, R. Asahi, T. Takeuchi, H. Sato, O.Y. Kontsevoi, and A.J. Freeman, *Z. Kristallogr.* 224 (2009) 17.]

1.76 ± 0.04. Therefore, NiAl B2-compound is also excluded from the family of the **e/a** = 3/2 compounds. The valency of Ni in NiAl is deduced to be 0.52, provided that the valency of Al is three. It is important to note that the valency of Ni is determined to be 0.53 ± 0.07 within the accuracy of the present analysis, regardless of whether the partner element to Ni is either Al or Zn, and that it is definitely higher than 0.15 for Ni_2Zn_{11} gamma-brass discussed in Chapter 8, Section 8.2. We believe that the present Hume-Rothery plot can now resolve the dilemma raised by Hume-Rothery [20] and Raynor [21] in the 1950s.

Finally, we show the DOS and the Hume-Rothery plot for MnZn B2-compound in Figures 10.10a,b, respectively [18]. The Fermi level falls inside the Mn-3d band. There is no pseudogap at the Fermi level. The **VEC** reaches (7 + 12)/2 = 9.5 at the Fermi level. A straight line can be drawn in Figure 10.10b by connecting data points above about +10 eV and near the bottom of the valence band, where the variance is small. From the intercept at the Fermi level, the square of the Fermi diameter for MnZn B2-compound is determined to be 1.77 ± 0.05 in units of $(2\pi/a)^2$. The $(e/a)_{total}$ value is calculated to be 1.23 ± 0.03 from the relation

$$\left(e/a\right)_{total} = \frac{8\pi k_F^3}{3N}$$

where the number of atoms per unit cell, N, is equal to 2 and k_F is in units of

$$\frac{2\pi}{a}$$

Thus, MnZn B2-compound cannot be designated as an $e/a = 3/2$ compound. The effective e/a value of Mn turned out to be +0.46, if the valency of two is assigned to Zn.

In this way, positive e/a values were deduced for all TM elements studied, regardless of whether the partner element is Al or Zn. This is obviously in conflict with negative valencies proposed by Raynor [5].

10.4 $(e/a)_{TM}$ VALUE FOR THE TRANSITION METAL ELEMENT IN THE PERIODIC TABLE

The value of $(e/a)_{total}$ has been deduced from the Hume-Rothery plot for a number of alloys, including B2-compounds, gamma-brasses, and 1/1-1/1-1/1 approximants. As described above, the value of $(e/a)_{TM}$ for the TM element was then derived by assuming a nominal valency for its partner element, which has been selected from polyvalent metals like Mg, Al, Si, and Zn in the periodic table. The results are summarized in Figure 10.11. The

Li 1.0[A]	Be											B	C	N
Na	Mg (2.0)											Al (3.0)	Si (4.0)	P
K	Ca	Sc	Ti	V 0.23[B]	Cr	Mn 0.46[C]	Fe 0.7[D] 0.68[E]	Co 0.26[F]	Ni 0.15[G] 0.53[H]	Cu 0.97[I]	Zn (2.0)	Ga	Ge	As
Rb	Sr	Y	Zr	Nb	Mo	Tc	Ru 0.68[J]	Rh	Pd 0.07[K]	Ag 0.96[L]	Cd	In	Sn	Sb
Cs	Ba	La	Hf	Ta	W	Re	Os	Ir	Pt	Au	Hg	Tl	Pb	Bi

A: AgLi-B2 and Ag_5Li_8 gamma-brass, B: Al_8V_5 gamma-brass, C: MnZn-B2, D: Fe_2Zn_{11} gamma-brass, E: Al-Cu-Fe-Si approximant, F: Co_2Zn_{11} gamma-brass, G: Ni_2Zn_{11} gamma-brass, H: NiZn-B2 and NiAl-B2, I: CuZn-B2, Cu_5Zn_8 and Cu_9Al_4 gamma-brasses, J: Al-Cu-Ru-Si approximant, K: $Pd_2 Zn_{11}$ gamma-brass and L: AgMg-B2. The number in the bracket is the e/a value *a priori* given from the valency of the element.

FIGURE 10.11 $(e/a)_{TM}$ derived from the Hume-Rothery plot for transition metal elements in the periodic table [from U. Mizutani, R. Asahi, T. Takeuchi, H. Sato, O.Y. Kontsevoi, and A.J. Freeman, *Z. Kristallogr.* 224 (2009) 17]. Numbers in the bracket refer to nominal valency corresponding to the number of outermost electrons in the form of a free atom.

e/a value for Li and the noble metals agrees with the nominal valency and is apparently independent of their surroundings such as a crystal structure, constituent partner elements, and their composition. However, the value for the TM element, particularly Ni, is found to be dependent on its surroundings: $(e/a)_{Ni} = 0.15$ for Ni_2Zn_{11} gamma-brass but $(e/a)_{Ni} = 0.53$ for NiAl and NiZn B2-compounds. A striking difference in the electronic structure between them must be responsible for this: a pseudogap apparently pushes a whole Ni-3d band well below the Fermi level in Ni_2Zn_{11} gamma-brass (see Figure 8.7b), whereas its absence plus higher Ni concentration leaves the Fermi level in the middle of the Ni-3d antibonding subband in both NiZn and NiAl B2-compounds (see Figures 10.8a and 10.9a).

It is important to investigate, at this stage, whether the value of $(e/a)_{TM}$ listed in Figure 10.11 is physically more plausible than that, say, proposed by Raynor [5]. For this purpose, we point to the e/a dependences of the number of atoms per unit cell, N, and e/uc for group (I) and (II) gamma-brasses shown in Figures 8.30a,b, respectively. We find that N decreases while e/uc increases with increasing e/a for all gamma-brasses studied in a quite universal manner. A universal behavior is particularly evident among the data in both (a) and (b) for Cu-Cd, Ni-Zn, and Co-Zn gamma-brasses, where an increase in the number of vacancies with increasing e/a is significant. We consider that such a consistent behavior is obtained, thanks to correct determination of the value of $(e/a)_{TM}$ for Ni and Co.

One might still wonder whether a choice of a particular e/a scheme like Raynor's negative valencies is simply a matter of a shift in an electron concentration scale. It would become more convincing if some evidence can be provided for the soundness of the Hume-Rothery plot method. Figure 10.12 shows the melting temperature of several B2-compounds as functions of the two different sets of e/a values: (a) one derived from Raynor's negative valencies for TM elements and (b) the other derived from the Hume-Rothery plot, as described in Section 10.3 [22]. There is no correlation in (a) but the data fall on a straight line in (b). An increase in $(e/a)_{TM}$ means an increase in electron density outside the MT sphere. We consider that the higher an average electron concentration outside the MT sphere, the stronger is the bonding strength between the neighboring atoms and, in turn, the higher is the melting temperature.

10.5 PHYSICS BEHIND THE MATCHING CONDITION

The universality of the Hume-Rothery stabilization mechanism should be best studied by testing the matching condition $2k_F = |\mathbf{G}|$ given by Equation

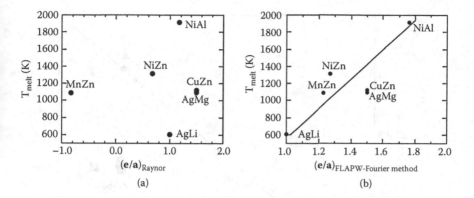

FIGURE 10.12   Melting temperature of several B2-compounds as a function of (a) $(e/a)_{Raynor}$ derived using Raynor's negative valencies [from G.V. Raynor, *Prog. Met. Phys.* 1 (1949) 1] and (b) $(e/a)_{TM}$ derived from the Hume-Rothery plot based on the FLAPW-Fourier method [from U. Mizutani, T. Noritake, T. Ohsuna, and T. Takeuchi, *Phil. Mag.* 90 (2010) 1985].

4.1. Indeed, the matching condition has been discussed quite frequently in the field of quasicrystals and approximants for the past two decades. In 2005, Tsai [8] and Ishimasa [10] reviewed the matching condition by analyzing their own experimental data. Under an implicit assumption that the Hume-Rothery stabilization mechanism works for quasicrystals containing TM elements, Tsai discussed the stability of quasicrystals by plotting the reciprocal lattice vector K_p against the Fermi diameter $2k_F$. His results [8] are reproduced in Figure 10.13, where $(e/a)_{total}$ is first evaluated by using Raynor's negative valency for the TM element and then the Fermi diameter $2k_F$ is calculated by inserting it into the relation

$$\left(e/a\right)_{total} = \frac{8\pi k_F^3}{3N}$$

As emphasized in preceding sections, there are indications that the negative valencies proposed by Raynor are physically problematic. Moreover, the reciprocal lattice vector K_p appears to be always selected from one of the major x-ray diffraction peaks in such a way that the matching condition is best fulfilled.* Ishimasa [10] also tested the $2k_F = K_p$ relation

---

* Both K_p and $|G|$ represent the magnitude of the reciprocal lattice vector to satisfy the matching condition. In this monograph, K_p is used when it is selected from the measured x-ray diffraction peak, while $|G|$ is used when determined from the FLAPW–Fourier analysis.

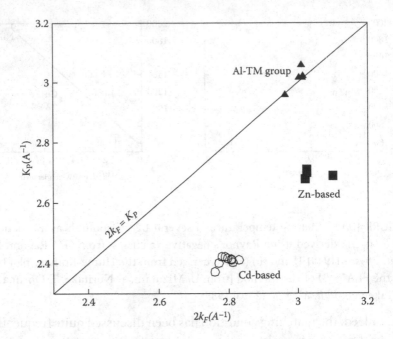

FIGURE 10.13   Reciprocal lattice vector K_p derived from one of the major x-ray diffraction peaks versus the Fermi diameter $2k_F$ for the Al-TM, Zn- and Cd-based quasicrystals. Here the Fermi diameter was calculated by assuming Raynor's negative valencies for transition metal elements involved. [From A.P. Tsai, *The Science of Complex Alloy Phases*, edited by T.B. Massalski and P.E.A. Turchi, TMS (The Minerals, Metals & Materials Society, 2005), pp. 201–214.]

for Zn-Mg-Sc, Zn-Fe-Sc, Zn-Ni-Sc, and Zn-Ga-Mg-Sc quasicrystals by assigning empirical **e/a** values proposed by Haworth and Hume-Rothery [13] for the TM element. We consider the construction of such $2k_F$ versus K_p diagram to be far from being satisfactory as a true test of the matching condition. A rigorous test of the matching condition should be made by determining both the $2k_F$ and the *critical* reciprocal lattice vector $|\mathbf{G}|$ from first-principles FLAPW band calculations.

Prior to the test of the matching condition based on the FLAPW-Fourier analysis, two remarks need be addressed first. In Chapter 8, Sections 8.2 and 8.3, dealing with group (II) gamma-brasses, we stressed that, though the $|\mathbf{G}|^2 = 18$ electronic states play a key role, its neighboring $|\mathbf{G}|^2$ states also contribute to lowering the electronic energy. The number of critical $|\mathbf{G}|$ values would become inherently more than two for alloys containing a large amount of the TM element, where the d-states-mediated-splitting occurs. In Chapter 9, Sections 9.2 and 9.3 dealing with RT-type 1/1-1/1-

1/1 approximants, we emphasized the necessity of taking into account, at least, two sets of $|\mathbf{G}|^2$ s, since more than two form factors are closely distributed near the Fermi level. A superposition of the Brillouin zones associated with more than two different $|\mathbf{G}|^2$ s near the Fermi level would effectively form a more or less spherical Brillouin zone net and its interaction with electrons near the Fermi level gives rise to sparse energy dispersion relations in almost all directions in the reciprocal space and, in turn, gives rise to a pseudogap across the Fermi level in the DOS in CMAs like Al-Mg-Zn and Al-Li-Cu 1/1-1/1-1/1 approximants. It is further claimed in Chapter 9, Section 9.3.3 that the most *critical* $|\mathbf{G}|^2$ can be still extracted as the one participating most significantly in the formation of a FsBz-induced pseudogap: $|\mathbf{G}|^2 = 46$ for Al-Li-Cu, and $|\mathbf{G}|^2 = 50$ for Al-Mg-Zn.

In the remainder of this section, we will work out the matching condition, using the data only for group (I) gamma-brasses and RT-type Al-Mg-Zn and Al-Li-Cu 1/1-1/1-1/1 approximants, all of which are well characterized by a FsBz-induced pseudogap. The $2k_F$ versus $|\mathbf{G}|$ diagram is shown in Figure 10.14a. Both $|\mathbf{G}|$ and $2k_F$ for Cu_5Zn_8, Cu_9Al_4, Ni_2Zn_{11} and Pd_2Zn_{11} gamma-brasses are unanimously determined from the FLAPW-Fourier method and from the Hume-Rothery plot, as discussed in Chapters 7 and 8.* As mentioned above, the most *critical* $|\mathbf{G}|$ is deduced to be $\sqrt{50}$ and $\sqrt{46}$ for Al-Mg-Zn and Al-Li-Cu approximants, respectively. On the other hand, the value of $2k_F$ can be safely determined by inserting valencies of one, one, two, two and three for Cu, Li, Mg, Zn, and Al, respectively, into the relation

$$k_F = \left[\frac{3N}{8\pi} (e/a)_{total} \right]^{1/3}$$

where N is 160 for both approximants and k_F is in units of

$$\frac{2\pi}{a}$$

The data for group (I) gamma-brasses and the two RT-type approximants discussed above are plotted in Figure 10.14a, where data points are found to fall in a narrow region satisfying the condition $2k_F = |\mathbf{G}|$ and centered

---

* Ni_2Zn_{11} and Pd_2Zn_{11} gamma-brasses are discussed in Section 8.2.2.1 by classifying them into group (II). It is, however, concluded that they should be better classified into group (I).

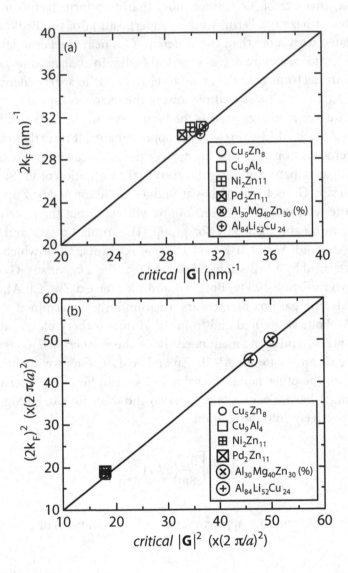

FIGURE 10.14 Test of the matching condition given by Equation 4.1 for group (I) gamma-brasses and RT-type 1/1-1/1-1/1 approximants [23]. (a) $|G|$ represents the *critical* reciprocal lattice vector causing a FsBz-induced pseudogap. Both $|G|$ and the Fermi diameter $2k_F$ are in the units of $(nm)^{-1}$. (b) $|G|^2$ and $(2k_F)^2$ are normalized with respect to $(2\pi/a)^2$, where a is the lattice constant.

at 31 $(nm)^{-1}$ [23]. This means that, regardless of e/a and the size of the unit cell, electrons near the Fermi level always possess the Fermi diameter equal to about 31 $(nm)^{-1}$ and interfere with the set of lattice planes associated with the critical $|G|$ equal to 30 $(nm)^{-1}$. This may be taken as a clear demonstration of the Hume-Rothery stabilization mechanism for CMAs characterized by a purely FsBz-induced pseudogap at the Fermi level.

Figure 10.14b is likewise constructed by plotting $(2k_F)^2$ against $|G|^2$ in units of $(2\pi/a)^2$, i.e., by expressing them as nondimensional quantitites, using the same data as in Figure 10.14a [23]. The data for group (I) gamma-brasses fall exactly at $|G|^2 = 18$. The value of $(2k_F)^2$ also falls in the very vicinity of 18. This confirms the validity of the matching condition given by Equation 4.1. In RT-type Al-Mg-Zn and Al-Li-Cu approximants having larger unit cells, the sets of *critical* lattice planes responsible for the FsBz-induced pseudogap are those of {543}, {550}, and {710} with $|G|^2 = 50$ and the {631} with $|G|^2 = 46$, respectively. The larger the size of the unit cell, the larger is the magnitude of the *critical* reciprocal lattice vector. Hence, Figure 10.14b may be more convenient than Figure 10.14a to confirm which set of lattice planes participates in an interfering event with electron waves at the Fermi level and to cause a FsBz-induced pseudogap there.

Figure 10.15 shows x-ray diffraction spectra for the $Al_{15}Mg_{44}Zn_{41}$ quasicrystal and $Al_{25.5}Mg_{39.5}Zn_{35}$ 1/1-1/1-1/1 approximant [24], together with the Brillouin zones causing a pseudogap at the Fermi level in both cases [23]. In (a), an arrow indicates a conventional way to select an appropriate K_p from the diffraction spectrum. In the case of the Al-Mg-Zn approximant, electrons at the Fermi level interact most significantly with the Brillouin zone in (b) composed of totally 84 zone planes: 48-fold {543}, 12-fold {550}, and 24-fold {710} zone planes (see Chapter 9, Section 9.2.3 and also [25]). The 48-fold {631} zone planes become the most important in the Al-Li-Cu approximant (see Chapter 9, Section 9.3.3, and [26]). The Brillouin zone bounded by 60-fold (222100) zone planes for the Al-Mg-Zn quasicrystal is illustrated in (c).

The depth of a pseudogap is known to depend on the magnitude of the form factor in the framework of the NFE model. The three different zone planes {543}, {710}, and {550} in the approximant have their individual form factors and make different contributions to the formation of a pseudogap, though the total number of zone planes is higher than that in the quasicrystal. Instead, 60-fold (222100) zone planes in the quasicrystal are equivalent and, hence, would be more effective to produce a deeper pseudogap across the Fermi level. Indeed, the measured electronic specific

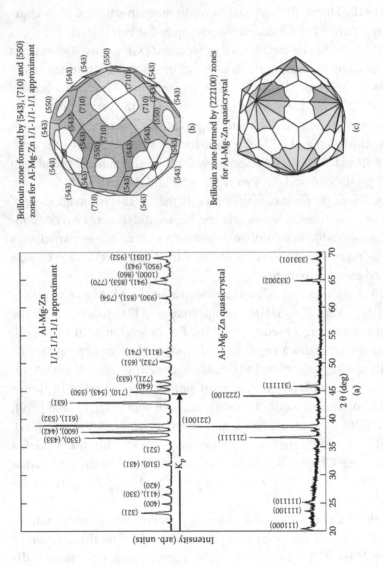

FIGURE 10.15 (a) x-ray diffraction spectra for $Al_{25.5}Mg_{39.5}Zn_{35}$ 1/1–1/1–1/1 approximant and $Al_{15}Mg_{44}Zn_{41}$ quasicrystal [24]. (b) Brillouin zone bounded by {543} + {710} + {550} zone planes for 1/1–1/1–1/1 approximant [23]. (c) Brillouin zone bounded by (222100) zone planes for quasicrystal [23]. Many arcs drawn onto the Brillouin zone planes illustrate contours due to the contact of free electron Fermi sphere with the radius $k_F = 15.7$ (nm)$^{-1}$ in the Al-Mg-Zn approximant and quasicrystal.

heat in the Al-Mg-Zn quasicrystal is smaller than that in its approximant [24]. We can say that an increase in the number of atoms in the unit cell, N, increases the magnitude of the *critical* $|\mathbf{G}|$ or $|\mathbf{G}|^2$ in units of $(2\pi/a)$ or $(2\pi/a)^2$ (see Figure 10.14b) and, hence, the number of lattice planes having the *critical* lattice spacing $2d$ in the real space or the number of equi-distant zone planes in reciprocal space. This, in turn, increases the number of directions in the reciprocal space, along which the interference with electrons near the Fermi level occurs. This makes it possible to form a deeper pseudogap and to lower the electronic energy more efficiently to stabilize such a complex structure in the framework of the Hume-Rothery stabilization mechanism.

10.6 UNIVERSAL TEST OF THE MATCHING CONDITION

Now we try to extend the test of the matching condition $2k_F = |\mathbf{G}|$ to the data for alloys including those characterized by an orbital hybridization-induced pseudogap, where the FLAPW-Fourier analysis and the Hume-Rothery plot have been already performed. Rather than working on $2k_F$ versus $|\mathbf{G}|$ diagram in reciprocal space, we construct wavelength λ_F versus $2d$ diagram in real space by converting $2k_F$ into λ_F through the relation $\lambda_F = 2\pi / k_F$ and the *critical* reciprocal lattice vector $|\mathbf{G}|$ into the *critical* lattice spacing $2d$ for the set of relevant lattice planes. Alloys studied are classified into three families: the first includes group (I) gamma-brasses and RT-type 1/1-1/1-1/1 approximants, where the *Hume-Rothery stabilization mechanism* operates (Chapters 7 and 9 and Section 10.5), the second covers those, in which orbital hybridization effects and d-states-mediated-FsBz-interactions dominate, and the third B2-compounds, where no pseudogap appears at the Fermi level.

All the numerical data including λ_F and $2d$ are summarized in Table 10.1. The data for group (I) gamma-brasses and the two RT-type approximants obeying the Hume-Rothery stabilization mechanism are plotted in Figure 10.16a. This is the real space version of Figure 10.14a. Obviously, all the data well satisfy the matching condition $\lambda_F = 2d$ and fall in a narrow range of $2d = 0.40 \pm 0.01$ nm, regardless of the size of the unit cell.

The data for d-states-mediated-splitting systems, which include group (II) gamma-brasses and MI-type Al-Cu-TM-Si (TM = Fe, Ru) approximants, are plotted in Figure 10.16b. It is clear that the data no longer fall in a narrow region at $2d = 0.40 \pm 0.01$ nm but are scattered over a

TABLE 10.1   Test of the Matching Condition $\lambda_F = 2d$ for Various Intermetallic Compounds

Phase	Alloys	Lattice Constant (nm)	Pseudogap at E_F	$(e/a)_{total}$	$2k_F$ $(nm)^{-1}$	λ (nm)	Set of Lattice Planes	2d (nm)
A	AgLi	0.3168	×	0.9	24.3	0.51	{110}	0.45
	AgMg	0.3302	×	1.48	26.9	0.47	{110}	0.47
	CuZn	0.29539	×	1.48	30.1	0.42	{110}	0.42
	NiZn	0.29083	×	1.27	29.0	0.43	{110}	0.41
	NiAl	0.2882	×	1.76	32.7	0.38	{110}	0.41
	MnZn	0.3070	×	1.23	27.2	0.46	{110}	0.43
B	Cu_5Zn_8	0.8878	○	1.60	30.4	0.41	{330}+{411}	0.42
	Cu_9Al_4	0.8707	○	1.60	31.0	0.41	{330}+{411}	0.41
	Al_8V_5	0.9223	◎	1.94	31.2	0.40	{330}{411}	0.43
							{332}	0.39
	Ag_5Li_8	0.9907	×	1.00	23.2	0.54	{321}	0.46
C	$Al_{30}Mg_{40}Zn_{30}$ (%)	1.435	○	2.3	30.9	0.41	{543}+{710}+{550}	0.41
	$Al_{84}Li_{52}Cu_{24}$	1.389	○	2.05	30.7	0.41	{631}	0.40
	$Al_{108}Cu_6Fe_{24}Si_6$	1.248	◎	2.59	35.6	0.35	{543}+{710}+{550}	0.35

Note: A: B2-compounds, B: gamma-brasses, C: 1/1-1/1/1-1/1 approximants. ○: purely FsBz-induced pseudogap systems, ◎: orbital-hybridization-induced pseudogap systems and ×: systems having no pseudogap across the Fermi level.

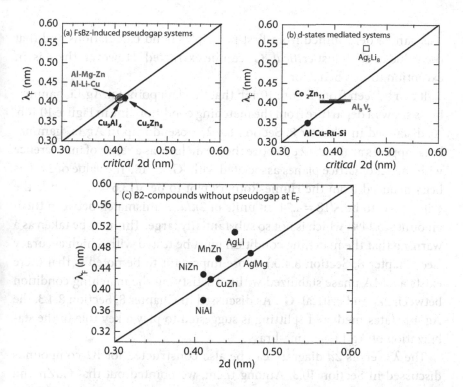

FIGURE 10.16 λ_F versus $2d$ for (a) group (I) Cu_5Zn_8 and Cu_9Al_4 gamma-brasses and RT-type $Al_{30}Mg_{40}Zn_{30}$ and $Al_{52.5}Li_{32.5}Cu_{15}$ 1/1-1/1-1/1 approximants, (b) d-states-mediated-splitting systems and (c) B2-compounds. λ_F: electron wavelength at the Fermi level and $2d$: *critical* lattice spacing for the set of lattice planes responsible for the formation of a pseudogap in (a) and (b). In (c), $2d$ represents the lattice spacing of the set of {110} lattice planes of B2-compounds.

range of the lattice spacing from 0.35 to 0.46 nm.* In the case of Co_2Zn_{11} and Al_8V_5 gamma-brasses, we emphasized in Chapter 8, Sections 8.2 and 8.3, that there exist, at least, two *critical* $|G|^2$s of 18 and 22 due to the d-states-mediated-FsBz interactions. Accordingly, the range of *critical* $2d$ is widened. Though the data points are still limited, we are inclined to believe that the matching condition generally holds even for orbital hybridization-induced pseudogap systems, provided that $2k_F$ and *critical*

---

* The *critical lattice* spacing $2d$ for $Al_{30}Mg_{40}Zn_{30}$ and $Al_{68}Cu_7Fe_{17}Si_8$ 1/1-1/1-1/1 approximants is calculated to be 0.40 and 0.35 nm by inserting their lattice constants $a = 1.4355$ and 1.248 nm into the relation $2d = 2a/\sqrt{50}$, respectively. A smaller lattice constant in the MI-type approximant yields a shorter *critical* lattice spacing than in the RT-type 1/1-1/1-1/1 approximant. Nevertheless, the value of **e/uc** becomes almost the same between the two families of approximants (see Figure 10.17a).

$|G|$s can be determined from first-principles band calculations and that the value of the most *critical* $|G|$ can be extracted. However, there is an exception to this behavior.

It can be seen from Figure 10.16b that the data point for Ag_5Li_8 gamma-brass shows a departure from the matching condition line in Figure 10.16b. As discussed in Chapter 8, Section 8.4.2, a pseudogap in Ag_5Li_8 gamma-brass appears at about +2 eV above the Fermi level as a result of interference with the set of lattice planes associated with $|G|^2 = 18$. The value of $2k_F$ has been deduced from the Hume-Rothery plot to be $\sqrt{13.6} = 3.66$ while the critical $|G|$ to be $\sqrt{18} = 4.24$ in units of $2\pi/a$. A mismatch between them amounts to 14%, which is not so substantially large. This may be taken as a warning that the matching condition must be tested with a high accuracy (see Chapter 8, Section 8.4.3.). More important to be noted is that there exists a CMA phase stabilized without satisfying the matching condition between $2k_F$ and critical $|G|$. As discussed in Chapter 8, Section 8.4.3, the Ag-4d-states-mediated-splitting is suggested to play a key role in the stabilization of Ag_5Li_8 gamma-brass.

The λ_F versus $2d$ diagram can be also constructed for B2-compounds discussed in Section 10.3. Among them, we pointed out that CuZn and AlMg B2-compounds satisfy the Hume-Rothery rule with $e/a = 3/2$ but that others do not. Since B2-compound contains only two atoms in the unit cell, the only relevant reciprocal lattice vector near the Fermi level must be $|G|^2 = 2$ corresponding to the set of {110} lattice planes. The corresponding lattice spacing $2d$ is calculated from the {110} lattice planes, though it is not *critical* because the FsBz interaction involved is too weak to cause a noticeable pseudogap near the Fermi level (see small van Hove singularity due to {110} zones in the bcc-Cu DOS in Chapter 5, Section 5.6). It happens that data points for CuZn and AgMg B2-compounds fall on the matching condition line $\lambda_F = 2d$. We have essentially constructed an "*effective Fermi sphere*" representing the momentum distribution of electrons outside the MT sphere, which is located within ±5% from the {110} zone planes of the Brillouin zone of a bcc lattice.* To the best of our knowledge, the *effective Fermi sphere* is a new concept. Its physical impli-

---

* It may be worthwhile mentioning that, even though the mismatch between $2k_F$ and $|G|$ is only ±5%, discussing the stabilization in terms of the matching condition is meaningless for B2-compounds because of the absence of a pseudogap at the Fermi level.

cation will be discussed in parallel with the **e/a** value deduced from the Hume-Rothery plot.

10.7 ORBITAL HYBRIDIZATIONS VERSUS FsBz INTERACTIONS

We are aware that the FsBz interactions can occur only for systems, in which diffraction intensities consist of either a finite or infinite number of δ-functions, while orbital hybridization effects can occur even in disordered systems like in amorphous alloys and even in liquid state, where the Bragg condition is lost (see Chapter 4, Section 4.2). As emphasized in Chapter 8, Section 8.3, the orbital hybridization between the V-3d and Al-3p states in Al_8V_5 gamma-brass results in the formation of a pseudogap at about +0.5 eV above the Fermi level in its total DOS. This would naturally contribute to lowering the electronic energy, since electrons are exclusively filled into the V-3d bonding subband. An orbital hybridization-induced pseudogap also exists in the energy range from +1.5 to +2.0 eV in AlV B2-compound [27]. Therefore, we consider it difficult to discuss a crystal structure-dependent phase stability solely in terms of orbital hybridization effects.

In this regard, we may recall discussions in Chapter 8, Sections 8.3, where it is shown that the $|\mathbf{G}|^2 = 18$ wave is the most important in forming bonding states due to the d-states-mediated-FsBz interactions in group (II) gamma-brasses like Co_2Zn_{11} and Al_8V_5. In Chapter 9, Section 9.4.2.1, we showed that a FsBz-induced pseudogap is inherently present but is hidden behind more prominent orbital hybridization-induced one in Al-Cu-Ru-Si 1/1-1/1-1/1 approximant. It should be recalled that the FsBz interactions and the d-states-mediated-FsBz interactions as well, are both structure- and **e/a**-sensitive. Indeed, Figure 10.16b demonstrated the validity of the matching condition for Co_2Zn_{11} and Al_8V_5 gamma-brasses in group (II) and MI-type Al-Cu-Ru-Si 1/1-1/1-1/1 approximant, in which the orbital hybridization predominates. This indicates that the d-states-mediated-FsBz interactions would potentially play a key role in the stabilization of CMAs characterized by an orbital hybridization-induced pseudogap.

In order to shed more light on the Hume-Rothery stabilization mechanism, we try to examine if the electron concentration parameter **e/a**, or more specifically, **e/uc** for CMAs tends to be fixed at a specific constant value, as imposed by the matching condition given by Equation 4.1, provided that the critical reciprocal lattice vector involved is the same. Relevant numerical data for various CMAs are listed in Table 10.2. It is

TABLE 10.2 $(e/a)_{total}$, N, **e/uc**, VEC, and **VE/uc** for Various CMAs

CMA		$(e/a)_{total}$	N	e/uc	VEC	VE/uc	Ref.
Cu_5Zn_8		1.615	52	84	11.615	604	CH.7
Cu_9Al_4	G	1.615	52	84	8.538	444	CH.7
Al_8V_5		1.94	52	100.9	3.77	196	CH.8
$Al_{30}Mg_{40}Zn_{30}$		2.30	160	368	5.3	848	CH.9
$Al_{52.5}Li_{32.5}Cu_{15}$		2.05	160	328	3.55	568	CH.9
$Al_{68}Cu_7Ru_{17}Si_8$	1/1	2.546	139	354	4.49	624	CH.9
$Al_{73.9}Re_{17.4}Si_{8.7}$		2.645	138	365	3.78	522	CH.9
Cd_6Ca		2.0	168	336	10.57	1776	[28]
Cd_6Yb		2.0	168	336	12.57	2112	[28]
$Al_{15}Mg_{43}Zn_{42}$	2/1	2.15	676–692	1453–1471	6.35	4293	[24,29–31]

Note: G: gamma-brasses, 1/1: 1/1-1/1-1/1 approximants, 2/1: 2/1-2/1-2/1 approximants.

recalled that gamma-brasses in groups (I) and (II) are characterized by the set of the *critical* lattice planes {330} and {411} with $|\mathbf{G}|^2 = 18$ and that both RT- and MI-type 1/1-1/1-1/1 approximants by the set of *critical* lattice planes {543}, {710}, and {550} with $|\mathbf{G}|^2 = 50$.

Included in Table 10.2 are also the data from literature: Cd_6Ca and Cd_6Yb and $Al_{15}Mg_{43}Zn_{42}$ 2/1-2/1-2/1 approximant. Both Cd_6Ca and Cd_6Yb 1/1-1/1-1/1 approximants are known to be described by the Tsai-type cluster and to contain 168 atoms per unit cell with space group $Im\overline{3}$ [28]. The atomic structure data for $Al_{15}Mg_{43}Zn_{42}$ 2/1-2/1-2/1 approximant is also available in literature [24,29–31].* It contains 676–692 atoms in its unit cell with space group $Pa3$. We can safely assume the set of {11 20}, {10 50}, {10 43}, and {865} lattice planes to be *critical* in $Al_{15}Mg_{43}Zn_{42}$ 2/1-2/1-2/1 approximant. The value of **e/uc** listed in Table 10.2 is simply calculated by taking the product of $(e/a)_{total}$, which is calculated by using $(e/a)_{TM}$ for the TM element in Figure 10.11 and nominal valencies for non-TM elements, and the number of atoms per unit cell, N. Similarly, the number of valence electrons per unit cell, **VE/uc**, is given by the product of **VEC** and N.

---

* The 2/1-2/1-2/1 approximant in Al-Mg-Zn alloy system was first discovered in 1995 by Takeuchi and Mizutani [24] at the composition $Al_{15}Mg_{43}Zn_{42}$. The atomic structure was later studied by Sugiyama et al. [29] at the composition $Al_{16}Mg_{42}Zn_{42}$ and by Lin and Corbett [30] at the composition $Al_{12.6}Mg_{31.8}Zn_{55.6}$. Lin and Corbett claimed the possession of 692 atoms per unit cell. More recently, Kreiner [31] reported that the 2/1-2/1-2/1 approximant exists at composition very close to $Al_{15}Mg_{43}Zn_{42}$ but not at $Al_{12.6}Mg_{31.8}Zn_{55.6}$ and that it contains 676 atoms per unit cell. We chose N = 676–692 for $Al_{15}Mg_{43}Zn_{42}$. Fortunately, however, uncertainties in N cause no serious error on a log-log scale in Figure 10.17.

Using all these data, we plotted **e/uc** as a function of **VE/uc** in Figure 10.17a on a log-log scale. It can be seen that the value of **e/uc** is fairly well kept constant within the scatter of a few percentiles for each family of CMAs characterized by the same *critical* $|G|^2$. The data for group (I) gamma-brasses are well centered at **e/uc** = 52 × (21/13) = 84. In the case of group (II) Al_8V_5 gamma-brass, the value of **e/uc** is slightly off from this characteristic value (see Table 10.2). This is naturally understood, since Al_8V_5 gamma-brass is excluded from the family obeying the **e/a** = 21/13 rule and is characterized by critical reciprocal lattice vectors $|G|^2$'s over 18 to 22 (see Chapter 8, Section 8.3).* We may say from the data in Figure 10.17 (a) that all gamma-brasses including Al_8V_5 obey the matching condition with $|G|^2$ = 18 to 22.† More surprising is that all 1/1-1/1-1/1 approximants fall in the neighborhood of **e/uc** = 350, including not only RT- and MI-type ones but also Tsai-type Cd_6Ca and Cd_6Yb. Note that all of them belong to space group $Im\overline{3}$. Thus, the data in Figure 10.17a strongly indicate that $|G|^2$ =18, 50, and 125 can be assigned as being critical for families of gamma-brasses, and 1/1-1/1-1/1 and 2/1-2/1-2/1 approximants, respectively.

To strengthen above arguments, we plot the value of **e/uc** as a function of the *critical* $|G|^2$ in Figure 10.17b on a log-log scale. All the data are found to fall on a straight line with the slope of approximately 1.5, being well consistent with the relation

$$(e/uc) \propto \left[\left(2k_F \right)^2 \right]^{3/2} = \left[|G|^2 \right]^{3/2}$$

derived from the matching condition. Obviously, the assignment of a *critical* $|G|^2$ to each family of CMAs is proved to be correct. The discussions above in relation to Figure 10.17 are also supported by the theory based on the free electron model by Mott and Jones in 1936 [32]. Its importance will be emphasized in Conclusions, Chapter 11.

Before ending this section, three remarks may be added. Firstly, it is claimed that all 1/1-1/1-1/1 approximants, irrespective of RT-, MI-, and Tsai-type clusters involved, create a FsBz-induced pseudogap through

---

* The value of **e/uc** for Al_8V_5 is deduced to be 100.9 (= 1.94 × 52), as listed in Table 10.2. This is larger than the maximum number of electrons per unit cell, i.e., $(e/uc)_0$ = 90 (see Section 7.6.).

† The value of $2k_F$ for Al_8V_5 has been deduced from the Hume-Rothery plot to be $\sqrt{21}$ = 4.58 in units of $2\pi/a$ (see Section 8.3.2). A mismatch, defined as $(2k_F - |G|)/2k_F$, amounts to +7.4 and −2.4% when *critical* $|G|$s are $\sqrt{18}$ = 4.24 and $\sqrt{22}$ = 4.69, respectively.

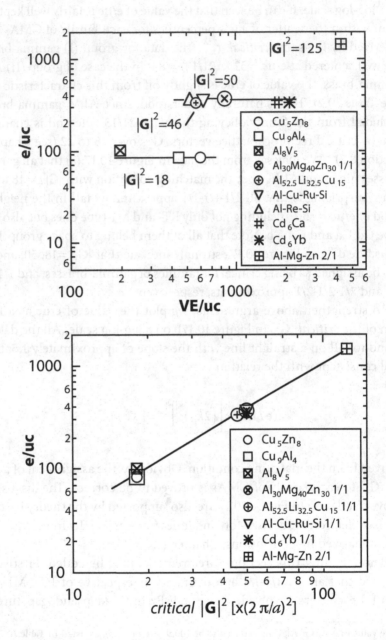

FIGURE 10.17 (a) **e/uc** as a function of **VE/uc** for various CMAs on a log-log scale. (b) **e/uc** as a function of *critical* $|G|^2$ for various CMAs on a log-log scale.

interference with the set of {543}, {710}, and {550} lattice planes. Secondly, the matching condition is functioning for all CMAs, as long as they are characterized by a sizable pseudogap across the Fermi level. In other words, we can state that systems obeying the Hume-Rothery stabilization mechanism cover not only those stabilized by forming a purely FsBz-induced pseudogap like group (I) gamma-brasses and RT-type 1/1-1/1-1/1 approximants but also those stabilized by producing a d-states-mediated-FsBz-induced pseudogap, which includes group (II) gamma-brasses and MI-type approximants. Finally, the $(e/a)_{TM}$ value listed in Figure 10.11 is proved to be physically well acceptable.

Judging from the arguments above, we may say that the data for all quasicrystals, if both $2k_F$ and *critical* $|G|$ s were to be rigorously determined, would have more perfectly satisfied the matching condition $2k_F = |G|$ than those shown in Figure 10.13.

10.8 ROLES OF **VEC** AND **e/a** IN DESIGNING NEW CMAS

It is now interesting to consider how one should differentiate the role of the two electron concentration parameters **VEC** and **e/a** to design a new CMA characterized by a pseudogap across the Fermi level. Let us once again direct our attention to the Seebeck coefficient in Fe_2VAl doped with three different types of elements (see Section 10.1) [4]. It has been well established that late TM elements like Co, Pt, etc., early TM elements like Ti, Zr, Mo, W, etc. and non-TM elements like Si and Ge, etc. can be substituted for Fe, V, and Al, respectively, to vary **VEC**. The formation of a master curve when plotted against **VEC** is already shown in Figure 10.4. The **VEC** for the host Fe_2VAl is simply calculated as

$$VEC = \frac{2*8+1*5+1*3}{4} = 6.0 \qquad (10.1)$$

It is clear that the data on the Seebeck coefficient fall on a universal curve only if **VEC** is chosen as an electron concentration parameter, and that the data can be sharply divided into the p- and n-type regimes corresponding to a positive and negative Seebeck coefficient across **VEC** = 6.0, respectively.

The LAPW band calculations have been performed for the Fe_2VAl intermetallic compound and its total DOS is reproduced in Figure 10.18 [33]. Its electronic structure is characterized by a deep pseudogap across the Fermi level, though Fe_2VAl contains only four atoms per unit cell and cannot be

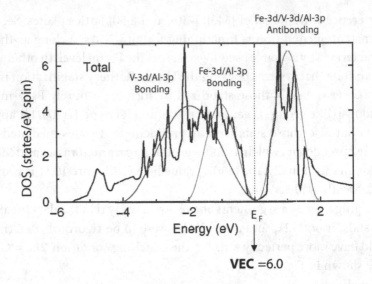

FIGURE 10.18 The total DOS for Fe_2VAl [31]. The Fermi level corresponds to **VEC** = 6.0. Gaussian curves are roughly drawn as an eye guide to approximate the TM-3d/Al-3p (TM = V and Fe) bonding and antibonding subbands.

regarded as a CMA. The pseudogap is apparently caused by splitting of both Fe-3d and V-3d states into bonding and antibonding subbands as a result of orbital hybridizations between these d-states and Al-3p states. The thermoelectric power is known as one of electron transport phenomena determined by the DOS at the Fermi level [1]. Hence, the emergence of a master curve indicates that a rigid-band model holds well, regardless of the choice of atomic species for three types of dopants. Indeed, thanks to the validity of the rigid-band model, Nishino [4] could design a number of p- and n-type thermoelectric materials based on Fe_2VAl. This clearly indicates that the **VEC** can be used as a powerful parameter, when a rigid-band model holds.

Among various quaternary alloys Nishino's research group has so far synthesized, we selected those containing TM elements, whose $(e/a)_{TM}$ is available from Figure 10.11. Included are $(Fe-Mn)_2VAl$, $(Fe-Pd)_2VAl$, $Fe_2V(Al-Si)$, and $Fe_2V(Al-Ge)$. The Seebeck coefficient for these systems is shown in Figure 10.19 (a) as a function of $(e/a)_{total}$, which is calculated by taking a composition-weighted average of $(e/a)_{TM}$ for the TM element and nominal valencies for polyvalent elements. The corresponding **VEC** dependence of the Seebeck coefficient is shown in Figure 10.19b. It is clear that the universal curve collapses, when plotted against $(e/a)_{total}$, and that

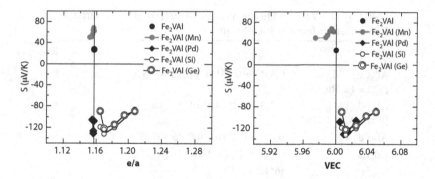

FIGURE 10.19   (a) Seebeck coefficient as a function of $(e/a)_{total}$, which is calculated using $(e/a)_{TM}$ in Figure 10.11, $(e/a)_{Al} = 3$ and $(e/a)_{Si} = (e/a)_{Ge} = 4$. (b) Seebeck coefficient as a function of VEC for Fe_2VAl doped with Mn, Pd, Si, and Ge elements. The data on the Seebeck coefficient are due to courtesy by Prof. Y. Nishino.

the data are distributed in the very vicinity of $(e/a)_{total} = 1.16$ except for the data where Si and Ge are chosen as dopants. Hence, we can no longer distinguish the p- from the n-type with respect to $(e/a)_{total}$. However, it seems premature to conclude if the matching condition really holds true or breaks down in such structurally simple pseudogap system Fe_2VAl. More accumulation of $(e/a)_{TM}$ data in Figure 10.11 and the execution of the FLAPW-Fourier analysis for the Fe_2VAl are needed to confirm this behavior.

Table 10.3 lists the value of VEC for various quasicrystals in which orbital hybridizations between TM-d states and non-TM p-states result in a deep pseudogap across the Fermi level (see Chapter 9, Figures 9.11, 9.14, and 9.16). Quasicrystals listed are divided into several families, depending on characteristic features of their valence band structures. We see that the VEC is essentially constant for a given family of quasicrystals. For example, a common rigid-band cannot be assumed for $Al_{63}Cu_{25}TM_{12}$ (TM = Fe, Ru, and Os) and $Al_{70}Pd_{20}TM_{10}$ (TM = Mn, Tc, and Re), since VECs involved are different. However, a common value of $(e/a)_{total}$ is likely assigned to them, as inferred from the data in Figure 10.17 (a). Indeed, values of $(e/a)_{total}$ for $Al_{63}Cu_{25}TM_{12}$ (TM = Fe, Ru, Os) and $Al_{70}Pd_{20}TM_{10}$ (TM = Mn, Re) quasicrystals turn out to be 2.22 and 2.16, respectively, if $(e/a)_{TM}$ listed in Figure 10.11 is employed. It is almost at the value around 2.2.*

---

* In this evaluation, we assumed $(e/a)_{TM}$ for Pd to be 0.07, which is obtained for Pd_2Zn_{11} gamma-brass, where the Pd-4d band is immersed well below the Fermi level (see Figure 8.2). A slightly larger $(e/a)_{TM}$ would be obtained, if the Hume-Rothery plot is made for other Pd-compounds, in which the Pd-4d band is located closer to the Fermi level.

TABLE 10.3   VEC for Quasicrystals Characterized by the Orbital Hybridization-Induced Pseudogap

	VEC	Ref.
$Al_{63}Cu_{25}TM_{12}$ (TM = Fe, Ru, Os)	0.63*3 + 0.25*11 + 0.12*8 = 5.60	8
$Al_{70}Pd_{20}TM_{10}$ (TM = Mn, Tc, Re)	0.7*3 + 0.2*10 + 0.1*7 = 4.8	8
$Al_{70}Pd_{20}V_5Co_5$	0.7*3 + 0.2*10 + 0.05*5 + 0.05*9 = 4.8	8
Cd_6Ca	(6*12 + 2)/7 = 10.57	8
Cd_6Sr	(6*12 + 2)/7 = 10.57	8
In_4Ag_2Ca	(4*13 +2 *11 + 2)/7 = 10.85	8
Cd_6Yb	(6*12 + 16)/7 = 12.57	8
$Ag_{47}Ga_{38}Yb_{15}$	0.47*11 + 0.38*13 + 0.15*16 = 12.51	8
In_4Ag_2Yb	(4*13 + 2*11 + 16)/7 = 12.85	8
$Zn_{84}Mg_8TM_8$ TM = Ti, Zr, Hf	0.84*12 + 0.08*2 + 0.08*4 = 10.56	10
$Zn_{74}Ni_{10}Sc_{16}$	0.74*12 + 0.1*10 + 0.16*3 = 10.36	10
$Zn_{74}Co_6Sc_{16}$	0.78*12 + 0.06*9 + 0.16*3 = 10.38	10
$Zn_{74}Fe_7Sc_{16}$	0.77*12 + 0.07*8 + 0.16*3 = 10.28	10
$Zn_{75}Mn_{10}Sc_{15}$	0.75*12 + 0.1*7 + 0.15*3 = 10.15	10
$Zn_{72}Cu_{12}Sc_{16}$	0.72*12+0.12*11+0.16*3 = 10.44	10
$Zn_{74}Ag_{10}Sc_{16}$	0.74*12+0.10*11+0.16*3 = 10.46	10
$Zn_{74}Au_{11}Sc_{16}$	0.74*12 + 0.11*11 + 0.16*3 = 10.57	10
$Cu_{48}Ga_{34}Mg_3Sc_{15}$	0.48*11 + 0.34*13 + 0.03*2 + 0.15*3 = 10.21	10

Tsai [8, 9] employed Raynor's negative valency scheme for TM elements and designed a new quasicrystal by substituting 5 at.%V and 5 at.%Co for 10 at.%Mn in $Al_{70}Pd_{20}Mn_{10}$ with the hope that $(e/a)_{total}$ is kept unchanged.* If $(e/a)_{TM}$ in Figure 10.11 is used, the value of $(e/a)_{total}$ turns out to be 2.16 and 2.13 for $Al_{70}Pd_{20}Mn_{10}$ and $Al_{70}Pd_{20}V_5Co_5$, respectively, being again kept unchanged. Moreover, this is a condition leading to a constant VEC equal to 4.8. Hence, his successful synthesis of the quaternary quasicrystal does not necessarily lend support to the validity of Raynor's negative valency scheme. The usefulness of the parameter VEC in quasicrystals would more effectively come into play, if electronic properties like the Seebeck coefficient in Figure 10.4 are available and plotted as a function of VEC. The parameter $(e/a)_{total}$, if it is evaluated using $(e/a)_{TM}$ listed in Figure 10.11, may be also useful in design of a new CMA characterized

---

* Raynor did not discuss the valency of V atom [5]. Tsai [8, 9] apparently assigned its valency as −5.66 by extrapolating Raynor's negative e/a tendency to increase by one in negative direction with decreasing the atomic number by one.

by a pseudogap at the Fermi level, since we may tune compositions and a combination of constituent elements so as to keep it unchanged.

10.9 SUMMARY

The value of $(e/a)_{TM}$ for most of 3d-TM elements in the periodic table can be deduced from the Hume-Rothery plot and is summarized in Figure 10.11. Its value is always a small and positive number less than unity in sharp contrast to negative valencies proposed by Raynor in 1949. Its soundness was confirmed in various ways such as the $(e/a)_{total}$ dependence of the melting temperature for B2-compounds shown in Figure 10.12, and the e/uc behavior in Figure 10.17.

The matching condition given by Equation 4.1 was studied by using the Fermi diameter $2k_F$ deduced from the Hume-Rothery plot and critical $|G|$s from the FLAPW-Fourier method for gamma-brasses, 1/1-1/1-1/1 approximants and B2-compounds. The matching condition holds true for all CMAs, as long as they are characterized by a sizable pseudogap across the Fermi level. It is emphasized that a FsBz-induced pseudogap, including the d-states-mediated one, is both structure- and e/a-sensitive through the Brillouin zone specific to a given structure. The data shown in Figures 10.16 and 10.17 can be taken as a strong evidence for this. All 1/1-1/1-1/1 approximants with space group of either $Im\overline{3}$ or $Pm\overline{3}$, which include RT-, MI-, and Tsai-types, form a FsBz-induced pseudogap through interference with the set of {543}, {710}, and {550} lattice planes. It is claimed that systems obeying the Hume-Rothery stabilization mechanism cover not only those stabilized by forming a purely FsBz-induced pseudogap like in group (I) gamma-brasses and RT-type 1/1-1/1-1/1 approximants but also those stabilized by forming a d-states-mediated-FsBz-induced pseudogap like in group (II) gamma-brasses and MI-type approximants.

Finally, we discussed how the role of the two electron concentration parameters VEC and e/a should be differentiated in a possible design of a new CMA characterized by a pseudogap across the Fermi level. The VEC can be used as a useful parameter to design a new alloy as long as a rigid-band model holds, whereas the parameter e/a or e/uc calculated from Figure 10.11 is also useful, since it plays a different role from VEC and points to the value imposed by the matching condition (see Chapter 7, Figure 7.18, and Figure 10.17).

REFERENCES

1. R.M. Bozorth, *Ferromagnetism*, (D van Nostrand, New York, 1951); U. Mizutani, *Introduction to the Electron Theory of Metals* (Cambridge University Press, Cambridge, 2001).
2. C.H. Cheng, K. Gupta, E.C. van Reuth, and P.A. Beck, *Phys. Rev.* 126 (1962) 2030.
3. B.T. Matthias, *Phys. Rev.* 97 (1955) 74.
4. Y. Nishino, *The Science of Complex Alloy Phases*, edited by T.B. Massalski and P.E.A. Turchi, TMS (The Minerals, Metals & Materials Society, 2005), pp. 325–344.
5. G.V. Raynor, *Prog. Met. Phys.* 1 (1949) 1.
6. A.P. Tsai, A. Inoue, and T. Masumoto, *Jpn. J. Appl. Phys.* 27 (1988) L1587; A.P. Tsai, A. Inoue, Y. Yokoyama, and T. Masumoto, *Mater. Trans. Jpn. Inst. Met.* 31 (1990) 98.
7. Y. Yokoyama, A.P. Tsai, A. Inoue, T. Masumoto, and H.S. Chen, *Mater. Trans. Jpn. Inst. Met.* 32 (1991) 421.
8. A.P. Tsai, *The Science of Complex Alloy Phases*, edited by T.B. Massalski and P.E.A. Turchi, TMS (The Minerals, Metals & Materials Society, 2005), pp. 201–214.
9. A.P. Tsai, *Sci. Technol. Adv. Mater.* 9 (2008) 013008.
10. T. Ishimasa, *The Science of Complex Alloy Phases*, edited by T.B. Massalski and P.E.A. Turchi, TMS (The Minerals, Metals & Materials Society, 2005), pp. 231–249.
11. G. Trambly de Laissardière, D. Nguyen Manh, L. Magaud, J.P. Julien, F. Cyrot-Lackmann, and D. Mayou, *Phys. Rev.* B52 (1995) 7920.
12. E.S. Zijlstra and S.K. Bose, *Phil. Mag.* 86 (2006) 717.
13. J.B. Haworth and W. Hume-Rothery, *Phil. Mag.* 43 (1952) 613.
14. U. Mizutani, T. Takeuchi, and H. Sato, *Prog. Mat. Sci.* 49 (2004) 227.
15. T. Ishimasa, S. Kashimoto, and R. Maezawa, *Mater. Res. Soc. Symp. Proc.* 805 (2004) 3–14.
16. W. Hume-Rothery, *J. Inst. Metals* 35 (1926) 295.
17. U. Mizutani, R. Asahi, H. Sato, T. Noritake, and T. Takeuchi, *J. Phys.: Condens. Matter* 20 (2008) 275228.
18. U. Mizutani, R. Asahi, T. Takeuchi, H. Sato, O.Y. Kontsevoi, and A.J. Freeman, *Z. Kristallogr.* 224 (2009) 17.
19. W. Hume-Rothery, *Atomic Theory for Students of Metallurgy* (The Institute of Metals, London, 1962), p. 135.
20. W. Hume-Rothery, *J. Inst. Metal,* 9 (1961–62) 42.
21. V. Raynor, *The Electronic Structure of the Transition Metals*, The report of a conference held at Abingdon in November 1952 (Inter-Service Metallurgical Research Council Basic Properties of Metals Committee).
22. U. Mizutani, T. Noritake, T. Ohsuna, and T. Takeuchi, *Phil. Mag.* 90 (2010) 1985.
23. U. Mizutani, *MATERIA* (in Japanese) 46, No. 2 (2007) 77.
24. T. Takeuchi and U. Mizutani, *Phys. Rev.* B 52 (1995) 9300.

25. H. Sato, T. Takeuchi, and U. Mizutani, *Phys. Rev.* B 64 (2001) 094207.
26. H. Sato, T. Takeuchi, and U. Mizutani, *Phys. Rev.* B 70 (2004) 024210.
27. U. Mizutani, R. Asahi, H. Sato, and T. Takeuchi, *Phys. Rev.* B 74 (2006) 235119.
28. C.P. Gomez and S. Lidin, *Phys. Rev.* B 68 (2003) 024203.
29. K. Sugiyama, W. Sun and K. Hiraga, *J. Alloys and Compounds* 342 (2002) 139.
30. Q. Lin and J.D. Corbett, *Proc. Natl. Acad. Sci U.S.A.*, 2006 September 12; 103(37) 13589.
31. G. Kreiner, presented at Quasicrystal Meeting, Stuttgart, Germany, January 30, 2009.
32. N.F. Mott and H. Jones, *The Theory of the Properties of Metals and Alloys* (Oxford University Press, England, 1936).
33. R. Weht and W.E. Pickett, *Phys. Rev.* B 58 (1998) 6855.

CHAPTER 11

Conclusions

In 1948, Hume-Rothery wrote one of his famous books entitled *Electrons, Atoms, Metals and Alloys* in the form of a dialogue between a "Young Scientist" and an "Older Metallurgist" to help readers in metallurgical industries be stimulated by a deliberate raising of questions [1]. The present author has decided to adopt this admirable approach to help his readers learn what is new in this monograph and what still remains unsolved and is left over for future work.

Young Scientist: I know that the Hume-Rothery electron concentration rule has received basic support in terms of the free electron theory advanced by Mott and Jones in 1936, which could provide a reason why the electron per atom ratio **e/a** values can favor the respective fcc, bcc, and gamma-brass phases in the noble metal alloys. I understand that the argument was that a spherical Fermi surface touches the respective Brillouin zone planes. Why did you write a new book on this topic?

Older Metallurgist: Yes, you may well ask. However, we are well aware that the free electron model was too simple to explain the various electronic properties of "realistic" metals and alloys. A quasicrystal was discovered in 1984. People started to recognize the importance of a pseudogap at the Fermi level in the DOS that contributes to stabilizing such complex metallic compounds. A pseudogap can never be worked out within the free electron model. As mentioned in Chapter 2, Section 2.3, the electronic energy of a given system can be lowered most efficiently, if the Fermi level of an alloy is located in the range of the DOS pseudogap.

As noted at the end of Chapter 10, Section 10.5, an increase in the number of atoms in the unit cell accompanies an increase in the number of Brillouin zone planes (or the number of relevant equivalent lattice planes), making the Brillouin zone "more spherical." This, in turn, increases the number of directions in the reciprocal space, along which the interference with electrons near the Fermi level occurs. Hence, the larger the number of atoms per unit cell, the deeper is the pseudogap formed at the Fermi level, and the more efficient is the lowering of the electronic energy tending to stabilize a given structure. This is the gist of the Hume-Rothery stabilization mechanism for CMAs. Hence, the need for performing first-principles electronic structure calculations for CMAs containing more than 50 atoms per unit cell has become urgent to resolve some of the yet unsettled questions regarding the Hume-Rothery electron concentration rule, particularly, for CMAs. This is the present book's aim.

YS: Why did you place so much emphasis on the difference between the two different electron concentration parameters, **e/a** and **VEC**?

OM: As discussed in Chapter 3, Section 3.2, Jones fully ignored the Cu-3d band, and calculated the DOS for fcc- and bcc-Cu in the NFE model. By having done this, he could automatically select an **e/a** value as an electron concentration parameter and directly link his results with the Hume-Rothery electron concentration rule, which is also expressed in terms of **e/a**. As emphasized in Chapter 1, Sections 1.2 and 3.2, however, the discovery of the neck in the Fermi surface contour in pure Cu by Pippard in 1957 eliminated the applicability of the Jones approach. Indeed, his failure obviously stemmed from neglecting the Cu-3d band. To elucidate the physics behind the Hume-Rothery electron concentration rule, I have emphasized the need for performing first-principles band calculations, where the Cu-3d band Jones ignored is fully taken into account. However, an integrated DOS now gives rise to the **VEC** instead of **e/a**. So, a problem has newly arisen how to extract from first-principles band calculations the electron concentration parameter **e/a**, which is essential in the Hume-Rothery electron concentration rule.

In collaboration with my colleagues, we have developed a very useful technique named the *FLAPW-Fourier method* to extract from first-principles FLAPW band calculations the *critical* reciprocal lattice vector $|G|$ responsible for the formation of a pseudogap. The essence of the FLAPW-Fourier method relies on the fact that the FLAPW wave function outside the MT sphere is expanded into plane waves with respect to the reciprocal lattice vector allowed to a given lattice (see Chapter 7, Sections 7.3 and 7.4). By extending the FLAPW-Fourier method, we have also developed the *Hume-Rothery plot*, which allows us to extract the value of **e/a** from first-principles band calculations. Interestingly, the above approach made it also possible to derive a new set of effective **e/a** values for transition metal (TM) elements in the periodic table. They turn out to be slightly positive numbers across the 3d-series, as summarized in Figure 10.11.

YS: You stressed in Chapter 5 that the total-energy difference ΔU_{total} between the two competing phases is crucially important in the discussion of their relative stability, and pointed out that in the phase competition between fcc- and bcc-Cu, the largest contribution to ΔU_{total} is not the valence-band structure energy difference ΔU_v but the $\Delta U_{pot.outside\,MT}$ (see Table 5.3). This is rather surprising, since many of us had implicitly understood ΔU_v to be the largest, as Jones did. Nevertheless, you took into account only the **VEC** dependence of ΔU_v within the rigid-band model and judged the **VEC** dependence of $\Delta U_{pot.outside\,MT}$ to be essentially negligible in the α/β phase transformation (Chapter 5, Section 5.6). I see the reason for your approach, since only the term U_v among those in Equation 5.2 can be pursued as a function of the **VEC** through the FsBz interactions, which are reflected as the van Hove singularities on the competing DOSs. However, in the remaining chapters, you have apparently ignored the contribution from terms other than U_v in Equation 5.2 and discussed an "absolute stability" for CMAs solely in terms of U_v without considering competing phase(s). Is this reasonable?

OM: I appreciate your difficulty here. A discussion about the relative stability between two competing phases even at absolute zero gets to be rather involved. As mentioned in connection with stability of Ag$_5$Li$_8$ gamma-brass in Chapter 8, Section 8.4.3, I argued that one has to find from the phase diagram an appropriate metastable phase as a counterpart competing with a given CMA compound. A candidate could be selected by examining neighboring phases having a slightly different composition. However, to make first-principles band calculations feasible, the two competing phases so chosen must be perfectly ordered at the same composition. So, a construction of such two ordered compounds having the same composition is almost impossible. In other words, a discussion about relative stability between fcc- and bcc-Cu as in Chapter 5 must be regarded as a very rare and ideal case. As stated at the end of Chapter 5, Section 5.7, α and β-phases in the Cu-Zn alloy system are competing with each other within only a few kJ/mol in the total-energy difference. Even in such an ideal case, we realized the discussion on relative stability to be a formidable task, particularly since van Hove singularities are extremely small for structurally simple phases (see Chapter 5, Figure 5.10).

That is why I emphasized that such a severe condition is more relaxed when dealing with CMAs. As shown in Figure 2.7, the presence of a deep pseudogap across the Fermi level in the DOS can lower the valence-band structure energy by 10 to 60 kJ/mol. As repeatedly emphasized, most CMAs are indeed characterized by a pseudogap at the Fermi level. People would further say that a quasicrystal having a deep pseudogap at the Fermi level is not competing with a hypothetical phase characterized by the free electron-like monotonic band, but is actually competing with an approximant having also a pseudogap. The situation can become again very delicate. Hence, I simply assumed throughout the volume that the presence of a pseudogap at the Fermi level is significant enough to stabilize such a CMA only through the reduction in the valence-band structure energy U_v without worrying about other terms in Equation 5.2 and any relative stability with possible competing phases. So, there is much work still to be done here.

YS: Why did you spend so many pages for gamma-brasses rather than quasicrystals among various CMAs?

OM: As stated in the Introduction, gamma-brasses have played a considerable role in the development of modern solid-state physics. More practically speaking, the execution of the FLAPW band calculations is of vital importance to extract the FsBz interactions, which play a key role in the Hume-Rothery electron concentration rule. The size of the unit cell must be large enough to produce a sizable pseudogap at the Fermi level but is still small enough to perform efficiently the FLAPW band calculations. Gamma-brasses containing 52 atoms per unit cell are ideally suited for this purpose. Moreover, there are more than 20 binary alloy systems that crystallize into the gamma-brass structure with space group of either $I\bar{4}3m$ or $P\bar{4}3m$. This provides a unique opportunity to make systematic studies within a given family of CMAs. It is also worthwhile mentioning that research on both Cu_5Zn_8 and Cu_9Al_4 gamma-brasses dates back to the 1920s, being initiated by Westgren, Phragmén, and Bradley and developed later by Hume-Rothery, Mott, and Jones in the 1930s. Indeed, we could extract an essence of the physics behind the Hume-Rothery electron concentration rule by performing first-principles band calculations for a series of gamma-brasses with and without containing TM elements.

YS: Does the ratio 21/13 carry a special meaning for the stabilization of Cu_5Zn_8 and Cu_9Al_4 gamma-brasses?

OM: As stated in Chapter 1, Section 1.3, Westgren and Phragmén were the first in 1928 to mention that the ratio 21/13 can be commonly assigned to both compounds, provided that valencies of Cu, Zn, and Al are assumed to be unity, two, and three, respectively: (5 * 1 + 8 * 2)/13 = 21/13 and (9 * 1 + 4 * 3)/13 = 21/13. However, we can deduce only a fractional number close to 21/13 from the *Hume-Rothery plot* based on the FLAPW band calculations (Chapter 7, Section 7.4). If it is expressed as a numerical value like 1.615, one cannot judge whether both compounds can be said to be really characterized by the same *e/a* value. To emphasize the physics behind it, or the existence of a common FsBz interaction between them, we should better use the ratio 21/13 rather than 1.615. Instead, the value of **VEC** is no longer the same between them: (5 * 11 + 8 * 12)/13 = 151/13 = 11.615 and (9 * 11 + 4 * 3)/13 = 111/13 = 8.538. In this case, the use of a complicated ratio like 151/13 would make little sense.

YS: Can you explain why Cu_5Zn_8 gamma-brass crystallizes into the structure with space group of $I\bar{4}3m$, whereas Cu_9Al_4 gamma-brass into the structure with space group $P\bar{4}3m$?

OM: The x-ray diffraction spectra are very similar between the Cu-Zn and Cu-Al gamma-brasses (see Appendix 2, Figures A2.1 and A2.3). The latter can be distinguished from the former only by the existence of (221), (300), and (210) diffraction lines associated with the CsCl-type super-lattice structure. There is a difference in atomic size ratio between these two alloy systems: $r_{Cu} / r_{Zn} = 1.278/1.394 = 0.92$ and $r_{Cu} / r_{Al} = 1.278/1.432 = 0.89$. It may be worthwhile mentioning the work by Brandon et al. [2]. They discussed the difference in the ordering schemes in the gamma-brass structure in terms of packing efficiencies at the respective radius ratios of the two constituent elements and in terms of maximizing the number of unlike-atom neighbors while minimizing the number of like-atom neighbors in the structure.

YS: You said at the beginning of Chapter 7 that any alloy with an off-stoichiometric composition would not remain stable at absolute zero, since the configurational entropy remains finite. This must be a consequence of the "third law of thermodynamics." Hence, a handling of off-stoichiometric alloys requires the arguments about phase stability at finite temperatures. You emphasized the role of vacancies in the case of gamma-brasses in Chapter 7, Sections 7.6 and 7.7. The number of quenched-in vacancies and the degree of chemical disorder would depend on a thermal history of a sample studied. I wonder if a reliable discussion can be made on phase stability of an off-stoichiometric alloy and a finite solid solution range as well.

OM: I fully agree with your comments. Indeed, the discussions in Chapter 7, Sections 7.6 and 7.7, are concerned with phase stability over the solid solution range at finite temperatures and are not as accurate as those made in Sections 7.2 to 7.4 for ordered Cu_5Zn_8 and Cu_9Al_4 gamma-brasses at absolute zero. I had to assume the rigid-band model to discuss the solid solution range in group (I) gamma-brasses. The rigid-band model was also employed for the discussion on the α/β-phase transformation in the Cu-Zn alloy system in Chapter 5. Its justification is not self-evident.

It is true that an off-stoichiometric alloy can exist only as a metastable phase because of the presence of chemical disorder. A phase diagram would eventually lose any finite solid solution range around an ordered compound, if appropriate annealing is done for an infinitely long time. In the case of the Cu-Zn alloy system, only pure Cu and Zn at the both ends, and Cu_5Zn_8 gamma-brass and CuZn B2-compound would remain stable at absolute zero. A thermally stable quasicrystal can exist only at a very specific composition without containing any disorder. However, any quasicrystals and CMAs so far synthesized in laboratories would contain a certain degree of quenched-in disorder, no matter how sharp the diffraction spots are. Perfectly ordered compounds may exist only in our imagination. In a practical approach, one has to decide on a maximum tolerable presence of such disorder, above which the system could be no longer treated as a "perfectly ordered compound."

YS: So you said that even if there are vacancies and chemical disorder at high temperatures, a given structure may settle down to have a perfect chemical order corresponding to an expected stoichiometry at absolute zero. Since all the structures are observed only at "real" temperatures, we really don't know if they would be there at absolute zero. I know that you can only do your first-principles band calculations at absolute zero, but this avoidance of dealing with entropy and off-stoichiometry still worries me.

OM: Nobody can surely say that both Cu_5Zn_8 and Cu_9Al_4 gamma-brasses are stable as ordered compounds at absolute zero. At any real temperatures, one always has to worry about quenched-in-vacancies or chemical disorder even in ordered compounds. Its presence inevitably involves statistical averaging over all possible randomly distributed atom configurations. In the present monograph, I limited myself to the assumption that ordered compounds like Cu_5Zn_8 and Cu_9Al_4 gamma-brasses exist as a stable phase at absolute zero to take full advantage of using first-principles band calculations. No such works at absolute zero have been reported in the past for the H-R rules.

As I mentioned in Chapter 5, alpha and beta phases are competing within only a few kJ/mol in the valence-band structure energy difference at absolute zero. Such a small difference in the electronic energy arising from extremely small van Hove singularities on the DOS (Figures 5.10 and 5.11) would be smeared out, as soon as statistical averaging is made for randomly distributed atom configurations at off-stoichiometric compositions. Nevertheless, we have the H-R rule with $e/a = 1.4$ for noble metal alloys. As you say, this empirical rule was established at the working temperatures. This is a big dilemma. It is beyond our present ability to discuss the α/β phase competition by incorporating factors like the entropy term at finite temperatures into the free-energy difference, though it is unlikely to expect any **e/a** dependent behavior in the entropy term. Clearly, there is a need to develop a theoretical tool to clarify accurately the effect of contributions arising from statistical averaging of atom configurations on the DOS and to calculate the electronic structure and the total-energy for an off-stoichiometric alloy with the same accuracy as that available now for an ordered compound.

YS: What is remaining as a future work regarding the Hume-Rothery electron concentration rule?

OM: Hume-Rothery himself had hoped to eventually enable engineers and industry researchers to design a new functional alloy by fully playing with the critical alloying parameters, i.e., atomic size factor, electronegativity difference, and electron concentration without relying on highly sophisticated calculations. However, all of us are now well aware that this was too optimistic. To design a new alloy, one must first consult the equilibrium phase diagram, because a material we handle is available only at "real" temperatures. Of course, the recent development of computer science concerning first-principles calculations is extremely powerful and will definitely serve more and more important roles in the future. The development of a new functional material can be made in a more efficient way with the assistance of first-principles calculations. For example, the evaluation of the total-energy in first-principles band calculations like WIEN2k certainly provides us with valuable information about whether a new material designed is liable to be stabilized or not, before its synthesis is attempted in laboratory.

Nevertheless, I consider the availability of an **e/a** table like that shown in Figure 10.11 to be of very useful importance as an assisting tool in alloy designing in the future. I trust that its usefulness and power has been demonstrated in many cases in the present monograph: you can find examples directly from the **e/a**, or **e/uc**, plots such as in Figures 7.18, 8.30, 8.32b, 10.12b, 10.17a,b, 10.19a, etc., and others indirectly from the matching condition such as in Figures 10.14 and 10.16. However, there still remain many missing data in Figure 10.11. Data for remaining TM elements could be added. In Chapter 10, Section 10.4, we emphasized that a different set of $(e/a)_{TM}$ values must be assigned to some TM elements like Ni, depending on the location of its d-band relative to the Fermi level. Such "surrounding effect" on $(e/a)_{TM}$ must also be clarified. The "surrounding effect" on the **e/a** value for elements in the column IV and V in the periodic table should be also studied. All these works are likely to be done in near future by performing the Hume-Rothery plots for various compounds. A complete establishment of the **e/a** table in Figure 10.11 will be of great help in designing new materials.

I am hoping that the newly established **e/a** values can be coupled with the **VEC** and used to design new functional alloys and compounds. Future research will judge the effectiveness of this approach.

OM: Before ending our dialogue, I tell you how precisely the Mott and Jones theory proposed in 1936 [3] caught the essence of the Hume-Rothery electron concentration rule in spite of the use of the naive free electron model in the presence of the Brillouin zone having a vanishing energy gap. As discussed in connection with Figure 10.17 in Section 10.7, we could identify the critical $|G|^2$ in the units of $(2\pi/a)^2$ to be 18, 50, and 125 for the family of gamma-brasses, 1/1-1/1-1/1 and 2/1-2/1-2/1 approximants, respectively. Consider this within the free electron model. A critical $(e/a)_c$ was defined in Chapter 3, Section 3.1 as the electron concentration obtained when the Brillouin zone plane is inscribed by a Fermi sphere. Obviously, the Fermi energy and the energy at the center of the relevant Brillouin zone plane are given as $E_F = [3\pi^2(e/\mathbf{uc})/a^3]^{2/3}$ and $E_c = (\pi/a)^2|G|^2$ in the atomic units for cubic systems, respectively. The condition $E_F = E_c$ immediately leads to

$$\left(e/\mathbf{uc}\right)_c = \left(\frac{\pi}{3}\right)\left[|G|^2\right]^{3/2} \tag{11.1}$$

An insertion of $|G|^2$ = 18, 50, and 125 into Equation 11.1 results in $(e/\mathbf{uc})_c$= 80, 370, and 1464 for the family of gamma-brasses, 1/1-1/1-1/1 and 2/1-2/1-2/1 approximants, respectively.

TABLE 11.1 Critical $|G|^2$, N, $(e/a)_{total}$, e/uc, and $(e/uc)_c$ for CMAs

	critical $|G|^2$ [$\times (2\pi/a)^2$]	N	$(e/a)_{total}$	e/uc	$(e/uc)_c$	$[(e/uc) - (e/uc)_c]/$ $(e/uc)_c$ (%)
Cu_5Zn_8	18	52	1.615	84	80	5
$Al_{30}Mg_{40}Zn_{30}$ 1/1-1/1-1/1 approximant	50	160	2.30	368	370	−0.5
$Al_{68}Cu_7Ru_{17}Si_8$	50	139	2.58	358	370	−3.2
Cd_6Ca, Cd_6Yb	50	168	2.0	336	370	−9.2
$Al_{15}Mg_{43}Zn_{42}$ 2/1-2/1-2/1 approximant	125	692	2.15	1460	1464	−0.3

Note: Critical $|G|^2$: square of the reciprocal lattice vector responsible for causing a FsBz-induced pseudogap at the Fermi level, N: number of atoms per unit cell, $(e/a)_{total}$: electrons per atom ratio obtained by composition-weighted average of nominal valency for the non-TM element and $(e/a)_{TM}$ for the TM element listed in Figure 10.11, e/uc: number of electrons per unit cell given by the product of N and e/a, and $(e/uc)_c$: number of electrons per unit cell given by Equation 11.1.

As summarized in Table 11.1, the free electron model reproduces the data in Figure 10.17a within the accuracy of ±10% for the respective families. The discussion above clearly demonstrates that Mott and Jones laid the most basic theoretical foundations on the Hume-Rothery electron concentration rule.

REFERENCES

1. W. Hume-Rothery, *Electrons, Atoms, and Alloys* (Metal Industry, The Louis Cassier Co., Ltd., 1948).
2. J.K. Brandon, H.S. Kim, and W.B. Pearson, *Acta Cryst.* B35 (1979) 1937.
3. N.F. Mott and H. Jones, *The Theory of the Properties of Metals and Alloys* (Oxford University Press, England, 1936).

Appendix 1: Atomic Size Ratio Rule

The atom size ratio of constituent elements must be close to unity to allow the formation of an alloy. For example, the ratio of the Goldschmidt radii of two constituent atoms is in between 0.8 and 1.2 for favorable alloy formation. Its theoretical foundation based on the elasticity theory was advanced in 1950s by Friedel [1] and Eshelby [2]. The size difference is defined as $\varepsilon = (r_B - r_A)/r_B \cong (r_B - r_A)/r_A$, where r_i ($i = A$ or B) refers to the Goldschmidt radius of the element A and B. Both Friedel and Eshelby evaluated the elastic energy of an alloy on the basis of the sphere-in-hole model, in which a spherical solute atom is inserted into a hole in the matrix. For example, Eshelby [2] deduced the following criterion for a tolerable size difference $|\varepsilon|$ to allow the formation of a primary solid solution:

$$|\varepsilon| < \left(\frac{kT_m}{\mu V_a} \frac{\gamma}{3} \right)^{1/2} \qquad (A1.1)$$

where T_m is the melting point, μ is the shear modulus, γ is the total volume change per solute atom and V_a is the volume per atom. The value of γ turns out to be about 1.61, since

$$\gamma = 3\frac{1-\sigma}{1+\sigma}$$

holds for homogeneous isotropic systems, where σ is the Poisson ratio nearly equal to 0.3. Eshelby found $|\varepsilon|$ to be around 0.15 by estimating $kT_m/\mu V_a$ to be 0.042 and took this as a possible interpretation for the Hume-Rothery 15% size rule [2].

More recently, Egami and Waseda [3] proposed the minimum solute concentration c_{min}^B in an A-B binary alloy system to obtain an amorphous phase by rapid quenching and revealed that c_{min}^B is inversely proportional to the atomic volume mismatch between the solute atom B and the solvent atom A:

$$c_{min}^B \times \frac{\left(V_B - V_A\right)}{V_A} = 0.1 \qquad (A1.2)$$

where V_A and V_B represents the volumes of the atom A and B, respectively. Equation A1.2 is immediately rewritten in terms of the atom size ratio r_B / r_A:

$$c_{min}^B = 0.1 \times \left(\frac{1}{\left(r_B / r_A\right)^3 - 1}\right) \qquad (A1.3)$$

where $r_A < r_B$ is assumed. A minimum solute concentration (%) is plotted as a function of the atom size ratio in Figure A1.1. The value is about 30%, when $r_B / r_A = 1.1$ but is sharply decreased to less than 5%, when the ratio is increased to 1.45. Remember that this is the criterion for the formation

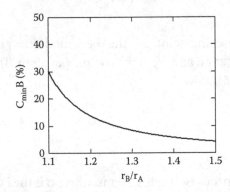

FIGURE A1.1 Atomic size ratio r_B/r_A dependence of the minimum solute concentration (%), beyond which an amorphous phase is favorably formed in an A-B binary alloy system. [From T. Egami and Y. Waseda, *J. Non-Cryst. Solids* 64 (1984) 113.]

of an amorphous phase extending outside a primary solid solution. This is, therefore, wider than the 15% size rule but points to an enhanced instability of a primary solid solution with an increasing size ratio.

REFERENCES

1. J. Friedel, *Adv. Phys.* 3 (1954) 446.
2. J.D. Eshelby, *Solid State Physics*, edited by F. Seitz and D. Turnbull (Academic Press, New York, 1956), vol. 3, pp. 115–119.
3. T. Egami and Y. Waseda, *J. Non-Cryst. Solids* 64 (1984) 113.

Appendix 2: Crystal Structures of Gamma-Brasses

Detailed studies of the atomic structure of gamma-brasses are important in gaining a deeper insight into the electronic structure and the stability of complex metallic alloys (CMAs). The present author in collaboration with T. Noritake, Toyota Central Research and Development Laboratories, Inc., studied the atomic structure of a number of binary gamma-brasses in a systematic manner by using a laboratory x-ray diffractometer with Cu-Kα radiation (Rigaku, RINT-TTR), unless otherwise stated. All samples, for which x-ray diffraction spectra are displayed in Section A2.1, were synthesized by the present author between 2005 to 2009. The measured x-ray diffraction spectra for representative gamma-brasses are shown with a brief comment. In particular, the data for the Li-Pb alloy clearly show the absence of gamma-brass phase at the composition $Li_{10}Pb_3$ and those for Mn-In alloys indicate the difficulty in synthesizing a single-phase gamma-brass phase.

In Section A2.2, we compiled previous works on the atomic structure of gamma-brasses with I-, P-, F-, and R-cells available in literature in the chronological order.

A2.1 X-RAY DIFFRACTION SPECTRA FOR A SERIES OF BINARY GAMMA-BRASSES

The powder diffraction spectra for 62 at.%Zn-Cu and 62 at.%Zn-Ag gamma-brasses are shown in Figure A2.1. All diffraction lines are indexed in terms of the Miller indices, whose sum is even. Both compounds are indeed identified as being typical of I-cell gamma-brasses with space

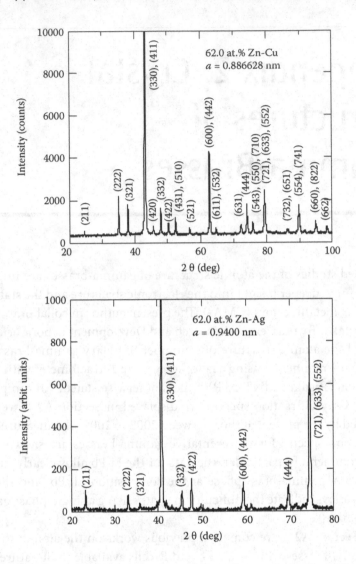

FIGURE A2.1 x-ray powder diffraction spectra taken with Cu-Kα radiation for 62.0 at.%Zn-Cu and 62.0 at.%Zn-Ag gamma-brasses.

group $I\bar{4}3m$. The most intense (330) and (411) diffraction line is characteristic of the cubic gamma-brass. The diffraction spectrum for 60 at.%Cd-Cu gamma-brass is shown in Figure A2.2. All the diffraction lines can be indexed with space group $I\bar{4}3m$. In contrast to the data for Cu-Zn and Ag-Zn gamma-brasses shown in Figure A2.1, the (310) diffraction line is unusually strong. This would reflect its unique atomic arrangements in the 26-atom cluster (see Chapter 7, Section 7.6 and Section A2.2.1.4).

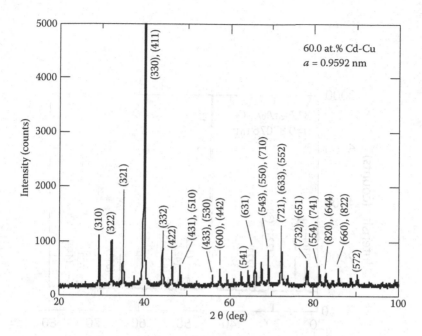

FIGURE A2.2   x-ray powder diffraction spectra taken with Cu-Kα radiation for 60.0 at.%Cd-Cu gamma-brass.

Diffraction spectra for 31.5 at.%Al-Cu and 31.0 at.%Ga-Cu gamma-brasses are shown in Figure A2.3 as typical examples for P-cell gamma-brass. The existence of (210), (221), and (300) diffraction lines, which are underlined, are characteristic of P-cell gamma-brass with space group $P\bar{4}3m$, since the sum of the Miller indices in them is odd. The (221) plus the (300) diffraction line associated with $|\mathbf{G}|^2 = 9$ is only weakly observed in the Cu-Ga gamma-brass.

The preparation of a single-phase Al_8V_5 gamma-brass was found to be difficult, since it is formed as a line compound through the peritectic reaction between Al-rich liquid phase and Al-V bcc phase at 1670°C in the Al-V phase diagram [1]. The sample was melted in a BN crucible using an induction furnace. Figure A2.4 shows the x-ray diffraction spectrum for Al_8V_5 gamma-brass after the heat-treatment at 1050°C for 24 h. All the diffraction lines were seemingly indexed without any trace of impurity phases. However, we realized that diffraction lines of V-rich AlV bcc phase taken with Cu-Kα radiation happens to be so close to those of gamma-brass that its presence could be hardly detected. Indeed, we were often disturbed by the segregation of V-rich Al-V bcc phase in the gamma-brass matrix, the existence of which was detected in micrographs taken

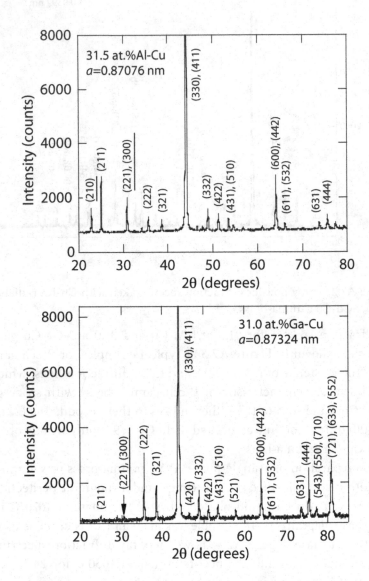

FIGURE A2.3 x-ray powder diffraction spectra taken with Cu-Kα radiation for 31.5 at.%Al-Cu and 31.0 at.%Ga-Cu gamma-brasses.

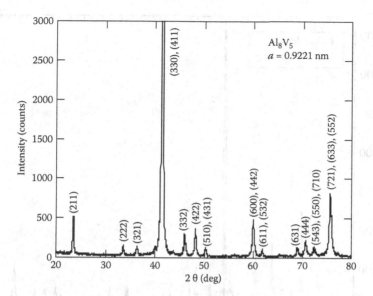

FIGURE A2.4   x-ray powder diffraction spectra taken with Cu-Kα radiation for Al_8V_5 gamma-brass.

with SEM (scanning electron microscope). It may be clever to synthesize a single-crystal for the structure analysis, as reported by Brandon et al. [2, A2.2.1.12], rather than trying to make a polycrystalline Al_8V_5 gamma-brass ingot.

According to the Mn-In phase diagram [1], Mn_3In compound is formed as a result of the peritectic reaction between In-rich liquid and Mn-rich β-Mn phase at 910°C. We melted an appropriate amount of Mn and In with the ratio of Mn:In = 3:1 in an alumina crucible using an induction furnace and annealed the resulting ingot at 900°C for 336 h. The SEM micrograph revealed a mixture of Mn_3In gamma-brass and α-Mn phase. To reduce the impurity phase, melt-spinning technique was employed to form rapidly quenched ribbon samples with a few 10 μm in thickness and 5–10 mm in length. Ribbons were subsequently annealed at 890°C for 72 h. As shown in Figure A2.5, α-Mn phase still remained as an impurity phase. The growth of a tiny single crystal is probably a clever way to synthesize a single-phase Mn_3In gamma-brass (see Section A2.2.2.10).

The gamma-brass phase can be relatively easily prepared in both Ni-Zn and Co-Zn alloy systems. Figure A2.6 shows two diffraction spectra for 17.0 and 25.0 at.%Ni-Zn gamma-brasses. The diffraction lines for the Ni-poor sample are quite sharp and can be perfectly indexed in terms of space group $I\bar{4}3m$, while those for the Ni-rich one are fairly broad.

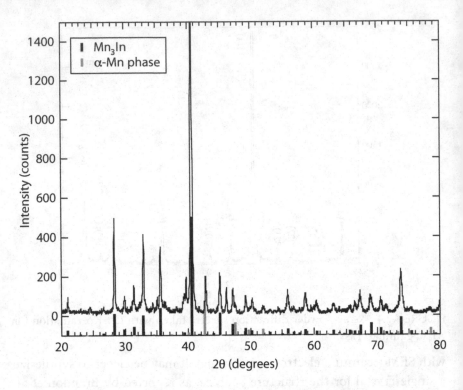

FIGURE A2.5 X-ray powder diffraction spectra taken with Cu-Kα radiation for Mn₃In gamma-brass. The ribbon sample was annealed at 890°C for 72 h.

We originally thought this as a sign of the difficulty in producing a high-quality gamma-brass phase in Ni-rich composition range. However, it was soon realized that this is caused by the formation of a long-period super-lattice, as reported by Morton [4]. As shown in the insert, we can observe three successive sub-peaks on smaller angle side of the main peak due to {330} and {411} lattice planes. This, along with TEM observations [5], was taken as evidence for the formation of a long-period super-lattice structure in Ni-rich Ni-Zn gamma-brasses.

In contrast to Ni-Zn gamma-brass, a single-phase gamma-brass was easily synthesized over 15 to 30 at.%Co in the Co-Zn gamma-brass alloy system without any formation of a long-period super-lattice structure [5]. The diffraction spectra for 15.0 and 29.0 at.%Co-Zn samples corresponding to both low and high ends of its composition range are shown in Figure A2.7. All the diffraction lines are indexed using space group $I\bar{4}3m$. Unfortunately, the Co-Zn gamma-brass phase is incorrectly referred to as possessing space group $P\bar{4}3m$ in its phase diagram [1].

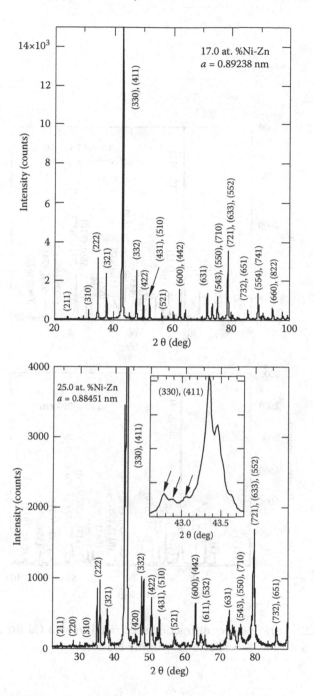

FIGURE A2.6   x-ray powder diffraction spectra taken with Cu-Kα radiation for 17.0 at.%Ni-Zn and 25.0 at.%Ni-Zn gamma-brasses.

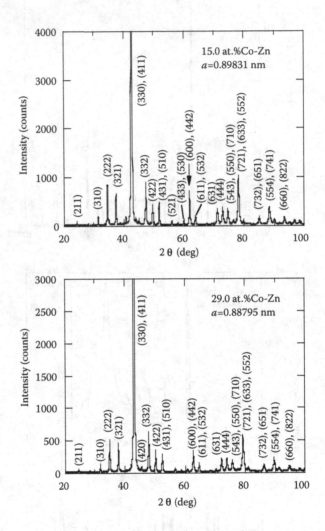

FIGURE A2.7   x-ray powder diffraction spectra taken with Cu-Kα radiation for 15.0 at.%Co-Zn and 29.0 at.%Co-Zn gamma-brasses.

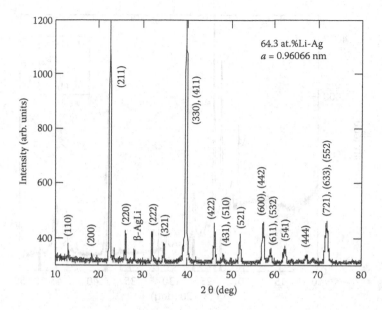

FIGURE A2.8   x-ray powder diffraction spectra taken with Cu-Kα radiation for 64.3 at.%Li-Ag gamma-brass.

Three alloys with 64, 70, and 73 at.%Li in the Ag-Li alloy system were synthesized in a molybdenum crucible using an induction furnace in a glove-box. The preparation of a high-quality Ag-Li gamma-brass was difficult because of volatility and reactivity of lithium. The ingot was annealed at 458 K for 10 days. Only the 64 at.%Li-sample was brittle enough to grind into powders in a glove-box. Figure A2.8 shows the diffraction spectrum for the 64.3 at.%Li-Ag sample [6]. The diffraction peaks confirm the formation of gamma-brass with space group $I\bar{4}3m$. In the phase diagram [1], this phase is described as γ_3 with space group "$P\bar{4}3m$?" A small peak at 2θ = 28.1° was attributed to β-phase AgLi phase precipitated as an impurity phase. More important to be noted is the appearance of a huge (211) peak, which is comparable to the most intense (330) and (411) peak in magnitude. This is apparently caused by Ag atoms densely distributed over the set of {211} lattice planes and is suggested to serve as a crucial role in its stabilization (see Chapter 8, Section 8.4.3). More detailed structure analysis of this alloy was made, using the powder diffraction spectrum taken with the wavelength of 0.050226 nm at the beam-line BL02B2, Synchrotron Radiation Facility, Spring-8, Japan [6].

We attempted to synthesize $Li_{10}Pb_3$ gamma-brass, which is listed as a compound containing 52 atoms in its unit cell with space group $P\bar{4}3m$

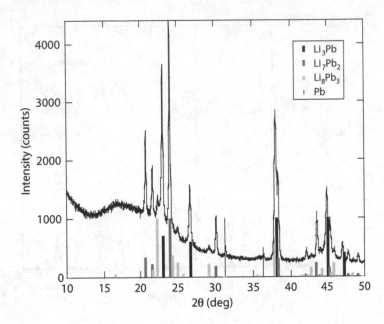

FIGURE A2.9 x-ray powder diffraction spectra taken with Cu-Kα radiation for 23.0 at.%Pb-Li alloy.

in the Li-Pb phase diagram [1]. An appropriate amount of 99+%Li sheet and 99 %Pb shots was melted for 30 s in a molybdenum crucible using an induction furnace in Ar-gas circulating glove-box. The ingot thus obtained was annealed at 700°C for 10 h. As shown in Figure A2.9, almost all diffraction lines were indexed in terms of Li_3Pb, Li_7Pb_2, Li_8Pb_3, and Pb. No evidence for the growth of gamma-brass phase was confirmed. This lends strong support to the work by Zalkin and Ramsey [7,8]. We conclude the absence of $Li_{10}Pb_3$ gamma-brass phase in the Li-Pb alloy system.

REFERENCES

1. H. Okamoto, *Phase Diagrams for Binary Alloys* (ASM Internationals, Materials Park, OH, 2000).
2. J.K. Brandon, W.B. Pearson, P.W. Riley, C. Chieh, and R.Stokhuyzen, *Acta Cryst.* B 33 (1977) 1088.
3. J.K. Brandon, H.S. Kim, and W.B. Pearson, *Acta Cryst.* B35 (1979) 1937.
4. A.J. Morton, *Phys. Stat. Sol.* (a) 44 (1977) 205.
5. U. Mizutani, T. Noritake, T. Ohsuna and T.Takeuchi, *Phil. Mag.* 90 (2010) 1985.
6. T. Noritake, M. Aoki, S. Towata, T. Takeuchi, and U. Mizutani, *Acta Cryst.* B63 (2007) 726.
7. A. Zalkin and W.J. Ramsey, *J. Phys. Chem.* 60 (1956) 234.
8. A. Zalkin and W.J. Ramsey, *J. Phys. Chem.* 62 (1956) 689.

A2.2 GAMMA-BRASSES IN LITERATURE

A2.2.1 I-Cell Gamma-Brasses

The essence of each contribution to I-cell gamma-brasses is reviewed below:

A2.2.1.1 A. Westgren and G. Phragmén, Phil.Mag. 50 (1925) 311.

Carpenter pointed out for the first time the similarity of phase diagrams of Cu-Zn, Ag-Zn, and Au-Zn alloys. The structure of these three systems was studied by x-ray methods. The structure analogies are revealed such that, with increasing Zn concentration, fcc α-phase is followed by β-, γ-, and ε-phases before ending with Zn primary solid solution called η-phase. Owing to a significant difference in diffractive powers between constituent elements, β-phase in Ag-Zn and Au-Zn was safely proved to crystallize in the CsCl-structure. The CsCl-type structure was also suggested for Cu-Zn β-phase, though its identification was difficult because of negligibly small difference in diffractive power between Cu and Zn. They also found that all three γ-phases strikingly resemble each other and are also very similar to the phase in Cu-Al alloys. They mentioned that all gamma-brasses contain 52 atoms in the unit cubic cell. The chemical formula Cu_4Zn_9, Ag_4Zn_9, and Au_4Zn_9 were suggested, since these correspond to compositions within homogeneous gamma-brass ranges. Unfortunately, their chemical formula proved to be incorrect later, as evidenced by Bradley and Thewlis [A2.2.1.2].

A2.2.1.2 A.J. Bradley and J. Thewlis, Proc.Roy.Soc. (A) 112 (1926) 678.

The structure of the Cu-Zn gamma-brass was refined by using the data of Westgren and Phragmén [A.2.2.1.1]. There is a fundamental relationship between the structures of gamma-brasses (and also alpha-manganese) and a bcc lattice. Beta-brass is bcc with the lattice constant 0.2945 nm. The unit cell contains two atoms. The lattice constant of gamma-brass is 0.885 nm and each unit cell contains 52 atoms. The side of the unit cube is therefore exactly three times that of beta-brass and its volume is 27 times as great. If gamma-brass had 54 atoms in the unit cell, they might be arranged in exactly the same way as the atoms of beta-brass, namely, on a bcc lattice. In this case, gamma-brass would only give rise to those lines, which appear on beta-brass. The existence of extra lines for gamma-brass can be accounted for if we suppose two of 54 atoms to be omitted without greatly displacing the remaining 52 atoms. These considerations show that a possible structure of gamma-brass consists of a bcc arrangement with 1 atom in 27 removed, the remaining atoms being slightly displaced, but necessarily in such a manner that the cubic symmetry is preserved. The authors could conclude that "IT (inner tetrahedron)" and "CO (cubo octahedron)" atoms are Zn and "OT (outer tetrahedron)," and "OH (octahedron)" atoms are noble metals Cu, Ag, or Au (see Chapter 6, Figure 6.2). In this way they were led to conclude that there are 32 Zn atoms and 20 Cu atoms in the unit cell. Bradley and Thewlis concluded that the chemical formulas proposed by Westgren and Phragmén [A2.2.1.1] were incorrect and concluded that the true formulas should be Cu_5Zn_8, Ag_5Zn_8, and Au_5Zn_8.

A2.2.1.3 A.F. Westgren and G. Phragmén, Metallwirtschaft 7 (1928) 700; Trans. Farad. Soc. 25 (1929) 379.

The e/a = 21/13 rule for gamma-brass was first pointed out in 1928. The authors stated as follows: "When Cu, Ag, or Au is combined with a divalent metal (Zn, Cd, etc.), the homogeneity range of gamma-brass phase corresponds to the formula of the type Cu_5Zn_8, Ag_5Cd_8, etc., and if they are alloyed with a trivalent metal (Al, etc.), the concentration interval of the phase includes values corresponding to formula of the type Cu_9Al_4. In these phases, there are 21 valency electrons per 13 atoms." The regularities concerning the ratio of valency electrons to atoms are also mentioned in beta-phase with the ratio 3:2 and in close-packed hexagonal structure (ε-phase) with the ratio 7:4. Note that their proposal on Cu_9Al_4 was made before the detailed structure refinement by Bradley in 1929 (see A2.2.2.2).

A2.2.1.4 A.J. Bradley and C.H. Gregory, Phil. Mag. 12 (1931) 143.

A considerable resemblance was expected in two gamma-brasses Cu_5Zn_8 and Cu_5Cd_8. As a first step, cadmium atoms simply replaced Zn atoms and an attempt was made to get an agreement between calculated and observed intensities by adjusting the parameters. However, this turned out to be impossible. They were led to a very definite but unexpected conclusion. Although the positions occupied by the 52 atoms are similar in Cu-Zn and Cu-Cd, the distributions of the atoms are quite different. The best agreement was obtained by putting 16 Cu atoms into IT and OT, whereas 4 Cu and 32 Cd atoms are put into OH and CO in a random way.

A2.2.1.5 R.E. Marsh, Acta Cryst. 7 (1954) 379.

The refinement of crystallographic parameters for Ag_5Zn_8 gamma-brass was undertaken by least-squares analyses of powder and single-crystal x-ray diffraction photographs with Cu-Kα radiation. A small single crystal was isolated from the ingot and was oriented by Laue photography. Space group was confirmed to be $I\bar{4}3m$.

A2.2.1.6 O.v. Heidenstam, A. Johansson, and S. Westman, Acta Chem.Scand. 22 (1968) 653.

The atomic structure of Cu_5Zn_8, Cu_5Cd_8 and Cu_9Al_4 gamma-brasses was investigated by using neutron powder diffraction method. Both Cu_5Zn_8 and Cu_5Cd_8 have been reported to belong to space group $I\bar{4}3m$, while Cu_9Al_4 belongs to $P\bar{4}3m$. The structure was described in terms of the two clusters, each comprising 26 atoms. The completely ordered Au_5Zn_8 type distribution is seen to give the best agreement with the observation. But somewhat randomized versions cannot be completely ruled out. Regarding Cu_5Cd_8, the model reported by Bradley and Gregory was found to be the best one.

A2.2.1.7 A. Johansson, H. Ljung, and S. Westman, Acta Chemica Scandinavica 22 (1968) 2743.

Single-phase gamma-brass Ni_xZn_{100-x} ($15.9 \le x \le 19.7$) alloys were studied. The neutron diffraction studies were made for the $x = 17.7$ sample. Its space group was identified as $I\bar{4}3m$. Among various models, Ni_2Zn_{11} with Ni on sites OT resulted in the best fit. In this model, there are no Ni-Ni contacts. In general, the minimum number of like-atom contacts is favored in the gamma-brass structure. Assuming Fe-Zn gamma-brass to be analogous to Ni-Zn gamma-brass, they carried out a refinement, but final values may be still less accurate.

A2.2.1.8 V.A. Edstrom and S. Westman, Acta Chemica Scandinavica 23 (1969) 279.

Single-phase Pd_xZn_{100-x} gamma-brasses ($x = 18.0$ and 20.0) were analyzed. The number of atoms per unit cell turned out to be N = 51.8 for $x = 18.0$.

A2.2.1.9 L. Arnberg and S. Westman, Acta Chemica Scandinavica 26 (1972) 513.

The $Ir_{15.3}Zn_{84.7}$ sample was found to be a single-phase gamma brass. Both single crystal and powder data showed it to be bcc with space group $I\bar{4}3m$. The gamma-brass phase seems to be homogeneous between 15.3 and 15.7 at.%Ir. The lower composition corresponds almost exactly to the stoichiometric composition Ir_2Zn_{11}.

A2.2.1.10   L. Arnberg and S. Westman, Acta Chemica
Scandinavica 26 (1972) 1748.

The powder x-ray diffraction data were taken for Ag-Li alloy, using mono-chromatized Cu-Kα radiation with a focusing Guinier-Hagg camera of 80 mm diameter.

A2.2.1.11   J.K. Brandon, R.Y. Brizard, P.C. Chieh, R.K. McMillan,
and W.B. Pearson, Acta Cryst. B30 (1974) 1412.

The atomic structure of Cu_5Zn_8, Cu_5Cd_8, and Fe_3Zn_{10} was determined by using single-crystal x-ray diffraction techniques with Mo-Kα radiation. A good agreement with earlier results was confirmed in the refinements of Cu_5Zn_8 and Cu_5Cd_8 with support of the ordering in the former, and absence of any Cd on sites IT and OT in the latter. The refined structure of Fe_3Zn_{10} gave different results from those by Johansson et al. [6].

A2.2.1.12   J.K. Brandon, W.B. Pearson, P.W. Riley, C. Chieh, and
R. Stokhuyzen, Acta Cryst. B 33 (1977) 1088.

They studied both Al_8V_5 and Al_8Cr_5 gamma-brasses using the GE XRD-6 x-ray diffractometer with Mo-Kα radiation. It is found that Al_8V_5 has I-cell, whereas Al_8Cr_5 has R-cell. If an increase in electron concentration per unit cell is responsible for the stabilization of R-cell structure, it must be presumed that V absorbs some electrons provided by Al into its hybridized bands having mainly the d-character. If one electron per V atom were absorbed, this would allow three filled bands below the Fermi level and the electron concentration in the valence band would be 76 (= $20 \times (-1) + 32 \times 3$) per 52-atom cell for the composition Al_8V_5. However, it is unknown whether the absorption of electrons would be as great as this.

Note added by the present author: Their interpretation is not consistent with the data listed in Table 10.2 (see Chapter 8, Section 8.3.2, and Chapter 10, Section 10.7).

A2.2.1.13   T. Noritake, M. Aoki, S. Towata, T. Takeuchi, and
U. Mizutani, Acta Cryst. B 63 (2007) 726.

The atomic structure of $Ag_{36}Li_{64}$ gamma-brass was determined by analyzing the powder diffraction pattern taken with the use of synchrotron radiation beam with the wavelength of 0.050226 nm. It turned out that the compound contains 52 atoms in its unit cell with space group $I\bar{4}3m$ and that Li atom enters exclusively into sites IT and CO, whereas Ag atom into sites OT and OH in the 26-atom cluster. Small amounts of Li also exist in sites OT and OH, resulting in chemical disorder. They revealed that the volume of IT and CO shrinks, while that of OT and OH expands relative to that of the corresponding polyhedra in the original bcc lattice and that this is a universal feature found in other gamma-brasses like Cu_5Zn_8 and Al_8V_5, for which the structure data are available.

A2.2.2 P-Cell Gamma-Brasses

The essence of each of the following contributions to P-cell gamma-brasses is reviewed below:

A2.2.2.1 E.R. Jette, G. Phragmén, and A.F. Westgren, J. Inst. Metals, 31 (1924) 193.

The x-ray studies were made for the C16-type $CuAl_2$, $\gamma-$ and β-phases in the Cu-Al alloy system. They measured both lattice constant and density for 30.9, 37.0, and 43.9 at.%Al gamma-brasses and revealed the number of atoms per unit cell to decrease with increasing the Al concentration.

A2.2.2.2 A.J. Bradley, Phil. Mag. 6 (1929) 878.

The powder diffraction data were taken for $Cu_{100-x}Al_x$ (30.9<x<35.6) gamma-brasses. The structure was proved to be cubic. Moreover, some additional lines, which belong to planes with $h + k + l =$ odd, were observed. From this it was concluded that the unit cube is primitive and not body-centered in contrast to Cu-Zn gamma-brass. The structure of Cu-Al gamma-brasses was deduced to be cubic with 52 atoms per unit cell and could be best understood if the two subgroups of atoms are considered. There are 16 Al atoms and 36 Cu atoms in the "ideal" alloy, so that it is represented by the stoichiometric formula Cu_9Al_4 corresponding to a composition with 30.8 at.%Al.

A2.2.2.3 A.J. Bradley and P. Jones, J. Inst. Metals. 51 (1933) 131.

The two $Cu_{100-x}Al_x$ (x = 25.0 and 28.3) alloys obtained by quenching from temperatures 850 and 1000°C, respectively, turned out to be a single β'-phase with the CsCl-structure. It was also shown that gamma-brass phase exists between 30.9 and 35.6 at.%Al and the structure modification takes place, when Al concentration further increases. In the former, the atomic structure of the alternate clusters is identified. This is in agreement with earlier investigations and confirms the formula Cu_9Al_4 for the gamma-brass. The gamma-brass structure is modified beyond 35.6at.%Al. These alloys are no longer cubic but the departure from cubic symmetry is so slight that it is possible to take an assumed value for the lattice spacing, which is equal to the cube root of the volume of the unit cell.

A2.2.2.4 A.J. Bradley, H.J. Goldschmidt, and H.J. Lipson, J. Inst. Metals, 63 (1938) 149.

They studied the structure of totally 24 Cu-Al alloys over the composition range 30.9 to 50.2 at/%Al by using x-rays with Co-Kα radiation and revealed that an increase in Al content yields five phases including three gamma-brass structures denoted as γ, γ_1, and γ_2 plus $\zeta(Cu_4Al_3)$ and $\eta(CuAl)$ phases, all of which are derived from the same elementary body-centered cube. With increasing the Al content beyond Cu_9Al_4, the lattice constant gradually increases but the density decreases. The number of atoms per cell was found to be almost 52 within γ-phase over the range 31.3 to 35.2 at.%Al. In Al contents over 35.4 and 38.3 at.%, the lattice spacing becomes fairly constant but the density decreases steadily. This is referred to as γ_1-phase and the number of atoms per cell is distributed over 51.6 to 50.9, that is, more or less one vacancy is introduced into γ-phase. When Al content increases beyond 38.3 at.%, the lattice spacing begins to decrease as does the density. This is γ_2-phase and the number of atoms per cell is distributed over 50.5 to 49.7. Hence, there is a gradual elimination of atoms with increasing Al content. The lines become broader at low angles and are split into many components at higher angles. When the Al content reaches 35.5 at.%, there is a sudden change in symmetry.

A2.2.2.5 S. Westman, Acta Chem. Scand. 19 (1965) 1411.

The x-ray powder and single-crystal diffraction photographs were taken, using monochromatized Cu-Kα radiation. The composition of a single-phase sample was $Cu_{68.5}Al_{31.5}$ close to the stoichiometric $Cu_{69.2}Al_{30.8}$ or Cu_9Al_4. The lattice constant turned out to be a = 0.87023 nm. The structure was described in terms of two geometrically similar clusters of 26 atoms each and found to be consistent with that deduced by Bradley and Jones.

A2.2.2.6 O. v. Heidenstam, A. Johansson, and S. Westman, Acta Chem.Scand. 22 (1968) 653.

A refinement of the Cu_9Al_4 structure based on single-crystal diffractometer data was already carried out by Westman [A2.2.2.5]. However, some peculiarities remained at the end of the refinement, especially in the temperature factors of Al atoms. The present work is to repeat the refinement, using the old data with a program suited to the treatment of cubic symmetry. In their analysis, individual thermal parameter values lay within one standard deviation from the average value. This is taken as evidence that the atomic distribution deduced by their group must be quite reliable.

A2.2.2.7 R. Sokhuyzen, J.K. Brandon, PC. Chieh, and W.B. Pearson, Acta Cryst. B30 (1974) 2910.

Gamma-brass Cu_9Ga_4 is cubic with space group $P\bar{4}3m$. Single-crystal diffractometer data were obtained using the Zr-filtered Mo-Kα radiation. The structure was refined from 140 observed reflections. Three slightly different gamma phases exist in the Cu-Ga system in the composition range from 30 to 42.5 at.%Ga. The refinements were carried out starting with coordinates of Cu_9Al_4 reported by Westman [A2.2.2.5] and Heidenstam et al. [A2.2.2.6]. The chemical disorder exists on the sites CO in the clusters "a" and "b" in contrast to Cu_9Al_4.

A2.2.2.8 J.K. Brandon, R.Y. Brizard, W.B. Pearson, and D.J.N. Tozer, Acta Cryst. B33 (1977) 527.

Single crystals of Au_9In_4 and Ag_9In_4 were studied by using an automatic four-circle diffractometer with Mo-Kα radiation. The N = 52 atoms per unit cell was deduced for Au_9In_4 from the measured lattice constant and density. The starting structure model was that of Cu_9Al_4 determined by v. Heidenstam et al. [A2.2.2.6]. The refinement led to a structure, in which there was some slight mixing of In with Au on sites CO in the cluster "a" and some slight mixing of Au with In on sites CO in the cluster "b". Instead, accurate structure determination for Ag_9In_4 appears remote because of the small number of observable reflections with $h + k + l =$ odd resulting from a small difference in atomic scattering factors between In and Ag. Nevertheless, they have established without doubt that Ag_9In_4 has a primitive cubic cell rather than a bcc cell. Evidence for the primitive cell comes from the observation of 12 very weak $h + k + l =$ odd reflections. They also confirmed that Co-Zn gamma-brass has an I-cell (see Figure A2.7).

A2.2.2.9 L. Arnberg and S. Westman, Acta Cryst. A34 (1978) 399.

The crystal structure of Cu_9Al_4 has been further refined by using a $Cu_{35.4}Al_{16.6}$ or 31.9 at.%Al-Cu single-crystal with graphite-monochromatized Cu-Kα radiation. The atom position parameters do not differ much from those given by Heidenstam et al. [A2.2.2.6] but the precision is better by a factor of ~2 in the standard deviations. The anisotropy of thermal vibrations is discussed by plotting their directional dependence in the form of thermal ellipsoids. In the cluster "a," thermal vibrations of Al atoms on the IT and Cu atoms on the OT are directed toward the vacancies at 0, 0, 0 and 1/2, 1/2, 1/2. In the cluster "b," however, only remarkable oscillations along these directions are seen in Cu atoms on the IT, whereas Cu atoms on the OT vibrate isotropically.

A2.2.2.10 J.K. Brandon, H.S. Kim, and W.B. Pearson, Acta Cryst. B35 (1979) 1937.

Single-crystal x-ray diffraction experiments with Zr-filtered Mo-Kα radiation revealed that the alloy of composition Mn_3In has a gamma-brass type structure with space group $P\bar{4}3m$ and lattice constant $a = 0.942$ nm. They observed weak but easily visible reflections of type $h + k + l =$ odd, indicating that Mn_3In has a primitive cubic lattice.

Why the two different ordering schemes of Cu_9Al_4 and Mn_3In occur among P-cell gamma-brass structures can be understood in terms of packing efficiencies at the respective radius ratios of the two constituent elements and in terms of maximizing the number of In-Mn neighbors while minimizing the number of In-In neighbors in the structure. The location of the larger In atoms on sites OH and CO in the cluster "b" maximizes the packing fraction in I-cell instead of P-cell. Going from I-cell to P-cell reduces the number of contacts for the large In atoms with themselves and increases the number of unlike In-Mn contacts. This is the reason why Mn_3In has a P-cell rather than an I-cell.

A2.2.3 F-Cell Gamma-Brasses

The essence of each contribution to F-cell gamma-brasses is reviewed below.

A2.2.3.1 A. Westgren and G. Phragmén, Z. Metallkd. 18 (1926) 279.

The powder diffraction photographs were taken for three Cu-based gamma-brasses containing 61 at.%Zn, 31 at.%Al, and 20 at.%Sn. The strongest diffraction peak was observed at $\Sigma h^2 = 72$ for Cu-Sn gamma-brass, which was four times as large as $\Sigma h^2 = 18$ for both Cu-Zn and Cu-Al gamma-brasses. This indicates that the unit cell is twice as large as that in the prototype gamma-brasses. The measured density is 8.95 g/cm^3 and the lattice constant is deduced to be 1.791 nm, which is indeed almost twice that of Cu_5Zn_8 gamma-brass ($a = 0.887$ nm). The number of atoms in the unit cell is estimated to be 416. This is precisely eight times as large as those in already discussed Cu-Zn and Cu-Al gamma-brasses and confirmed that the unit cell of Cu_4Sn gamma-brass is constructed by doubly stacking the unit cell of the prototype gamma-brass along the x-, y-, and z-directions.

A2.2.3.2 A. Westgren and G. Phragmén, Z.anorg.Chem. 175 (1928) 80.
They reported the x-ray structure data for α-, β-, γ-, ε-, and η-phase Cu-Sn alloys. The β-phase was identified as the bcc structure containing 2 atoms in its unit cell with lattice constant a = 0.2972 nm. This could confirm Hume-Rothery's postulate that the structure of 15 at.%Sn is close to 16.6 at.%Sn corresponding to Cu_5Sn with e/a = 3/2 and is analogous to that in CuZn beta-brass.

There is a Cu-Sn phase very similar to those of γ-Cu-Zn and γ-Cu-Al phases. It crystallizes into the fcc structure containing 416 atoms with the lattice constant 1.791 nm. The chemical formula $Cu_{31}Sn_8$ (20.5 at.%Sn) was proposed to be the best, since the valence electron concentration becomes 21:13 in agreement with those of prototype Cu_5Zn_8 and Cu_9Al_4 gamma-brasses.

A2.2.3.3 J.D. Bernal, Nature, Lond. 122 (1928) 54.
The Cu_4Sn single-crystal is found to have a cubic structure with a face centered lattice of side 1.792 nm, thus confirming the powder photograph observation of Westgren and Phragmén [see Section A2.2.3.1]. With a cell of this large size, it is difficult to ascertain the number of atoms in the cell. However, its close relation to Cu_5Zn_8 gamma brass worked out by Bradley and Thewlis, which has a cell of almost exactly half the dimensions, 0.887 nm and gives intensities of reflections for the 50 corresponding planes of almost identical values, makes it almost certain that the total number of atoms in the cell is 52 × 8 = 416. Such a number cannot be made up from molecules of Cu_4Sn and the most probable values to fit with the density 8.95 are 328 atoms Cu and 88 atoms Sn, which makes the formula $Cu_{41}Sn_{11}$. Note that the chemical formula proposed by Westgren and Phragmén corresponds to 20.5 at.%Sn, whereas that by Bernal to 21.15 at.%Sn.

A2.2.3.4 J.S.L.Leach and G.V.Raynor, Proc.Roy.Soc. A224 (1954) 251.
The structure of $Cu_{12}Sn_2Al_2$ showed a complex diffraction pattern very similar in distribution and intensity of diffraction lines to that of the ordered gamma-brass $Cu_{41}Sn_{11}$. The lattice constant for $Cu_{74.77}Sn_{10.93}Al_{14.3}$ turned out to be 1.7783 nm, almost six times the side of the unit cube of the parent body-centered cubic structure (a = 0.2971 nm). The e/a value

of this compound is 1.6139 close to the value of 1.6 for Cu_4Sn or 85/52 (= 1.634) for $Cu_{41}Sn_{11}$.

A2.2.3.5 A. Johansson and S. Westman, Acta Chem.Scand. 24 (1970) 3471.

The gamma-brass-like phase with the approximate composition Pt_3Zn_{10} is described in terms of face-centered cubic structure with a lattice parameter of ~1.811 nm with space group $F\overline{4}3m$. The structure is formed by four different clusters "a", "b", "c" and "d", each with site-sets IT, OT, OH and CO. The sites OH in the cluster "a" are unoccupied, while those in the remaining three clusters "b", "c" and "d" are occupied by a mixture of Pt and Zn. In the cluster "b," either sites IT or OT are unoccupied and the two versions of this cluster are statistically distributed over the structure in equal numbers. The sites OT in all remaining clusters "a," "c," and "d" are filled with Pt, while the sites OH in clusters "b," "c," and "d" are occupied by a mixture of Pt and Zn.

A2.2.3.6 L. Arnberg, A. Johansson, and S. Westman, Acta Cryst. A31 (1975) S98.

The structure of gamma-brass-like $Cu_{41}Sn_{11}$ compound was reported. It is described as face-centered cubic with the lattice constant of ~1.798 nm with space group $F\overline{4}3m$. Four different clusters are centered at 0,0,0 etc, 1/2, 1/2, ½, etc., 1/4, 1/4, 1/4, etc., and 3/4, 3/4, 3/4, respectively. Sn atoms occupy one CO, one OH and one OT site of three different clusters. There are no Sn-Sn contacts in this structure model.

A2.2.3.7 M.H. Booth, J.K. Brandon, R.Y. Brizard, C. Chieh, and W.B.Pearson, Acta Cryst. B33 (1977) 30.

In cubic gamma-brasses with F-cell, four different 26-atom clusters, "a," "b," "c," and "d" pack together in an arrangement of a superstructure of the bcc lattice. Two single crystals $Cu_{41}Sn_{11}$ and Cu_9Sn_3Ni were selected from crushed portions of each alloy. Space group $F\overline{4}3m$ was chosen, since typical gamma-brass structures can be described in terms of the 26-atom clusters packed about point sites of $\overline{4}3m$ symmetry. In $Cu_{41}Sn_{11}$, the fractional occupancy of IT in the cluster "a" is consistent with the measured density. Fractional occupancies in Cu_9Sn_3Ni, which has a similar Sn distribution, occur on sites IT in the cluster "a" and also on sites CO in the cluster "c." Such additional vacancies found in Cu_9Sn_3Ni provide a mechanism to maintain valency electron concentration with fewer atoms per unit cell, since the percentage of low valency Cu atoms is smaller than in $Cu_{41}Sn_{11}$.

The ordering in $Cu_{41}Sn_{11}$ is accounted for on the basis that Sn atoms shall not be close neighbors. It can be also shown that this condition cannot be satisfied in P- or I-cells at this composition. This may justify the adoption of F-cell. Similar ordering occurs in Cu_9Sn_3Ni. A further consequence of the observed ordering of $Cu_{41}Sn_{11}$ in F-cell is an increase in the number of unlike Sn-Cu contacts. Maximizing the number of unlike contacts as a factor in the stability of the gamma-brass structures is important.

A2.2.4 R-Cell Gamma-Brasses

The essence of each contribution to R-cell gamma-brasses is reviewed below.

A2.2.4.1 A.J. Bradley and P. Jones, J. Inst. Metals. 51 (1933) 131.

This work was the first to point out that cubic gamma-brass phase exists only between 30.9 and 35.6 at.%Al in the Cu-Al alloy system and that the structure modification occurs when the Al concentration exceeds 35.6 at.%.

A2.2.4.2 A.J. Bradley and S.S. Lu, Z. Krist. 96 (1937) 20.

They showed Al_8Cr_5 gamma-brass to have a rhombohedral gamma-brass-like structure and determined the atom positions.

A2.2.4.3 A.J. Bradley, H.J. Goldschmidt, and H.J. Lipson, J. Inst. Metals, 63 (1938) 149.

They pointed out the occurrence of structural changes, as Al content increases beyond Cu_9Al_4. With increasing Al concentration, two derivative structures γ_1 and γ_2 of lower symmetry and fewer atoms per unit cell were mentioned. The cubic gamma-phase lasts up to 35 at.%Al, above which it is replaced by γ_1, which is further replaced by γ_2 above about 38 at.%Al.

A2.2.4.4 S. Westman, Acta Chem. Scand., 19 (1965) 2369.

The lattice constant is measured as a function of Al concentration over the range from 30.4 to 41.7 at.%Al covering both cubic and rhombohedral phases in the Cu-Al alloy system. The diffraction photographs and spectrogoniometer data identified space group $R3m$ for the 38.9 at.%Al sample.

A2.2.4.5 T. Lindahl, A. Pilotti, and S. Westman, Acta Chem. Scand. 22 (1968) 748.

The pseudo-cubic unit cell was found to possess $a = 0.94067$ nm and $\alpha = 90.413°$ for Cu-Hg gamma-brass. A combination with the measured density of 12.6 g/cm³ led them to conclude the chemical formula $Cu_{15}Hg_{11}$. A stoichiometric $Cu_{15}Hg_{11}$ may be derived from a hypothetical Cu_8Hg_5 structure by substitution of Hg for Cu in the only one-fold Cu position in the primitive rhombohedral cell.

The chemical formula Cr_9Al_{17} was proposed by substitution of Al for Cr in a one-fold position in the primitive cell. The pseudo-cubic cell was found to have its lattice constant $a = 0.91031$ nm and $\alpha = 90.326°$.

A2.2.4.6 T. Lindahl and S. Westman, Acta Chem. Scand. 23 (1969) 1181.

The Cu-Hg alloy they studied was described in terms of a rhombohedral structure with the pseudo-cubic lattice parameter of 0.94024 nm. It is alternatively described as an ordered version of the Cu_5Cd_8 structure-type, with Hg in one six-fold and two three-fold positions in the primitive unit of the bcc cell. The chemical formula Cu_7Hg_6 was proposed.

A2.2.4.7 J.K. Brandon, W.B. Pearson, P.W. Riley, C. Chieh, and R. Stokhuyzen, Acta Cryst. B33 (1977) 1088.

The gamma-brasses with R-cell with $\alpha<90°$ appear to have valence electrons in excess of 88 or 89 per 52-atom cell. They developed a hypothesis such that the distortion is stabilized by the valence-band structure energy and R-cell structure occurs at higher electron concentration than cubic gamma-brass.

A single crystal for the x-ray analysis was selected from the powder having 38.5 at.%Cr in Al. The x-ray analysis revealed space group $R3m$ (No. 160). In gamma-brasses with R-cell, the crystallographic site sets in I-cell are subdivided into IT1 + IT3, OT1 + OT3, OH3⁺ + OH3⁻ and CO6 + CO3⁺ + CO3⁻, where superscripts + and − are used to distinguish OH and CO positions with components along [111] and [$\overline{1}\,\overline{1}\,\overline{1}$] relative to the center of the cluster. IT1 and OT3 have components along [111], whereas IT3 and OT1 have components along [$\overline{1}\,\overline{1}\,\overline{1}$] from the center of the cluster. The body-centered rhombohedral cell with 52 atoms and $\alpha\sim90°$ allows convenient comparison of atomic coordinates with those in I-cell gamma-brasses.

In body-centered cubic gamma-brasses, the clusters of 26 atoms are roughly spherical in their overall shape. In rhombohedral Cr_5Al_8, where only one threefold symmetry axis along [111] exists, the ordering of atoms is such that smaller Cr atoms are predominant in the sites in that half of the cluster along the positive [111] direction. The larger Al atoms are predominant in the other half of the cluster. Hence, the observed atomic ordering, together with an expected collapse of the positive [111] half of

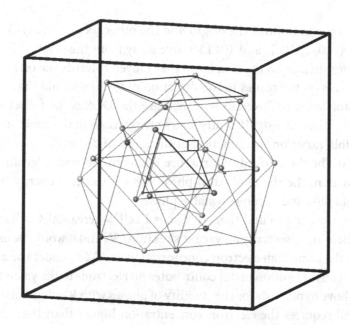

FIGURE A2.10 26-atom cluster "b" in R-cell. A symbol (□) indicates the vacancy on tetrahedral site IT1. The distorted inner tetrahedron and the distorted plane of the cubo-octahedron are drawn with black lines.

the cluster whenever an IT1 site is vacant, tends to give the clusters in Cr_5Al_8 a conical distortion with the point of the cone directed along the positive [111] direction. The situation, in which a rhombohedral distortion is induced as a result of an introduction of a vacancy into the IT1 site, is illustrated in Figure A2.10.

Because of the transition metals occurring in R-cell gamma-brasses, the valence electron concentration is hard to assess. The Al-Cu phase is the only one without a transition metal atom. If all sites are fully occupied, the Al-Cu phase would be homogeneous over the range 92.5 to 94.7 electrons per 52-atom cell. If in the Al-Cr phases the transition metal contributes no electrons to the valence band, they would contain approximately 93.6 to 97.5 electrons per 52-atom cell. Such evidence suggests that R-cell gamma-brasses probably occur at electron concentration higher than those at which cubic gamma-brasses are obtained. On this basis, it is possible to develop a hypothesis that R-cell phases are the results of a distortion, which is stabilized by the band structure energy.

Suppose that there is a rhombohedral distortion of the structure with $\alpha < 90°$, so that the six reciprocal planes (110), (011), $(\overline{1}\,\overline{1}0)$, $(\overline{1}0\overline{1})$, and

$(0\bar{1}\bar{1})$ move in towards the origin and the other six planes $(\bar{1}10)$, $(10\bar{1})$, $(0\bar{1}1)$, $(1\bar{1}0)$, $(\bar{1}01)$, and $(01\bar{1})$ move away from the origin. Overlap of the Fermi surface on the six planes that moved towards the origin would exert forces on the planes in the direction of the origin and thus stabilize the distortion. Such overlap will increase the DOS so that electrons are accommodated at a significantly lower energy than in the cubic structure. A possible variation as a function of the angle of distortion, α, of the free energy of the rhombohedral structure is shown. Beyond a certain degree of distortion, the rhombohedral phase has a lower free energy than the other phase(s) and becomes a stable phase.

It is surprising to find that Al_8V_5 has I-cell, whereas Al_8Cr_5 has R-cell, since there are essentially no vacant sites in Al_8V_5 and it would be expected to have the same high electron concentration as Al_8Cr_5 under the assumption that any transition metal contributes no electrons to the valence band. If the above hypothesis for the stability of phases with R-cell is correct, that is, R-cell requires the electron concentration higher than that for I-cell, Al_8V_5 should have the electron concentration lower than Al_8Cr_5. To reduce the electron concentration, only V, but not Cr, absorbs some electrons provided by Al into its hybridized bands, which have mainly d character.

Note added by the present author: See more discussion on the stability of Al_8V_5 gamma-brass in Chapter 8, Section 8.3.

A2.2.4.8 E.H. Kisi and J.D. Browne, Acta Cryst. B47 (1991) 835.
In this work, powders for neutron diffraction experiments were prepared by ball milling. The density was also measured. Neutron powder diffraction data were collected on the Debye-Scherrer-geometry high-resolution powder diffractometer, using $\lambda = 0.1376$ nm neutrons from a Ge monochromater. The starting model for refinement was that of stoichiometric Cu_9Al_4. It was found that $31.3 \sim 34.0$ at.%Al alloys were cubic with space group $P\bar{4}3m$ while 36.8 and 38.8 at.%Al alloys were rhombohedrally distorted with space group $R3m$.

The lattice constant increases with increasing Al content but at a decreasing rate in the cubic region. Instead, the lattice constant decreases with increasing Al content in the rhombohedrally distorted region. The distortion from cubic is smaller than that reported for other rhombohedral gamma-brasses. Measured densities show an accelerated decline in the rhombohedrally distorted region. There is a steady increase in the number of electrons per unit cell up to 34.0 at.%Al. However, within experimental error, the electron concentration per cell in the distorted region becomes constant at 88 in agreement with the theory of Jones.

(A) Al substitution in the cubic regime: The structure of stoichiometric Cu_9Al_4 gamma-brass is described as IT = Al, OT = Cu, OH = Cu, CO = Cu in the cluster "a", IT = Cu, OT = Cu, OH = Cu, CO = Al in the cluster "b". With an increase in Al content, significant amounts of Al substitution for Cu were found to occur on sites IT and OH in the cluster "b."

(B) Rhombohedrally distorted regime: The rhombohedral distortion was introduced into CO in the cluster "a," and sites IT and OH in the cluster "b" in the original cubic atom arrangements above. Excess Al was added to sites CO-3^+ in the cluster "a," whereas structural vacancies were introduced into sites IT-1, IT-3 and OH-3^- in the cluster "b." Among them, the site IT-1 in the cluster "b" is fully vacant in the 38.8 at.%Al sample, meaning the loss of one Cu atom. Further refinement failed because the large number of parameters involved prevented the use of chemical constraints to enforce the correct Al content. An emphasis is laid on the contraction and distortion around sites IT-1 and IT-3 and OH-3^- in the cluster "b" but that of sites CO-3^+ in the cluster "a" is small. It is therefore likely that the occupancy changes on the sites IT in the cluster "b" are primarily due to vacancies whilst on sites CO in the cluster "a" due to Al substitution.

The wide range of compositional stability for the gamma-brass structures in the Cu-Al system was postulated to occur by the formation of structural vacancies in order to keep the number of valence electrons per unit cell constant at 88 (see Chapter 7, Section 7.7). In the case of R-cell gamma-brass, Brandon et al. (1977) [see Section A2.2.4.7] suggested that a zone splitting occurs which maintains a Fermi surface-Brillouin zone boundary contact in a reduced cell containing 104 valence electrons per unit cell. Increasing the electron concentration would increase the rhombohedral distortion. However, the present work confirmed that this is not the case in Cu-Al alloys, since the rhombohedral distortion can maintain 88 valence electrons per unit cell.

Note added by the present author: It is possible to construct from the data in Section A2.2.4.8 the ordered R-cell Cu-Al gamma-brass. Ordered atom arrangements with space group $R3m$ are as follows: IT-1 = 1Al, IT-3

= 3Al, OT-1 = 1Cu, OT-3 = 3Cu, OH-3$^+$ = 3Cu, OH-3$^-$ = 3Cu, CO-6 = 6Cu, CO-3$^+$ = 3Cu, CO-3$^-$ = 3Cu in the cluster "a" and IT-1 = vacancy, IT-3 = 3Cu, OT-1 = 1Cu, OT-3 = 3Cu, OH-3$^+$ = 3Cu, OH-3$^-$ = 3Cu, CO-6 = 6Al, CO-3$^+$ = 3Al and CO-3$^-$ = 3Al in the cluster "b". This yields $Cu_{35}Al_{16}$ containing 51 atoms per cell as a result of the introduction of one vacancy on the site IT-1 at the cluster "b". The total electron concentration per unit cell, **e/uc**, is 83. The resulting **e/a** is 83/51 = 1.627. Alternatively, CO-3$^+$ = 3Al in the cluster "a" may be also chosen and otherwise the same as above. The chemical formula changes to $Cu_{32}Al_{19}$ and the value of **e/uc** is increased to 89 and **e/a** becomes 1.741. First-principles band calculations for these ordered structures will provide valuable information about the transformation into space group $R3m$.

Index

Printed in the United States
by Baker & Taylor Publisher Services

Printed in the United States
by Baker & Taylor Publisher Services